Communications in Computer and Information Science 3094

Series Editors

Rationale
The CCIS series is devoted to the publication of proceedings of computer science conferences. Its aim is to efficiently disseminate original research results in informatics in printed and electronic form. While the focus is on publication of peer-reviewed full papers presenting mature work, inclusion of reviewed short papers reporting on work in progress is welcome, too. Besides globally relevant meetings with internationally representative program committees guaranteeing a strict peer-reviewing and paper selection process, conferences run by societies or of high regional or national relevance are also considered for publication.

Topics
The topical scope of CCIS spans the entire spectrum of informatics ranging from foundational topics in the theory of computing to information and communications science and technology and a broad variety of interdisciplinary application fields.

Information for Volume Editors and Authors
Publication in CCIS is free of charge. No royalties are paid, however, we offer registered conference participants temporary free access to the online version of the conference proceedings on SpringerLink (http://link.springer.com) by means of an http referrer from the conference website and/or a number of complimentary printed copies, as specified in the official acceptance email of the event.

CCIS proceedings can be published in time for distribution at conferences or as post-proceedings, and delivered in the form of printed books and/or electronically as USBs and/or e-content licenses for accessing proceedings at SpringerLink. Furthermore, CCIS proceedings are included in the CCIS electronic book series hosted in the SpringerLink digital library at http://link.springer.com/bookseries/7899. Conferences publishing in CCIS are allowed to use our online conference service (Meteor) for managing the whole proceedings lifecycle (from submission and reviewing to preparing for publication) free of charge.

Publication process
The language of publication is exclusively English. Authors publishing in CCIS have to sign the Springer CCIS copyright transfer form, however, they are free to use their material published in CCIS for substantially changed, more elaborate subsequent publications elsewhere. For the preparation of the camera-ready papers/files, authors have to strictly adhere to the Springer CCIS Authors' Instructions and are strongly encouraged to use the CCIS LaTeX style files or templates.

Abstracting/Indexing
CCIS is abstracted/indexed in DBLP, Google Scholar, EI-Compendex, Mathematical Reviews, SCImago, Scopus. CCIS volumes are also submitted for the inclusion in ISI Proceedings.

How to start
To start the evaluation of your proposal for inclusion in the CCIS series, please send an e-mail to ccis@springer.com

Nurfadhlina Mohd Sharef · Azreen Azman ·
Tutut Herawan
Editors

Information Retrieval and Knowledge Management

7th International Conference, CAMP 2026
Putrajaya, Malaysia, August 12–13, 2026
Proceedings, Part II

Springer

Editors
Nurfadhlina Mohd Sharef
Universiti Putra Malaysia
Serdang, Malaysia

Azreen Azman
Universiti Putra Malaysia
Selangor, Malaysia

Tutut Herawan
Universiti Malaya
Kuala Lumpur, Malaysia

ISSN 1865-0929 ISSN 1865-0937 (electronic)
Communications in Computer and Information Science
ISBN 978-981-92-4304-4 ISBN 978-981-92-4305-1 (eBook)
https://doi.org/10.1007/978-981-92-4305-1

This Springer imprint is published by the registered company Springer Nature Singapore Pte Ltd.
The registered company address is: 152 Beach Road, #21-01/04 Gateway East, Singapore 189721, Singapore

Preface

The Seventh International Conference on Information Retrieval and Knowledge Management (CAMP 2026) was held on 12–13 August 2026 at the Bangi Resort Hotel, Malaysia. Organized by the Society of Information Retrieval and Knowledge Management Malaysia (PECAMP), CAMP serves as a platform for researchers, practitioners, and industry experts to share advances in information retrieval, knowledge management, computational linguistics, and data science.

The theme of CAMP 2026, "Beyond Retrieval: Information, Knowledge, and Intelligence," reflects the transformation of intelligent information systems from traditional retrieval mechanisms toward knowledge discovery, reasoning, prediction, and decision support. Recent advances in artificial intelligence, large language models, retrieval-augmented generation, multimodal learning, and knowledge-driven systems continue to reshape how information is transformed into actionable intelligence across diverse domains.

CAMP 2026 featured four technical tracks: Information Retrieval, Computational Linguistics, Knowledge Management, and Data Science. The conference received 113 submissions from researchers worldwide. Following a rigorous single-blind peer-review process conducted by an international program committee, 27 submissions were accepted for publication as regular papers, resulting in an acceptance rate of 24%.

This volume presents selected contributions that illustrate the convergence of information retrieval, knowledge management, computational linguistics, and data science in creating intelligent systems that support knowledge discovery and decision-making. These include retrieval-augmented and knowledge-enhanced artificial intelligence, multimodal learning, conversational systems, multilingual language technologies, machine learning, digital twins, intelligent decision support, and human-centered AI. The papers also demonstrate the application of information and knowledge technologies to address real-world challenges in healthcare, agriculture, education, cybersecurity, transportation, environmental sustainability, and public services. Collectively, the papers demonstrate how advances in information, knowledge, and intelligence can generate meaningful impact across research, industry, and society.

We sincerely thank all authors for their submissions and contributions. We also extend our appreciation to the Program Committee members, reviewers, track chairs, keynote speakers, special session organizers, sponsors, and organizing committee members, whose dedication and expertise ensured the success of CAMP 2026 and the quality of these proceedings. We also extend our sincere appreciation to Springer Nature as the publisher of the proceedings, for their invaluable support and guidance throughout this publication project.

We hope that the contributions presented in this volume will stimulate further research, foster interdisciplinary collaboration, and inspire future innovations in intelligent information systems.

June 2026

Nurfadhlina Mohd Sharef
Azreen Azman
Tutut Herawan

Organization

Honorary Chair

Tengku Mohd Tengku Sembok	PECAMP, Malaysia

General Co-chairs

Nurfadhlina Mohd Sharef	Universiti Putra Malaysia, Malaysia
Roslina Othman	IIUM, Malaysia

Conference Co-chairs

Shahrul Azman Noah	Universiti Kebangsaan Malaysia, Malaysia
Tutut Herawan	Universiti Malaya, Malaysia

Technical Program Chair

Azreen Azman	Universiti Putra Malaysia, Malaysia

Track Chairs

Mas Rina Mustaffa	Universiti Putra Malaysia, Malaysia
Shaiful Bakhtiar Rodzman	Universiti Teknologi MARA, Malaysia
Nor Liyana Mohd Shuib	Universiti Malaya, Malaysia
Sri Devi Ravana	Universiti Malaya, Malaysia

Publication Co-chairs

Mohd. Helmy Abdul Wahab	Universiti Tun Hussein Onn Malaysia, Malaysia
Haslizatul Fairuz Mohamed Hanum	Universiti Teknologi MARA, Malaysia

Aslina Baharum	Taylor's University, Malaysia
Suhaila Shahul Hameed	UniKL, Malaysia
Nurul Aiman Abdul Rahim	Taylor's University, Malaysia

Publicity Chair

Fakhrul Hazman Yusoff	UNITAR International University, Malaysia

Tutorials and Workshops Chairs

Azliza Yacob	University College TATI, Malaysia
Zirawani Baharum	UniKL, Malaysia

Finance Co-chairs

Masnizah Mohd	Universiti Kebangsaan Malaysia, Malaysia
Siti Syakirah Sazali	INTI International College, Malaysia

Sponsorship Chair

Dayang Nurfatimah Awang Iskandar	Universiti Malaysia Sarawak, Malaysia

Logistics Co-chairs

Azrul Hazri Jantan	Universiti Putra Malaysia, Malaysia
Alfian Abdul Halin	Universiti Putra Malaysia, Malaysia

Secretariat

Nurul Amelina Nasharuddin	Universiti Putra Malaysia, Malaysia
Nurul Sharefa Haneem Shahri	Universiti Putra Malaysia, Malaysia

Special Sessions Chairs

Annisa Ristya Rahmanti	Universitas Gadjah Mada, Indonesia
Nurfadhlina Mohd Sharef	Universiti Putra Malaysia, Malaysia
Brian Halubanza	Mulungushi University, Zambia
Musab Al Rawi	University of Technology and Applied Sciences, Oman
Vivi Indra Amelia Nasution	Universitas Terbuka, Indonesia
Shahrinaz Ismail	Universiti Sains Malaysia, Malaysia
Pradorn Sureephong	Chiang Mai University, Thailand
Suepphong Chernbumroong	Chiang Mai University, Thailand
Mohsin Iftikhar	Higher Colleges of Technology, UAE
Nidal Nasser	Alfaisal University, Saudi Arabia

Program Committee

Adhi Susilo	Universitas Terbuka Indonesia, Indonesia
Aggy Pramana Gusman	Universitas Putra Indonesia YPTK, Indonesia
Aldilas Achmad Nursetyo	ASEAN Biological Threats Surveillance Centre, Indonesia
Alfian Abdul Halin	Universiti Putra Malaysia, Malaysia
Amja Manullang	Middlesex University London, UK
Annisa Rahmanti	Universitas Gadjah Mada, Indonesia
Ardiansyah Dores	Universitas Mercu Buana, Indonesia
Ari Prayogo Pribadi	Middlesex University London, Indonesia
Asad Shah	University of Jeddah, Saudi Arabia
Aslina Baharum	Taylor's University, Malaysia
Azreen Azman	Universiti Putra Malaysia, Malaysia
Bagas Suryo Bintoro	Center of Health Behavior and Promotion, Indonesia
Brian Halubanza	Mulungushi University, Zambia
Dayang Nur Fatimah Awang Iskandar	Universiti Malaysia Sarawak, Malaysia
Devia Kartika	Universitas Putra Indonesia YPTK, Indonesia
Diva Kurnianingtyas	Universitas Brawijaya, Indonesia
Eissa Alshari	IBB University, Yemen
Fakhrul Hazman Yusoff	Sohar University, Oman
Farikin	Diponegoro University, Indonesia
Fatimah Khalid	Universiti Putra Malaysia, Malaysia
Firdaus Hafidz As Shidieq	Liverpool School of Tropical Medicine, UK

Hamdan Daniyal	Universiti Malaysia Pahang Al-Sultan Abdullah, Malaysia
Hanifah Wulandari	Universitas Gadjah Mada, Indonesia
Haslizatul Mohamed Hanum	Universiti Teknologi MARA, Malaysia
Javid Iqbal Thirupattur	Sunway University, Malaysia
Jing Fu	Chengdu University, China
Krittika Kantawong	University of Phayao, Thailand
Lailatul Qadri Zakaria	Universiti Kebangsaan Malaysia, Malaysia
Liu Ziyan	Universiti Putra Malaysia, Malaysia
Lucky LhauraVan FC	Universitas Lancang Kuning, Indonesia
Mas Rina Mustaffa	Universiti Putra Malaysia, Malaysia
Masnizah Mohd	Universiti Kebangsaan Malaysia, Malaysia
Masrah Azrifah	Universiti Putra Malaysia, Malaysia
Moch. Nafkhan Alzamzami	Institut Teknologi Sepuluh Nopember, Indonesia
Mohd Helmy Abd Wahab	Universiti Tun Hussein Onn Malaysia, Malaysia
Mohd Nordin Abd Rahman	Universiti Sultan Zainal Abidin, Malaysia
Mohsin Iftikhar	Higher Colleges of Technology, UAE
Muhammad Sam'an	Universitas Muhammadiyah Semarang, Indonesia
Muhd Eizan Shafiq Abd Aziz	Universiti Teknologi MARA, Malaysia
Musab Al Rawi	University of Technology and Applied Sciences, Oman
Myat Noe Win	University Malaya, Malaysia
Nabila Wardah Zamani	Universiti Putra Malaysia, Malaysia
Nasa Zata Dina	Universitas Airlangga, Indonesia
Nor Liyana Mohd Shuib	University of Malaya, Malaysia
Noraini Che Pa	Universiti Putra Malaysia, Malaysia
Nurfadhlina Mohd Sharef	Universiti Putra Malaysia, Malaysia
Nurul Amelina Nasharuddin	Universiti Putra Malaysia, Malaysia
Peer Azmat Shah	Higher Colleges of Technology, UAE
Pradorn Sureephong	Chulalongkorn University, Thailand
Pruet Putjorn	Mae Fah Luang University, Thailand
Puteri Suhaiza Sulaiman	Universiti Putra Malaysia, Malaysia
Qais Qassim	University of Technology and Applied Sciences, Oman
Rabiah Abdul Kadir	Universiti Kebangsaan Malaysia, Malaysia
Raihani Mohamed	Universiti Putra Malaysia, Malaysia
Razali Yaakob	Universiti Putra Malaysia, Malaysia
Rima Gema	Universitas Putra Indonesia YPTK, Indonesia
Roslina Othman	International Islamic Universiti Malaysia, Malaysia
Rosmayati Mohemad	University of Malaysia Terengganu, Malaysia
Rozi Nor Haizan Nor	Universiti Putra Malaysia, Malaysia

Ruqaya Subhi Tawfeeq	Al-Iraqia University, Iraq
Saffa Raihan Zainal Abidin	Universiti Teknologi MARA, Malaysia
Santichai Wicha	Mae Fah Luang University, Thailand
Shahrinaz Ismail	Universiti Sains Malaysia, Malaysia
Shahrul Azman Noah	Universiti Kebangsaan Malaysia, Malaysia
Shaiful Bakhtiar Rodzman	Universiti Teknologi MARA, Malaysia
Sharifah Sakinah Syed Ahmad	Universiti Teknikal Malaysia Melaka, Malaysia
Siti 'Aisyah Sa'dan	Universiti Teknologi MARA, Malaysia
Siti Syakirah Sazali	HELP University, Malaysia
Sonia Soans	University of Technology and Applied Sciences, Oman
Sri Devi Ravana	Universiti Malaya, Malaysia
Tutut Herawan	Universiti Malaya, Malaysia
Venkataraman Meenakshi	University of Texas at Austin, USA
Vivi Indra Amelia Nasution	Universitas Terbuka Indonesia, Indonesia
Xin Liu	Knowledge and Innovation Research Laboratory, Thailand
Yootthapong Tongpaeng	Mae Fah Luang University, Thailand
Yusmadi Yah Jusoh	Universiti Putra Malaysia, Malaysia
Zamri Abu Bakar	Universiti Teknologi MARA, Malaysia
Zulkifli Halim	Universiti Teknologi MARA, Malaysia

Contents

AI-Resilient Computer Science Education: Assessment Innovation, Curriculum Redesign, and Learning Analytics in the Era of Generative AI

Ethical AI in Education and Responsible Research: Frameworks, Governance, and Trustworthy AI for Knowledge-Driven Institutions

Data Governance and Knowledge Management: Higher Education Performance Analytics and Knowledge-Driven Decision Making

Transforming Data into Actionable Knowledge: AI and Retrieval for Intelligent Health Systems

AI Therapist Chatbot: Leveraging a Hybrid Approach

Shaidah Jusoh[1(✉)], Shu Qi Law[2], Khai Yi Leong[2], Ing Xian Jonas Lim[2], Kai Fan Woon[2], and Xuan Jing Bong[2]

[1] School of Artificial Intelligence and Robotics, Xiamen University Malaysia, Sepang, Malaysia
shaidah.jusoh@xmu.edu.my

[2] School of Computing and Data Science, Xiamen University Malaysia, Sepang, Malaysia

Abstract. There is a growing prevalence of psychological stress across all age groups. However, privacy concerns, lack of support, and social stigma remain significant barriers to seeking traditional therapy. This paper presents the design and development of a safe, emotionally aware AI therapist chatbot aimed at supporting mental health through conversational interaction. The proposed chatbot employs a hybrid approach, combining natural language processing (NLP) techniques with machine learning models. In this study, we adopt a multi-dataset strategy for emotion, intent, and response selection to ensure that outputs are supportive and contextually appropriate. A rule-based safety check module is integrated to detect crisis-related language and prevent inappropriate or harmful responses. The module comprises 26 rules, each representing a phrase that indicates a human user may be in danger. A large language model (Llama 3.1:8b) is utilized to rephrase responses for naturalness while maintaining safety constraints. Evaluation results demonstrate high accuracy (97%) in intent classification but highlight challenges in emotion detection due to overlapping vocabulary. Overall, the chatbot functions as a supportive companion, enhancing emotional wellness without replacing professional care.

Keywords: Chatbot · Mental Health Chatbot · AI Therapist Bot

1 Introduction

In recent years, mental health has become a worldwide issue, affecting people across all age groups. Widespread psychological stress is common in modern society due to rapid societal development, increasing competition, and the accelerating pace of life. For example, students face academic pressure and expectations regarding future career success. Most parents work away from home for whole days, leaving their children with little emotional support. If they also lack friends at school or experience bullying, they are highly susceptible to developing psychological problems or mental health illnesses. Working adults experience increasing workload stress, job instability, and financial responsibilities. In addition to workplace stress, adults face conflict and pressure in

This publication is supported by the Xiamen University Malaysia Research Fund (XMUMRF/2022-C9/IECE/0034)

N. Mohd Sharef et al. (Eds.): CAMP 2026, CCIS 3094, pp. 3–16, 2027.
https://doi.org/10.1007/978-981-92-4305-1_1

their relationships with their spouses and children. Conversely, older adults might face the problem of loneliness and emotional isolation, particularly the aged and unmarried people who have not interacted with society for a long time. These cumulative stressors may result in anxiety, depression, emotional stress, or other more severe psychological conditions unless addressed appropriately. Despite the increased awareness of mental health, many people remain unreceptive to professional help. This could be due to fear of invasion of personal privacy, fear of social stigma, and embarrassment about revealing sensitive feelings to others. Individuals who feel emotionally vulnerable might not want to undergo conventional face-to-face therapy because of embarrassment and fear of being judged by others; emotional problems often remain unexpressed and unresolved, increasing the risk of extreme emotional reactions and harmful behaviors in society. Even though conventional professional psychological services are a vital part of mental health care, they are not always available, affordable, and applicable to all. Many people require a safe, private, and non-judgmental space for them to express their emotions freely without fear of exposure or social judgment. In the era of artificial intelligence (AI), a system that can converse and interact with human users using spoken, written, and visual languages is a conversational AI system, a so-called chatbot.

Existing chatbots, especially general-purpose generative chatbots, such as ChatGPT, are not specifically designed to handle sensitive emotional interactions. These types of chatbots may give medical suggestions, while patients sometimes just need to be comforted. Sometimes, the chatbot provides unconditional support to users, regardless of right or wrong. This leads some loners to become increasingly extreme after constantly interacting with a generative chatbot, potentially resulting in them doing drastic things. For example, a person who had been bullied for a long time brought a gun to school and killed the bully. This study is one of the ongoing efforts for handling mental health issues.

In this study, we aim to design and develop a therapist chatbot that is safe, interpretable, and emotionally aware, providing conversational support. Rather than offering medical diagnoses or treatment, the chatbot is designed to recognize emotional expressions, respond appropriately, and ensure a secure space for users to express their emotions. This study focuses on the development of a therapist chatbot using natural language processing (NLP) techniques, machine learning methods, and a rule-based approach. The chatbot incorporates emotion recognition, intent classification, and controlled conversational response generation. Its primary function is to provide basic emotional assistance rather than clinical diagnoses or professional treatment, ensuring safe and responsible communication within the boundaries of emotions and conversation. Overall, the chatbot is developed to act as a supportive friend who provides comfort, understanding, and emotional validation and helps to enhance the emotional wellness of users. This paper is organized as follows: Sect. 2 presents topics related to the proposed study, Sect. 3 presents a methodology of the AI therapist chatbot, Sect. 4 presents a result analysis, and Sect. 5 discusses the results and their limitations; finally, Sect. 6 concludes the study.

2 Related Topics

2.1 Natural Language Processing

The ability of a chatbot to serve as a therapeutic agent is entirely limited by its underlying natural language (NL) understanding. Various studies indicate that the success of a mental health treatment largely relies on the accuracy of this initial input processing. Thus, NL processing is a main process for any mental health chatbot. It is responsible for filtering user input into a machine-processable format. User input, which is in the NL, must be processed well before the machine can express empathy. Standard preprocessing pipelines, as presented in Sect. III, involve a series of steps designed to strip noise while retaining the semantic signal. Tokenization breaks continuous text into discrete units, while stop-word removal eliminates high-frequency but low-information words (e.g., "and," "the"). However, in the field of mental health, researchers suggest that there is a need for more subtlety; the usage of self-referential pronouns (e.g., "I", "my") is a powerful indicator of depressive rumination, and the reduction process may degrade the psychological profile of the text as well as introduce noise in the psychometric data [5]. Further processes, such as stemming and lemmatization, return words to their basic forms (e.g., "suffering" becomes "suffer"), allowing the model to be tense-independent and to generalize across different grammatical tenses. The representation of text has also evolved. "Bag-of-words" (BOW), which discards word order and context, poses a limitation to a full understanding of a sentence. Current models, such as word embeddings (Word2Vec, GloVe) assign words to n-dimensional spaces where semantic similarity, represented by geometric proximity, allows a chatbot to make the inference that "sad" and "despondent" are not the same but do refer to a similar idea, even if they are distinct strings of characters.

2.2 Deep Learning Methods

Recurrent Neural Network (RNN) has become a starting point for many advanced machine learning models for text processing. This includes the Long Short-Term Memory (LSTM) network, which was traditionally considered the "gold standard" for sequential text processing. However, the training dynamic of these models has some limitations. According to [5], in mental health classification, typical LSTM models tend to overfit. Low accuracy (around 15–17%) can be observed in the training phase initially since the model has difficulty finding patterns from the noisy social media input. The authors in [5] demonstrate the specific advantages of using a bidirectional LSTM (BiLSTM) for classifying mental health text. While a standard LSTM processes text linearly (from start to finish), a BiLSTM processes the input in both directions simultaneously (e.g., left to right and right to left).

2.3 Transformer Revolution

The current state-of-the-art in mental health NLP is a transformer architecture, specifically Bidirectional Encoder Representations from Transformers (BERT). Unlike LSTM, BERT utilizes an "attention" mechanism, which assesses the significance of all words in a sentence by attending to every other word at once. The results of comparative studies demonstrate that transformer models are markedly superior. When evaluated against BiLSTMs or Convolutional Neural Networks (CNNs) for tasks such as depression detection or suicide risk assessment, fine-tuned BERT models consistently achieve higher F1 scores and accuracy metrics, often exceeding 90%. The pre-training of BERT on massive corpora of general text gives it a deep, innate understanding of syntax and semantics, which can then be "fine-tuned" on smaller, domain-specific mental health datasets. This transfer learning model initiated by BERT has given rise to a new hope, allowing highly precise models to be generated even when in-domain clinical data is scarce [3].

2.4 Hybrid Approach

Solely data-driven approaches, on the other hand, often lack "common sense" or a structured understanding of human psychology. To address this, [3] introduced emotion ontologies, which are hierarchical structures that sort and describe emotions and are incorporated into the chatbot architecture. Instead of just labeling an input as "Negative", the ontology allows the system to understand that "rage" is a high-intensity variant of "anger", or that "grief" is a complex emotional state related to "loss". With an NLP transformer for detection and an ontology for reasoning, chatbots were reported to respond appropriately to the emotional nuances and intensity of the user [3]. Other methods include leveraging dimen-sionality reduction algorithms such as probabilistic principal component analysis that map emotions into more generalized categories (positive, negative, ambiguous, and neutral). This is of particular benefit when dealing with large-scale, noisy data that could render the fine-grained emotion annotations less effective [6].

2.5 Safety as Critical Issues

While mental health chatbots show great potential, the ethical concerns surrounding generative AI in this field are troubling. The proficiency of large language models (LLMs) has given rise to what scholars term "deceptive empathy", instances where a model asserts, "I truly understand your pain". Brown University [1] highlighted how such fabricated empathy violates ethical practices.

Vulnerable users may develop pathologically dependent relationships, finding solace in the consistent support of a machine rather than navigating the friction inherent in human interactions. Furthermore, Stanford University (2025) [9] recently tested popular LLM-based therapy chatbots using vignettes of suicidal ideation. In several cases, the bots failed to detect the crisis or recognize the emergency. Moreover, poorly tuned versions produced harmful content related to self-harm methods or failed to intervene appropriately.

The devastating suicide of 14-year-old Sewell Setzer [7], who became emotionally attached to a chatbot character that failed to intervene during his time of crisis, serves as a grim case study. This tragedy illustrates the potentially fatal consequences of deploying general-purpose LLMs in mental healthcare. Without a "safety override" layer capable of detecting crisis-related keywords, these models risk becoming inadvertent accomplices to tragedy. Furthermore, AI models replicate the discrimination that exists in their training data. The Stanford study in 2025 [8] also found that chatbots showed biases against patients with symptoms of schizophrenia and alcohol use and were less "eager" to help such patients compared to those with depression. This implies that without careful tuning, AI could encode and perpetuate the systemic disparities that already exist in our healthcare. Another recent case regarding Replika shows the tension between user autonomy and corporate responsibility. Marketed as an AI companion, Replika became widely used for mental healthcare. But when the company swiftly removed 'erotic role-play' (ERP) in 2023 over child safety concerns, it catalyzed a mental health crisis for its users. Users in intense friendships and romantic relationships reported grief, loss, and distress, almost classically associated with breakups. This exposes the Autonomy-Control Paradox: the very features that made the bot effective for loneliness (intimacy and unrestricted interaction) were the ones deemed unsafe by regulators [2]

The arrival of AI in mental healthcare is an inflection point. The technologies powered by the transformative potential of Transformers and LLMs are achieving a linguistic conversance that can create genuine emotional rapport. However, the "Safety Gap" is increasing. Rapport-building exceeds the system's ability to facilitate safety. Deceptive empathy, crisis failure, and algorithmic stigma risks are not hypothetical; these are tangible harms that are already being experienced. To advance in this space in a safe manner, we may embrace *safety sandwich* architectures where the generative model is encased between two strict, rule-based safety classifiers that can trump the LLM in times of crisis.

3 Methodology

Our AI therapist chatbot is implemented as a web-based system using Streamlit, an open-source Python library that allows users to build an interactive dashboard and user interface using pure Python with minimal code. Therefore, we can focus on natural language processing techniques that are used in the chatbot without worrying about the additional web technologies, such as HTML, CSS, or external APIs. The primary consideration of this study is to provide health sensitivity because the users who use the chatbot might have different levels of emotional instability and hope to find comfort by using our chatbot. Thus, the proposed chatbot adopted emotional detection and intent classification as the core components before generating a response to avoid an inappropriate reply that may cause emotional distress.

Figure 1 demonstrates the overall flow of the proposed chatbot. The user input will be passed through a preprocessing pipeline to normalize informal conversational text. This ensures the user input can be transformed into clean text before emotional and intent classification, and this leads to better accuracy for the analysis. Next, a safety check will be performed to find harmful expressions in the user's input, such as death, self-injury, or suicide. In such cases, the chatbot will immediately display a warning message to encourage the user to seek help from qualified professionals because it exceeds the scope of the chatbot. The cleaned user's input will be sent to the emotion detection module to identify the user's emotional state and the intent classification module to determine the purpose of the message if it passes the safety check. The results are used in the response selection phase to retrieve responses, select an appropriate response, and display the response to the user. The following subsections present the details of the design and implementation process.

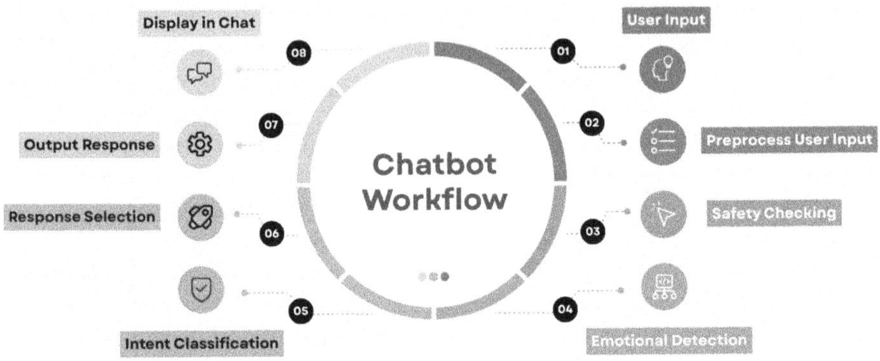

Fig. 1. Chatbot's workflow demonstrating the sequential stages of conversational processing.

3.1 Dataset Preparation

This study adopts a multi-dataset approach to support the development of a mental-health–oriented chatbot. Three different datasets were prepared and used in this study, namely, **Emotion**, **Intent**, and **Response**. Each dataset has a different role in the chatbot system, allowing it to recognize user emotions, understand user intentions, and provide suitable responses.

Emotion Dataset. The emotion dataset used in this study was collected from a publicly available Kaggle dataset titled Emotion Detection from Text [4]. This dataset consists of short text samples annotated with emotion-related labels and is widely used for emotion classification tasks. Each record in the dataset contains a text message and a corresponding emotion label. As the original dataset includes several fine-grained emotion categories, preprocessing steps were applied to clean and standardize the data before training the emotion classification model. These include converting to lowercase to ensure consistency across the dataset and to avoid treating the same word written in different cases as separate tokens. Next, informal expressions and abbreviations were normalized using a predefined slang dictionary. This helps reduce variation caused by casual typing while preserving the original meaning of the text. After normalization, noise removal was performed. Emojis were removed because the emotion classification in this project is based on textual features rather than visual symbols. Special characters, punctuation, and numbers were also removed using regular expressions, leaving only alphabetic characters and spaces. Any extra whitespace introduced during this process was normalized to keep the text clean and readable. Stopword removal was then applied using the NLTK English stopword list. However, negation words such as "not", "no", and "nor" were intentionally retained, as they play an important role in expressing emotional meaning. Removing these words could change the interpretation of a sentence, for example, turning "not happy" into "happy". Finally, lemmatization was applied using the WordNet lemmatizer to convert words into their base forms. This reduces vocabulary size and helps the model focus on the underlying meaning rather than grammatical variations. Through these steps, the emotion dataset was transformed into a cleaner and more consistent format suitable for emotion classification tasks.

Intent Dataset. The intent dataset was created to identify the purpose behind a user's message rather than the emotional state alone. Each data record consists of two attributes: text and intent label. Four intent categories were defined in this study: *advice*, *cheer up*, *expression*, and *affirmation*. A total of 320 text samples were constructed for the intent dataset. The data was manually generated to resemble realistic conversational inputs commonly found in mental-health or support-based chatbot interactions. Informal language and short expressions were intentionally used to better reflect how real users communicate. Each intent category contains an equal number of samples to ensure balanced representation during training. The intent dataset consists of short, well-structured conversational sentences that do not contain excessive noise such as emojis, URLs, or irrelevant symbols.

Response Dataset. The response dataset was designed to support the chat-bot's response selection mechanism. Each record contains three attributes: *emotion*, *intent*, and *response text*. The dataset covers all combinations of the eight emotion categories and four intent categories. For each emotion–intent combination, three different responses were prepared. In addition, an "any" intent category was also prepared as an option if the chatbot cannot confidently determine the user's intent. All responses were manually written to ensure they are appropriate, supportive, and safe for mental health–related interactions. A *retrieval-based approach* was adopted so that the chatbot selects responses from predefined templates rather than generating free-form text to prevent the risk of inappropriate or misleading replies. This structured data collection allows the chatbot to understand how the user feels, what the user wants, and how the system should respond in a controlled and emotionally appropriate manner. The response dataset was written in a controlled manner to ensure safety by making it clear and consistent, providing emotional appropriateness.

3.2 User Input Processing

The user input preprocessing module was designed to handle real-world chatbot messages, which are often informal, noisy, and unpredictable. Unlike the training datasets, user inputs may contain slang, emojis, repeated characters, links, or meaningless symbols. Therefore, a dedicated preprocessing pipeline was implemented to clean and standardize user messages before emotion detection and intent classification.

First, the text processing resources are prepared using NLTK. The script checks whether required resources (English stopwords and WordNet) are available and downloads them if needed. This makes the preprocessing module more reliable when running on different machines or environments. Stopwords are loaded from NLTK, and a small adjustment is made to preserve negation words such as "not", "no", and "nor", as removing them can change the meaning of a sentence. Next, a slang and abbreviation dictionary is used to normalize informal expressions commonly found in chats. This includes general chat abbreviations (e.g., "idk", "rn", "btw"), texting forms (e.g., "wanna", "gonna"), and common negation forms without apostrophes (e.g., "dont", "cant"). A few of local filler words (e.g., "ah", "lor") are also removed because they usually do not add useful semantic information in this task. The goal of this step is to reduce variation caused by informal typing while keeping the original meaning. To handle noisy real-world input, the pipeline removes website links, user mentions, and hashtags using regular expressions. Emojis are removed to simplify the text into a purely textual representation. The script also reduces repeated characters used for emphasis ("soooo" becomes "soo"), which helps avoid creating many slightly different word forms for the same meaning. After that, all non-alphabetic characters are removed, and extra whitespace is normalized so the final text remains clean and consistent. After cleaning, lemmatization and stopword filtering are applied to refine the text linguistically. Words are converted into their base forms using WordNet lemmatization, and common stopwords are removed while preserving negation words. Finally, a simple input validation step is applied. Empty inputs, meaningless symbol-only messages, and extremely low-signal noise inputs are filtered out. In addition, lightweight helper functions are included to detect common conversational patterns such as greetings, expressions of thanks, goodbyes, and questions. Although these functions are not required for the core cleaning process, they help improve the overall interaction flow of the chatbot. Overall, this preprocessing pipeline prepares user input in a consistent and cleaner form while retaining important meaning cues, making it suitable for downstream NLP tasks in a mental-health chatbot system.

3.3 Safety Check Module

The safety check module is our innovation in this study. It is designed to detect crisis-related language such as self-harm intent, suicidal thoughts, or extreme emotional distress. Its main purpose is to protect users by stopping the chatbot from generating normal responses when a high-risk message is detected. This ensures that potentially dangerous situations are handled carefully and responsibly before any further NLP processing occurs. This module uses a rule-based keyword matching technique. A predefined list of crisis-related keywords is created, including both single-word keywords (for example, suicide, die, kill) and multi-word phrases (for example, erase myself, done with life). Using both single words and phrases is important because crisis expressions can appear in different forms. Some users may use straightforward phrases, while others may express distress using shorter or incomplete statements. By checking both forms, this system increases the chance of detecting genuine crisis messages. Again, in a clearer view of the explanation, after preprocessing, the cleaned text is scanned to check the presence of these keywords. If any is detected, the safety check immediately returns a predefined crisis message encouraging the user to seek professional or emergency support. However, if no crisis-related term is found, the function will return "None", which allows the chatbot to continue with emotion and intent analysis. A rule-based approach is chosen for this module instead of a machine learning model because it is predictable, transparent, and reliable. Unlike TF-IDF-based models, which rely on word weighting and may overlook critical terms due to statistical effects, rule-based detection ensures that important crisis words are always checked directly. In this study, we created 26 rules, representing 26 possible phrases, for a safety check. Examples of phrases include "suicide", "done with life", and "unbearable". Although keyword matching cannot fully understand context, it provides a reliable safety layer that prioritizes user well-being over model complexity.

3.4 Emotion Classification Module

This emotion classification module identifies the emotional state expressed in the user's input text. By recognizing emotions such as happiness, sadness, anger, or others, the chatbot can adjust the tone and type of response it generates. This allows the system to understand the user's emotion for further response analysis. This module uses TF-IDF for feature extraction and logistic regression for classification. Words that appear frequently in a sentence but less frequently across all sentences receive higher weights. This helps the model focus on emotionally meaningful words while reducing the influence of common words that carry little emotional information. Hence, TF-IDF produces sparse feature vectors, which makes the approach efficient and suitable for real-time use. The feature extraction process also includes the use of n-grams, specifically unigrams and bigrams. Unigrams capture individual words while bigrams capture short word pairs. This allows the model to recognize emotional phrases such as "feel sad" or "not happy," which often convey emotion more clearly than single words. Once the TF-IDF features are extracted, a logistic regression classifier is applied to learn a linear decision boundary by assigning weights to each feature and computing class probabilities. The emotion with the highest probability is selected as the final prediction. This model performs well with sparse TF-IDF vectors that support multi-class classification and are fast to train and evaluate. However, it does not capture deep semantic meaning, which is well-suited for detecting surface-level emotional patterns in short chatbot messages and supports real-time deployment within the system.

3.5 Intent Classification

It is important to understand the user's intent for the chatbot to provide a meaningful response because users who express the same emotion may have different needs. For example, a user who feels sad may want to express feelings, seek emotional comfort, look for motivation, or ask for advice. By adding an intent classification module to the system, the chatbot can respond in a way that meets the user's expectations as much as possible. We used Term TF-IDF vectorization to transform raw text into a numerical vector to allow further processing using a machine learning model. It can measure the importance of a word in the user's input by finding the frequency in the intent dataset and assigning weights to each word. Additionally, unigrams and bigrams are considered to capture short phrases that are more indicative of user intent than individual words, such as "need help" or "feel sad". Different methods, such as fit_transform() and transform(), are used to fit the training and testing data to ensure consistency during the training and prediction stages.

After the TF-IDF vectorization, a multinomial Naïve Bayes classifier is deployed to predict the user's intent, as we have multiple intent types. In the training stage, the model learns the probability distribution of words across different intent categories and uses it to find probable intent in the user's input based on Bayes' theorem using the weight vector assigned by the vectorizer. Other than that, Laplace smoothing is applied by setting the smoothing parameter to alpha = 1.0. This can avoid zero probability when the words in the user's input do not appear in the training data because the dataset is small. Once the calculation is done, the intent with the highest probability will be chosen as the predicted intent. The vectorizer and classifier are saved into a local file using the joblib library to be loaded during runtime without retraining and enabling seamless integration into the chatbot system.

3.6 Response Selection

AI therapist chatbot selects a response from a response template to provide mental health support for different emotions and intents, rather than a generative response, to avoid inappropriate tone and cause mental breakdown or encourage risky behavior for the user. First, the system will filter potential candidates for the response based on the predicted emotion and intent from the response template using the predefined filter function. It will return a list of responses that match the predicted emotion and intent. A safety fallback condition is added in the function to return a more general open-ended response if the intent is empty. Similarly, the TF-IDF vectorizer will convert the user's input and possible candidates to numerical vectors based on the importance of words. After that, cosine similarity is used to compute the similarity between the user's input vector and the candidate response vector. Finally, the candidate response with the highest similarity will be chosen. Unlike the traditional rule-based chatbot, this approach returns the most suitable response by considering not only emotion and intent but also the context of the user's input and responses.

3.7 Response Generation

This system also adopted a large language model, which is Llama 3.1:8b, to improve the naturalness of the conversation between the chatbot and user. Due to its strong language generation capabilities and ease of local deployment without any external APIs, it allowed the chatbot to respond in a more empathetic and appropriate tone without the need to define multiple response templates. However, it is important to note that predicted emotion, predicted intent, and selected response will be given to the large language model to rephrase the selected final response through prompt design to preserve the original meaning and safety constraints, rather than provide a generative response. The combination of the system with a large language model can enhance the experience of the user while using the chatbot.

4 Result Analysis

The implemented AI therapist chatbot is demonstrated in Figs. 2 and 3. The hybrid approach was able to classify emotions and intention with accuracy values of 0.34 and 0.9685, respectively. These results demonstrate that despite only 34% of testing data being correctly classified, the evaluation result shows a very high accuracy on the test set of the intent classification model, which is 0.9685. These results show that the model can reliably predict user intent across all those intent classes. *CheerUp* and *Express* classes achieved perfect scores, which means the model classified all test samples in each of these categories correctly. The macro and weighted averages are both around 0.97, which shows that the model performs consistently across classes (Fig. 4).

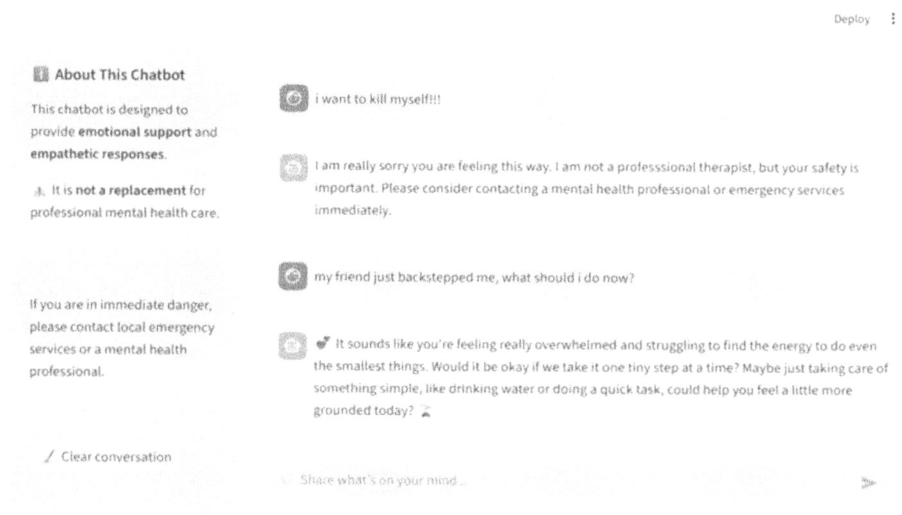

Fig. 2. Illustration of an empathetic chatbot interface designed to provide emotional support during user distress.

This low accuracy value for emotion classification is mainly due to the limitations of the TF-IDF feature representation and the logistic regression classifier.

TF-IDF assigns weights to words based on their frequency. It does not understand the emotional meaning or context of a sentence. As a result, different emotion categories often share similar vocabulary, such as sadness, anxiety, and neutral expressions, which makes them difficult for the model to differentiate clearly. The macro-average F1-score is 0.33, while the weighted-average F1-score is 0.36. These results indicate that the model performs better on emotion classes with more samples, while performance on minority classes is weaker. The report shows that the happy emotion has the highest performance, which may be due to clearer keywords and having more training samples. Meanwhile, the model performs poorly on emotions such as anger and tiredness, which may be due to fewer samples and higher lexical ambiguity.

Additionally, different emotions often use similar words; this makes them difficult to separate, which leads to misclassification. Overall, the challenge of emotion classification when using surface-level word patterns alone has been highlighted by these results.

The obtained confusion matrix also shows that most predictions fall on the diagonal axis, which means the model usually predicts the correct intent. This strong performance is also consistent with how the model is built. Naïve Bayes is especially suitable for this kind of task because intent messages often contain clear patterns and keywords like "What should I do?" for advice or "I feel..." for expression, so simple word-based probabilities can separate intents effectively. Overall, the intent model performs much better than the emotion model. With only four well-defined intent classes and clear language patterns, a TF-IDF with a Naïve Bayes approach is enough to achieve high accuracy in this system.

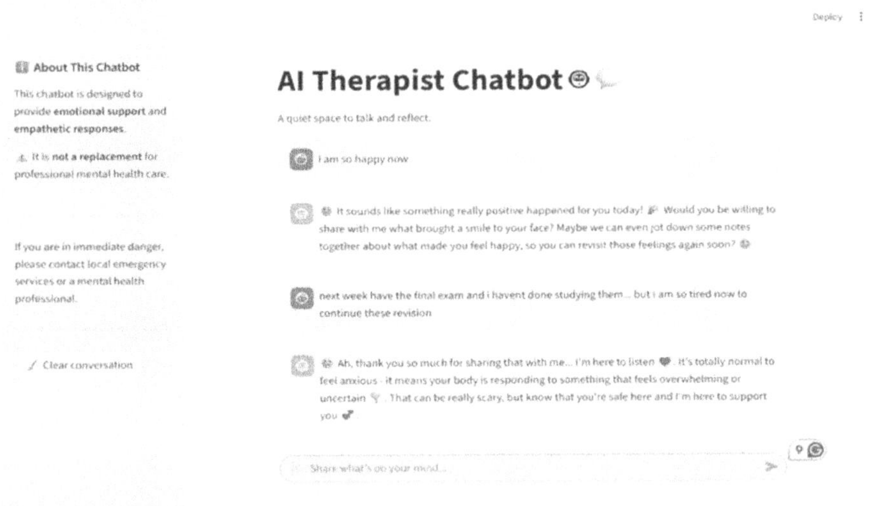

Fig. 3. Example of an AI therapist chatbot interface demonstrating empathetic dialogue. The system provides supportive responses to both positive and distress.

```
Accuracy: 0.96875

Classification Report:

              precision    recall  f1-score   support

      Advice       0.89      1.00      0.94        16
 Affirmation       1.00      0.88      0.93        16
     CheerUp       1.00      1.00      1.00        16
     Express       1.00      1.00      1.00        16

    accuracy                           0.97        64
   macro avg       0.97      0.97      0.97        64
weighted avg       0.97      0.97      0.97        64
```

Fig. 4. Accuracy report of intent classifications

5 Discussion

Despite the effectiveness of the system, some limitations remain. First, the emotion classification module relies on surface-level word patterns and does not understand deeper emotional context. As a result, emotions with similar vocabulary, such as sadness and anxiety, are often confused, especially when emotional expressions are subtle or indirect.

Besides, although class balancing was applied to reduce bias toward majority emotion classes, it cannot fully overcome the limitations of TF-IDF features. The model performance is still constrained by overlapping emotional language rather than the classifier itself.

Secondly, although the intent model achieved high accuracy, this performance is partly due to the use of the predefined and domain-specific dataset with clear intent categories. While this improves performance, the model may not generalize well to more diverse real-world conversations where intents are expressed in less predictable ways. Besides, the response engine uses the predefined dataset for chatbot responses, which also has limitations in not generalizing well with real-world conversations. Additionally, TF-IDF-based models rely heavily on word frequency and cannot generalize well to unseen expressions. Finally, although the system is efficient in combining multiple modules and utilizing a large language model for response rephrasing, the large language model rephrasing increases computational time, which may affect the scalability in larger deployments.

The limitations of this project can be addressed in future work through improving the capacity of the system to retrieve more comprehensive semantic and contextual information. The existing emotion classification is based on TF-IDF features, which emphasize the frequency of words on a surface level. The use of more complex and robust emotion representation approaches that can represent semantic similarity beyond exact lexical matches, including word embeddings, phrase-level representations, or language models, can be viewed as a way of better distinguishing between emotions that share common vocabulary and enhancing generalization to previously unseen expressions. This would enable the system to identify more implicit and subtle emotional cues other than lexical similarity. Moreover, the intent classification module that performs well on predefined and domain-specific data can be further improved by collaborating with professional mental health therapists to collect real-world case studies. By incorporating authentic user scenarios and therapist-approved responses into the database, the chatbot can better reflect real-life situations and provide more flexible and appropriate support.

6 Conclusion

Classical models of large language are generally powerful but often unpredictable and difficult to manage. When used in isolation, their outputs can be inconsistent and potentially risky in mental health applications. By combining large language models with emotion and intent detection, and constraining them to rephrase responses within fixed boundaries through safety checks, the system can provide safer and more controlled support for users. Although this chatbot is not designed to provide medical diagnoses or professional treatment, it may serve as an emotionally responsive conversational system that always accompanies users as a supportive companion. Additionally, the chatbot provides people with a private and accessible space to express their emotions, mitigate psychological stress, and receive emotional comfort. By offering this platform, mental health problems and emotional isolation in society can be significantly addressed and decreased to eliminate the possibility of extreme emotional consequences. In conclusion, this project aims to develop a therapist chatbot that utilizes statistical NLP methods to provide emotional support to people struggling with mental health problems.

References

1. Brown University: New study: AI chatbots systematically violate mental health ethics standards. Brown University (2025) https://www.brown.edu/news/2025-10-21/ai-mental-health-ethics
2. Ciriello, R., Hannon, O., Chen, A.Y., Vaast, E.: Ethical tensions in human-AI companionship: a dialectical inquiry into Replika. In: Proceedings of the Annual Hawaii International Conference on System Sciences (2024). https://doi.org/10.24251/hicss.2024.058
3. Elsaid, A.H., Khattab, M.A., Zawawi, N.S., Negied, N.K.: An ensemble learning approach for AI-powered Chatbot for mental health assessment and support. In: 2025 7th Novel Intelligent and Leading Emerging Sciences Conference (NILES), pp. 259–264 (2025). https://doi.org/10.1109/niles68063.2025.11232212
4. Gupta, P.: Emotion detection from text (2023). https://www.kaggle.com/datasets/pashupatigupta/emotion-detection-from-text
5. Kamdan, K., Fauziyah, N.G., Fadlullah, M.A., Hanif, D.A., Kharisma, I.L.: Early mental health detection and emotional states in teenagers through Chatbot systems using natural language processing (NLP). In: The 7th International Global Conference Series on ICT Integration in Technical Education & Smart Society, p. 64 (2025). https://doi.org/10.3390/engproc2025107064
6. Rahat, F.R. et al.: Data analysis using NLP to sense human emotions through Chatbot. In: Communications in Computer and Information Science, pp. 64–75 (2022). https://doi.org/10.1007/978-3-031-12641-3_6
7. Vaccaro, A., Antonelli, G., Fioravante, R.: Dealing with AI: loneliness, manipulations and suicides (2025). https://hbsp.harvard.edu/product/I00105-PDF-ENG. Case Study, Product Number I00105-PDF-ENG
8. Wells, S.: Exploring the dangers of AI in mental health care. Stanford HAI14 (2025). https://hai.stanford.edu/news/exploring-the-dangers-of-ai-in-mental-health-care
9. Wells, S., Haber, N., Moore, J.: New study warns of risks in AI mental healthtools. Stanford Report (2025). https://news.stanford.edu/stories/2025/06/ai-mental-health-care-tools-dangers-risks. Originally published by the Stanford Institute for Human-Centered AI

Automated SNOMED CT Mapping from Unstructured Indonesian Clinical Narrative Using Large Language Models: A Two-Level Information Retrieval Framework with Clinical Context Validation

Putu Gede Purwa Darmawan[1(✉)], Dian Kesumapramudya Nurputra[1], Annisa Ristya Rahmanti[1(✉)], and Nindya Widita Ayuningtyas[2]

[1] Department of Health Policy and Management, Faculty of Medicine, Public Health and Nursing, Universitas Gadjah Mada, Yogyakarta, Indonesia
putugedepurwadarmawan@mail.ugm.ac.id, annisaristya@ugm.ac.id
[2] Center for Health Policy and Management Faculty of Medicine, Public Health and Nursing, Universitas Gadjah Mada, Yogyakarta, Indonesia

Abstract. The national health interoperability platform SATUSEHAT mandates SNOMED CT adoption across Indonesian healthcare facilities, yet approximately 80% of electronic health record data remains in unstructured free-text narratives written in Bahasa Indonesia, making manual mapping labor-intensive. This study evaluated four large language models (LLMs): Gemini 3 Flash, GPT-4o-mini, LLaMA 3.1 8B, and Gemma 3 4B for automated SNOMED CT mapping from 395 fully anonymized pediatric outpatient anamnesis narratives selected using stratified random sampling. A gold standard of 1,207 clinical tuples was established through LLM-assisted sequential dual annotation with two independent expert reviews ($\kappa = 0.877$). This study defines a two-level Information Retrieval (IR) evaluation framework: Level 1 (Relaxed), assesses lexical code accuracy, and Level 2 (Strict Tuple), assesses code-plus-context accuracy with a Double Penalty mechanism for context mismatch. Gemini 3 Flash achieved the best overall clinical performance ($F1_2 = 84.08\%$, $\Delta F1 = 1.28\%$). GPT-4o-mini achieved the highest precision (84.22%) with perfect output calibration (1,207 entries), optimal for billing applications. Open-source models showed frequent errors in identifying clinical negation, with denied symptoms often misclassified as present findings. Gemma 3 4B showed the largest degradation ($\Delta F1 = 14.67\%$) with 181 context mismatches, of which 81% absent findings were incorrectly classified as present. LLaMA 3.1 8B exhibited extreme latency variability (CV = 112%, max 120.4 s) constituting an independent deployment disqualifier for real-time HMIS integration. This study demonstrates that single-level evaluation can systematically overestimate clinical performance by up to 14.67 percentage points, masking clinically significant context errors. A Human-in-the-Loop architecture is recommended for all models pending domain-specific fine-tuning for Indonesian clinical negation patterns.

Keywords: SNOMED CT · Large Language Models · Medical Entity Linking · Clinical Natural Language Processing · Information Retrieval · Negation

N. Mohd Sharef et al. (Eds.): CAMP 2026, CCIS 3094, pp. 17–31, 2027.
https://doi.org/10.1007/978-981-92-4305-1_2

Detection · Bahasa Indonesia · SATUSEHAT · Few-Shot Learning · Human-in-the-Loop

1 Introduction

Clinical terminology standardization is a prerequisite for semantic interoperability in modern healthcare systems. SNOMED CT (Systematized Nomenclature of Medicine Clinical Terms), encompassing over 350,000 active clinical concepts, is one of the core mandated terminology standards for clinical data exchange within Indonesia's national health interoperability platform, SATUSEHAT [1, 2]. This mandate is formalized through Minister of Health Regulation No. 24/2022 [3].

Despite the rapid adoption rates of EHR systems reaching 96% in some leading health systems [4], nearly 80% of healthcare data is still documented as unstructured free-text [5]. In Indonesian clinical practice, clinicians commonly use informal Bahasa Indonesia, characterized by non-standard abbreviations, regional expressions, telegraphic sentence structures, and symbolic negation such the minus sign (−) to indicate symptom absence. In addition, Indonesia does not yet have an official localized SNOMED CT edition [6]. These conditions make manual clinical concept mapping labor-intensive, inconsistent, and difficult to scale across healthcare facilities.

Large language models (LLMs) offer a transformative opportunity to automate clinical concept extraction and terminology mapping. Recent studies have shown GPT-4 correctly answers 86.6% of USMLE questions [7, 8], while preliminary evidence suggests that LLMs can substantially reduce manual effort in medical vocabulary mapping [9]. However, most SNOMED CT mapping studies are focused on English-language clinical text [10, 11] and prior evaluations only assess whether a SNOMED CT code is correctly identified, without validating the clinical context of the finding (present vs. absent) [12]. This evaluation gap, namely, context-blind assessment, is not merely a methodological limitation. When a model correctly identifies Seizure (SCTID 91175000) but incorrectly classifies it as Present despite the patient explicitly denying the condition, the error remains entirely undetected by conventional precision and recall metrics, yet carries direct clinical consequences including inappropriate prescriptions and misleading decision support. No existing framework imposes explicit penalties for such context inversions, leaving the true clinical reliability of SNOMED CT mapping systems systematically overestimated. A systematic search following Chang and Sung [10] found no prior study evaluating LLM-based SNOMED CT mapping from Indonesian clinical text. This study addresses this gap through: (1) the first systematic evaluation for SATUSEHAT; (2) a two-level IR evaluation framework with Double Penalty for context mismatch; and (3) a systematic comparison of commercial and open-source LLMs.

2 Background and Related Work

2.1 SNOMED CT and Clinical Terminology in Indonesia

SNOMED CT provides a polyhierarchical ontology organizing clinical concepts through 19 top-level hierarchies, of which the Clinical Findings hierarchy is exclusively targeted in this study, encompasses symptoms, signs, and diagnoses [1, 2]. Its SATUSEHAT implementation is governed by the Ministry of Health's official Terminology Appendix, a curated Indonesian subset. Implementation experiences at Indonesian tertiary hospitals reveal that manual SNOMED CT mapping requires interdisciplinary teams working for up to five years [13], underscoring the need for automated solutions.

2.2 LLMs for Medical Entity Linking

Medical entity linking, identifying clinical concepts in free text and mapping to ontology identifiers, has evolved from rule-based [14] through machine learning [15] to transformer-based methods [16]. ClinLinker [17] achieved 87,1% F1 for Spanish SNOMED CT linking; SNOBERT [18] demonstrated domain-specific pre-training benefits for English clinical text; Chen et al. [19] achieved accuracy of 90.5% and an F1-score of 64.5% for Chinese clinical terms using attention-based semantic matching. MapperGPT [20] showed GPT-class models can perform competitive entity linking without task-specific fine-tuning. No prior study has evaluated SNOMED CT mapping from Indonesian clinical text, nor developed frameworks explicitly penalizing clinical context errors [10].

2.3 Clinical Negation Detection in Indonesian Clinical Text

Negation detection is one of the most challenging and clinically important tasks in clinical NLP [21]. Indonesian clinical documentation expresses negation through explicit negators ("*tidak ada*", "*disangkal*"), symbolic notations ("-","(−)"), implicit constructions ("*dalam batas normal*"), and passive negation ("*disangkal pasien*"). These patterns are largely absent from existing multilingual LLM pre-training corpora and are rarely presented in standard NLP resources for Bahasa Indonesia [22]. Negation scope errors are consistently identified as the primary source of semantic inaccuracy in clinical entity extraction systems [21, 23].

2.4 Evaluation Frameworks for Clinical LLMs

Previous studies evaluating LLM performance for SNOMED CT mapping have predominantly adopted single-level lexical evaluation frameworks, measuring whether a model generates the correct terminology code without assessing the clinical context, specifically whether a finding is affirmed or denied. Prior studies evaluating SNOMED CT mapping including transformer-based pipelines [11], multilingual concept extraction frameworks [23], and GPT-class models such as MapperGPT [20] have uniformly adopted code-level accuracy metrics without distinguishing whether identified findings were affirmed or denied by the patient. A model that correctly identifies "Fever" (SCTID 386661006) but classifies it as present when the narrative states fever is absent introduces a clinically dangerous error indistinguishable from a true positive under standard recall metrics. Chang and Sung's scoping review [10] of 37 SNOMED CT-LLM integration studies identified this as a consistent methodological limitation. Critically, this gap reflects a broader pattern in clinical NLP evaluation where code-level accuracy serves as a proxy for clinical utility. Seo et al. [21] demonstrated that negation detection errors propagate systematically into downstream NLP tasks, yet no SNOMED CT evaluation framework has incorporated negation-aware penalties. The absence of such a penalty mechanism means that a model misclassifying 40% of negated findings as affirmed would appear indistinguishable from a model with perfect negation handling under single-level evaluation, a risk directly observed in this study.

2.5 Commercial vs. Open-Source LLMs in Healthcare

Cloud-based commercial LLM APIs require transmitting potentially re-identifiable patient data to third-party servers, creating compliance challenges under Indonesia's Personal Data Protection Law (UU No. 27/2022) and international frameworks [24, 25]. Locally deployable open-source models offer a privacy-preserving alternative but their performance gap for non-English clinical NLP remains insufficiently characterized [26]. This study provides the first systematic comparison for Indonesian clinical SNOMED CT mapping.

3 Methods

3.1 Study Design

This computer-based experimental study evaluated four LLMs for automated SNOMED CT mapping from Indonesian pediatric outpatient EHR narratives through four phases: (1) data collection and irreversible anonymization; (2) gold standard establishment with inter-rater reliability assessment; (3) LLM inference via few-shot prompt engineering; and (4) two-level IR evaluation. Ethical clearance was obtained from the Ethics Committee of the Faculty of Medicine, Public Health and Nursing, Universitas Gadjah Mada (Protocol No. KE/FK/1895/EC/2025).

3.2 Dataset

We extracted 395 fully anonymized clinical narratives (January–December 2025) from the outpatient EHR systems of Surya Husadha Hospital, Bali, Indonesia. From a population of 31,060 records, 395 record sample were retained after exclusion screening (administrative entries, non-clinical templates, referral-only records). Anonymization removed all direct identifiers prior to LLM processing. It should be noted that the dataset is restricted to pediatric outpatient narratives from a single private hospital network in Bali; findings may not generalize to adult or inpatient populations, or to clinical documentation styles from other Indonesian regions. Full dataset and gold standard characteristics are presented in Table 1.

Table 1. Comprehensive Dataset and Gold Standard Profile

Characteristic	Details
Data source	Surya Husadha Hospital (Denpasar, Nusa Dua, Ubung units), Bali
Collection period	January – December 2025

(continued)

Table 1. (*continued*)

Characteristic	Details
Total population	31,060 outpatient records
Final sample	395 fully anonymized narratives
Sampling method	Slovin's formula + stratified random sampling)
Patient population	Pediatric outpatient
Language register	Informal Bahasa Indonesia (mixed formal/colloquial, symbolic negation)
Narrative length	Median 60 characters (range 8–250)
Length distribution	Very short: 10.4% ǀ Short: 39.7% ǀ Medium: 31.9% ǀ Long: 18.0%
Top diagnostics	Respiratory (228) ǀ Asthma/Allergic (82) ǀ Gastrointestinal (51)
Gold Standard	
Total annotated tuples	1,207 clinical tuples
Unique codes	113 SNOMED CT codes (Clinical Findings hierarchy)
Inter-rater reliability	Cohen's $\kappa = 0.877$ (Almost Perfect Agreement)
Context distribution	77.5% Present (n = 936) / 22.5% Absent (n = 271)
Most frequent terms	Fever (20.5%) ǀ Cough (18.5%) ǀ Nasal discharge (16.1%)
High-Absent concepts	Bloody stool (78%) ǀ Seizure (71%) ǀ Fever (42%) ǀ Cough (15%)
Codes per patient	Median 3 (range 1–8; mean 3.06, SD 1.36)

3.3 Gold Standard and Inter-Rater Reliability

Gold standard development followed an LLM-assisted sequential annotation protocol to improve efficiency without sacrificing clinical validity. Two clinical experts participated in the annotation process: Expert 1, a general physician with health informatics expertise, and Expert 2, an academic health informatician with international SNOMED CT certification. LLM outputs were used solely as an initial candidate list to reduce annotation burden; Expert 1 performed a critical review of each candidate directly against the raw narrative text, conducting comprehensive curation that included deletion of incorrect entities, addition of omitted findings, and correction of SNOMED CT mappings using the SATUSEHAT Terminology Appendix as the primary reference (SNOMED CT International Browser as secondary), and assigned clinical context labels (Present or Absent) for each entity. Expert 2 then acted not merely as a validator but as a fully independent second annotator, conducting a direct critical review of the raw notes to add omitted entities, remove incorrect ones, and correct mapping ensuring absolute clinical consistency and terminological accuracy. To prevent anchoring bias, Expert 2's review was performed directly on the source text, not on Expert 1's annotations. The SNOMED CT scope was restricted to the Clinical Findings hierarchy per SATUSEHAT guidelines.

Of 1,207 total annotated tuples, 1,133 (93.87%) received direct agreement. The remaining 74 (6.13%) were resolved through facilitated consensus, yielding 100% final agreement. Inter-rater reliability was quantified as Cohen's $\kappa = (0.9387–0.50)/(1–0.50) = 0.877$, indicating Almost Perfect Agreement [27], commensurate with comparable SNOMED CT entity linking studies ($\kappa = 0.72–0.89$) [17, 18]. The gold standard comprised 1,207 tuples across 113 unique SNOMED CT codes (77.5% Present, 22.5% Absent).

3.4 LLM Selection and Configuration

Four LLMs were selected to span a commercial-to-open-source deployment spectrum: (1) Gemini 3 Flash (Google, 2025), via official API; (2) GPT-4o-mini (OpenAI, 2024), via OpenAI API; (3) LLaMA 3.1 8B (Meta AI, 2024), open-source model deployed locally simulating an on-premise hospital server; (4) Gemma 3 4B (Google, 2025), open-source compact multilingual model deployed locally.

All models received identical few-shot prompt templates structured with: (i) a context declaration establishing the model as an experienced medical coder; (ii) a task specification instructing extraction of clinical finding entities with SNOMED CT code and Present/Absent context assignment; and (iii) three concrete few-shot examples demonstrating the expected structured output format. No model-specific prompt optimization was applied and all models were executed using default inference settings

3.5 Pre-Processing Pipeline

Raw model outputs were standardized before evaluation using two preprocessing steps: (1) Clinical Deduplication, repeated symptom expressions within the same narrative (e.g., "*batuk batuk*") were normalized into a single entity; (2) Non-Finding Exclusion, entities corresponding to laboratory values or temporal descriptors were excluded.

The full pipeline is illustrated in Fig. 1.

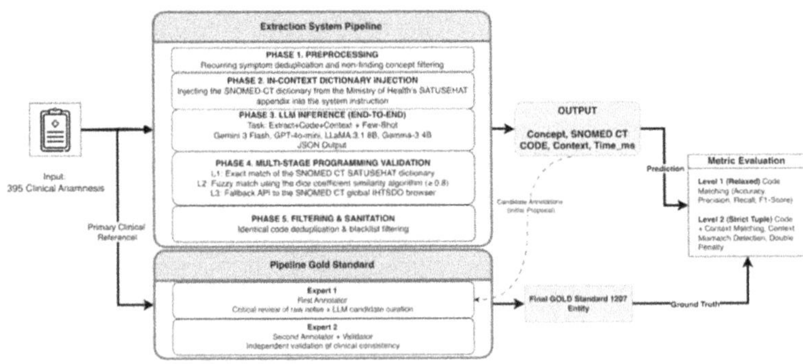

Fig. 1 Extraction pipeline Gold Standard pipeline, and two-level evaluation framework for SNOMED CT mapping.

3.6 Two-Level Evaluation Framework

3.6.1 Level 1: Relaxed (Terminology Extraction)

Level 1 treats mapping as a per-patient set operation comparing model-generated SNOMED CT codes against gold standard codes regardless of clinical context. True Positives (TP_1) are codes present in both sets; False Positives (FP_1) are model-generated codes absent from the gold standard (hallucinations); False Negatives (FN_1) are gold standard codes not generated by the model (omissions). Standard Precision, Recall, F1-Score, and Accuracy metrics are computed from these counts.

3.6.2 Level 2: Strict Tuple (Clinical Intent)

Level 2 evaluates each code-matched pair for context accuracy. A tuple (Patient, Code, Context) is TP_2 only when code AND context exactly match the gold standard. Context mismatches incur a Double Penalty, analogous to dual-penalization approaches in information retrieval evaluation [15] where partial matches satisfying one criterion while violating another are penalized on both dimensions:

$$\begin{aligned} &\textit{Context Mismatch} \\ &\qquad \Rightarrow FP^2 + 1(\textit{misleading clinical claim})\ AND\ FN_2 \\ &\qquad + 1(\textit{target not fulfilled}) \end{aligned} \tag{1}$$

$$FP_2 = \textit{Hallucinations} + \textit{Context Mismatches}; \tag{2}$$

$$FN_2 = \textit{Omissions} + \textit{Context Mismatches} \tag{3}$$

Precision, Recall, F1-Score, and Accuracy at Level 2 are computed using the same formulas as Level 1 applied to these redefined FP_2 and FN_2 counts. This formulation ensures $TP_2 + FN_2$ always equals total gold standard entities, guaranteeing R_2 reflects true clinical data integrity. $\Delta F1 = F1_1 - F1_2$ quantifies performance degradation attributable to context errors. All computations used Python 3.12 with pandas and custom evaluation scripts.

Table 2. Illustrative Comparison of Level 1 and Level 2 Evaluation on the Same Clinical Record

Narrative	Clinical entity	SCTID	Model output context	Gold standard context	L1 verdict	L2 verdict
"batuk pilek ada sesak – demam –"[a]	Cough	49727002	Present	Present	TP_1	TP_2
	Nasal discharge	64531003	Present	Present	TP_1	TP_2
	Dyspnea	267036007	Present	Absent	TP_1	CM → FP_2 + FN_2
	Fever	386661006	Present	Absent	TP_1	CM → FP_2 + FN_2

Under L1, all four codes are correctly identified (F11 = 100% for thisrecord). Under L2, two context mismatches (Dyspnea and Fever classified asPresent instead of Absent) each incur Double Penalty, reducing F12 to 50%.

4 Results

This section reports experimental results across two dimensions. First, model output profile and response time characteristics are examined to assess operational suitability for real-time HMIS deployment. Second, performance is evaluated at both levels of the IR framework, Level 1 (Relaxed) measuring lexical code accuracy and Level 2 (Strict Tuple) measuring clinical context accuracy, with Delta-F1 quantifying the clinical cost of context-blind evaluation for each model. All four models were evaluated against the same 1,207-tuple gold standard across 395 narratives under identical conditions

4.1 Model Output Profile

The four LLMs showed substantial differences in output volume and generation patterns. GPT-4o-mini generated exactly 1,207 entries, demonstrating perfect output calibration. Gemini 3 Flash produced 1,329 (+10.1%), Gemma 3 4B 1,290 (+6.9%), and LLaMA 3.1 8B 1,634 (+35.4%), indicating progressive over-generation. Mean response times: Gemma 3 4B 5.1 s, GPT-4o-mini 6.3 s, Gemini 3 Flash 10.2 s, LLaMA 3.1 8B 13.4 s. LLaMA 3.1 8B exhibited extreme variability (range 0.58–120.4 s, CV = 112%), exceeding the 30-s HMIS API timeout threshold, constituting an independent deployment disqualifier irrespective of F1 performance (Fig. 2).

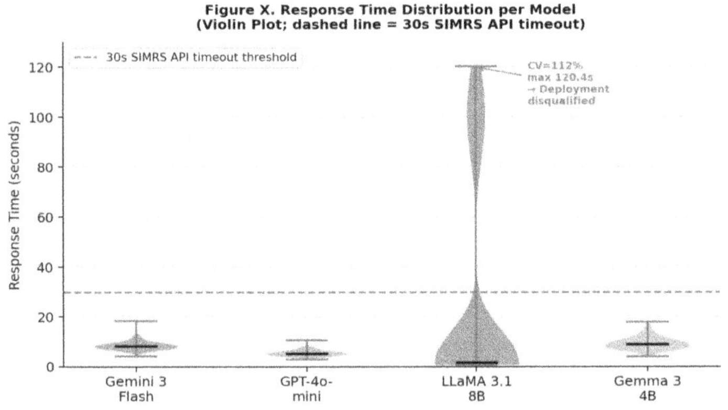

Fig. 2. Response Time Distribution per Model

4.2 Level 1 and Level 2 Performance

The two-level evaluation framework revealed substantial performance differences across models and evaluation levels, with ΔF1 serving as the primary indicator of clinical context reliability.

Table 3. Comprehensive performance metrics for all LLMs at Level 1 (Relaxed) and Level 2 (Strict Tuple)

Model	Level	TP	FP	FN	CM	P (%)	R (%)	F1 (%)	Acc (%)	ΔF1 (%)
Gemini 3 Flash	L1[a]	1,070	232	135	–	82.18	88.80	85.36	74.46	
	L2[b]	1,054	248	151	16	80.95	87.47	84.08	72.54	1.28
GPT-4o-mini	L1	998	187	207	–	84.22	82.82	83.51	71.7	
	L2	950	235	255	48	80.17	78.84	79.49	65.97	4.01
Gemma 3 4B	L1	940	322	265	–	74.48	78.01	76.20	61.56	
	L2	759	503	446	181	60.14	62.99	61.53	44.44	14.67
LLaMA 3.1 8B	L1	927	602	278	–	60.63	76.93	67.81	51.3	
	L2	830	699	375	97	54.28	68.88	60.72	43.59	7.08

[a]L1: Level 1 (Relaxed); [b]L2: Level 2 (Strict Tuple)

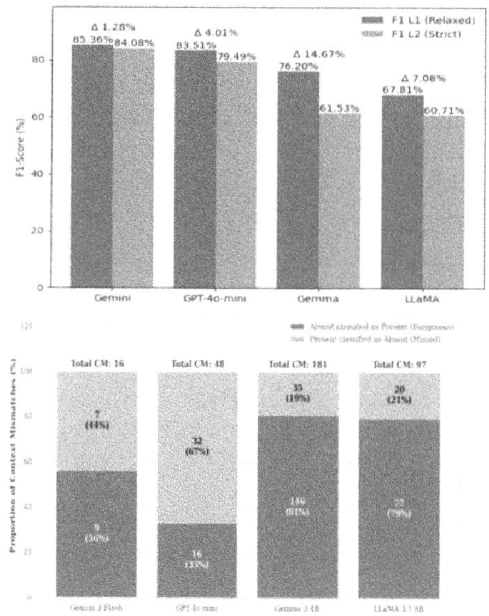

Fig. 3. (a). Performance Comparison Across Two Evaluation Levels and (b) Context Mismatch Direction per Model

Table 3 presents comprehensive metrics for all models at both levels including absolute TP, FP, FN counts, standard performance metrics, and Context Mismatch count enabling direct inspection of the Double Penalty mechanism applied at Level 2. At Level 1, Gemini 3 Flash achieved the highest $F1_1$ (85.36%) and R_1 (88.80%); GPT-4o-mini achieved the highest P_1 (84.22%); LLaMA 3.1 8B recorded the lowest P_1 (60.63%) with 602 false positives. At Level 2, ΔF1 ranged from 1.28 percentage points (Gemini

3 Flash) to 14.67 percentage points (Gemma 3 4B). Context Mismatch counts were 16 (Gemini 3 Flash), 48 (GPT-4o-mini), 181 (Gemma 3 4B), and 97 (LLaMA 3.1 8B).

Figure 3 illustrate F1 degradation magnitude (a) and context mismatch directionality respectively (b).

5 Discussion

5.1 Clinical Importance of Context-Aware Evaluation

The main finding of this study is that the predominant approach in existing literature, which evaluates only whether the correct SNOMED CT code is identified, may provide an overconfident estimate of clinical utility while masking potentially dangerous errors [10, 17, 18]. This finding has direct patient safety implications, as context misclassification (e.g., absent-as-present) may lead to inappropriate clinical decision-making. Performance degradation from Level 1 to Level 2 ranged from 1.28 to 14.67 percentage points across models, a margin with direct patient safety implications. The Double Penalty mechanism is mathematically justified by its property that $TP_2 + FN_2$ always equals total gold standard entities, ensuring R_2 genuinely reflects clinical data completeness. A context mismatch represents two simultaneous failures: a false positive assertion (misinformation generated) and a false negative omission (true clinical information not captured), violating both non-maleficence and beneficence principles [12].

To illustrate concretely: consider the clinical narrative 'batuk pilek ada sesak - demam -' (cough present, nasal discharge present, dyspnea absent, fever absent). A model generating SNOMED CT codes for all four entities achieves perfect Level 1 performance, all codes are correctly identified (F11 = 100%). However, if the model classifies Dyspnea (267036007) and Fever (386661006) as Present rather than Absent, Level 2 evaluation reveals two context mismatches, each incurring a Double Penalty (Table 2). The model receives 2 TP2 but 2 FP2 and 2 FN2, F12 drops to 50% for this record. In clinical deployment, this error would appear as two affirmed findings in the patient record, potentially triggering unwarranted clinical investigation or pharmacological intervention, while standard evaluation would have reported this record as perfectly coded

5.2 Commercial Models: Gemini 3 Flash and GPT-4o-Mini Performance Profiles

Gemini 3 Flash and GPT-4o-mini demonstrate clinically deployable performance with distinct profiles aligned to different use-cases. Gemini 3 Flash achieved $F1_2 = 84.08\%$ ($\Delta F1 = 1.28\%$, 16 context mismatches). Its balanced mismatch direction (56% Absent-as-Present vs. 44% Present-as-Absent) indicates random rather than systematic error, a safer profile for clinical deployment. Combined with the highest Recall ($R_1 = 88.80\%$), it is optimally positioned for Clinical Decision Support System (CDSS) applications [11].

GPT-4o-mini's highest P_1 (84.22%) and exact output calibration (1,207 entries = gold standard) make it uniquely suited for BPJS billing code generation, where false positives carry financial and legal consequences [13]. Its exact output volume also eliminates structural anomalies in HMIS database integration. However, its systematic under-detection bias (67% of context mismatches: Present-as-Absent) requires monitoring in complete-symptom-documentation applications. Taken together, the two models present a decision framework: Gemini 3 Flash for clinical completeness-driven applications and GPT-4o-mini for precision-driven billing both still requiring Human-in-the-Loop verification.

5.3 Open-Source Models and Indonesian Clinical Negation Gap

Both open-source models showed weaker performance in handling clinical negation, particularly when denied symptoms were incorrectly recorded as present findings. Gemma 3 4B showed the largest performance degradation (ΔF1 = 14.67%, n = 181 context mismatches, 81% Absent-as-Present). LLaMA 3.1 8B recorded 97 context mismatches (79% Absent-as-Present), the lowest precision (P_1 = 60.63%), and the highest over-generation (1,634 outputs; +35.4%). These findings reveal a systematic deficiency in Indonesian clinical negation processing, consistent with the broader low-resource clinical NLP literature [21, 22], and represent its first documentation for Indonesian text.

The gap is driven by three intersecting linguistic features: (1) symbolic telegraphic negation (−) the trailing minus sign (*"demam –"*) is pervasive in Indonesian outpatient settings yet virtually absent from general-domain corpora; (2) colloquial abbreviations (*"nafsu mami N", "btk"*) lacking canonical NLP lexicon forms; and (3) passive negation constructions (*"disangkal pasien"*) requiring multi-token dependency resolution. Critically, failure mechanisms differ: LLaMA 3.1 8B ignores negation markers entirely; Gemma 3 4B misinterprets implicit negation constructs. This distinction requires model-specific remediation strategies. LLaMA 3.1 8B's extreme latency variability (CV = 112%) additionally constitutes an independent deployment disqualifier.

5.4 Human-in-the-Loop Implementation Imperative

Even Gemini 3 Flash produced 16 context mismatches across 395 patients (4.1% affected). At SATUSEHAT's scale of 11,864+ connected facilities, such errors would accumulate significantly. A Human-in-the-Loop (HITL) architecture, LLM as coding assistant, medical records personnel as verifier, is strongly recommended for initial deployment [30, 31]. HITL verification should be prioritized for: (1) clinical narratives >100 characters (2) SNOMED concepts with high Absent proportions (Seizure 71%, Bloody stool 78%); and (3) complex negation patterns using symbolic notation ('−', '(−)') or passive constructions. A phased rollout is recommended: initiate at 5–10 national referral hospitals before broader SATUSEHAT-wide deployment. In parallel, investment in an Indonesian Medical NLP Benchmark a publicly available, annotated corpus covering diverse clinical document types, regional language variations, and SNOMED CT hierarchies beyond Clinical Findings would provide the research community with a standardized evaluation resource, accelerating the development of domain-specific models tailored to Indonesian healthcare.

5.5 Qualitative Analysis of Model Errors

Three primary error typologies were identified. Hallucinations occurred when models extracted concepts not present in the narrative, for example, given *"tidak ada lendir dan darah"* (no mucus or blood in stool), GPT-4o-mini generated code 8765009 (hematemesis), failing to synthesize negation scope across both tokens. Omissions stemmed from idiosyncratic abbreviations; *"nafsu mami N"* (normal appetite) caused all open-source models to miss concept (161825005). Context Mismatches, the most clinically dangerous category, resulted from failure to parse symbolic negation: in *"batuk pilek ada sesak – demam –"*, both GPT-4o-mini and LLaMA 3.1 8B classified dyspnea (267036007) as Present, entirely failing to parse the trailing minus sign as a negation marker.

5.6 Comparison with Related Work

Our Level 1 F1-scores (Gemini 85.36%, GPT-4o-mini 83.51%) are competitive with state-of-the-art non-English SNOMED CT entity linking: ClinLinker [17] achieved ≈87% for Spanish using fine-tuned bi-encoders; Chen et al. [19] achieved ≈83% for Chinese using task-specific attention mechanisms. Achieving comparable performance through zero domain-specific fine-tuning purely via few-shot prompting in a low-resource language demonstrates exceptional LLM generalization [29]. This finding extends Matentzoglu et al.'s [20] findings on LLM entity linking generalization to the underexplored domain of Southeast Asian clinical NLP. GPT-4o-mini (6.3 s) and Gemini 3 Flash (10.2 s) also compare favorably with Dagli et al. [28], who reported 34–39 s per record for complex spinal surgery notes.

5.7 Optimization Strategies and Future Research Directions

The findings highlight five interrelated research priorities. First, a publicly accessible Indonesian Medical NLP Benchmark dataset with comprehensive negation pattern coverage is the most foundational prerequisite. Second, open-source model fine-tuning must be model-specific: LLaMA 3.1 8B requires explicit negation token coverage in instruction-tuning data; Gemma 3 4B requires broader implicit and symbolic negation surface form coverage, a single strategy cannot address both. Third, prospective evaluation of HITL deployment is needed to assess real-world verification workload, claim accuracy, and clinician acceptance. Fourth, the two-level IR framework should be extended to other SNOMED CT hierarchies (Procedures, Substances, Observable Entities).

Fifth, and most immediately actionable, is empirical testing of In-Context Dictionary Injection (ICDI), a resource-efficient prompt-engineering strategy targeting the specific error typologies documented here without full fine-tuning. ICDI comprises three mechanisms: (1) Colloquial Synonym Expansion, augmenting the SATUSEHAT dictionary with local abbreviation metadata (e.g., *"btk"* → Cough SCTID 49727002) to suppress omission errors; (2) Explicit Negation Anchoring, adding rigid prompt rules mandating Absent classification for entities preceded by *"tidak"*, *"disangkal"*, or followed by '–' or '(–)', directly targeting the Absent-as-Present bias in Gemma 3 4B (81%) and LLaMA 3.1 8B (79%); and (3) Dynamic Contextual Retrieval (Mini-RAG), inserting only the 10–20 most contextually relevant SNOMED codes per narrative to mitigate context-window overload driving LLaMA's over-generation. Controlled ablation studies evaluating each mechanism against the gold standard established here would provide a scalable roadmap for resource-limited Indonesian healthcare institutions.

6 Limitations

This study has several limitations. First, as noted in Sect. 3.2, the dataset was limited to pediatric outpatient anamnesis narratives from a single private hospital network in Bali, which may limit generalizability to adult/inpatient populations or other regions in Indonesia where documentation styles and language use may differ. Second, while inter-rater reliability was excellent ($\kappa = 0.877$), the gold standard was developed using an LLM-assisted sequential annotation process in which model outputs were used as initial candidates lists before expert review. A fully de-novo annotation process with multiple independent annotators would further strengthen gold standard validity. Third, SNOMED CT code matching was based on exact code matching. In cases where models selected semantically similar but non-identical concepts, these were counted as incorrect, which may underestimate practical clinical performance. Fourth, the evaluation was restricted to the Clinical Findings hierarchy following SATUSEHAT implementation guidelines. Other hierarchies such as procedures, medications, and substances were not included. Fifth, model performance was assessed at a single point in time and may change as commercial and open-source models continue to receive updates. Finally, API cost analysis, a critical factor for resource-limited settings, was not performed and warrants future investigation.

7 Conclusions

Automated SNOMED CT mapping from unstructured Indonesian clinical narratives using LLMs is technically feasible and has strong potential to support terminology standardization within SATUSEHAT interoperability system. Gemini 3 Flash provides the best overall clinical utility ($F1_2 = 84.08\%$, $\Delta F1 = 1.28\%$), while GPT-4o-mini is optimal for precision-critical billing applications ($P_1 = 84.22\%$). Open-source models exhibit critical deficiencies in Indonesian clinical negation detection that must be addressed through targeted fine-tuning before deployment.

Our primary methodological contribution is the two-level IR evaluation framework with Double Penalty for context mismatch, exposing a systematic overconfidence of up to 14.67 percentage points in single-level evaluations. The $\Delta F1$ metric provides a novel, interpretable measure of model clinical context consistency applicable to any clinical NLP task where affirmed and denied conditions must be distinguished. Future priorities include: (1) development of Indonesian clinical NLP benchmark datasets with comprehensive negation pattern coverage; (2) model-specific fine-tuning for open-source LLMs; (3) prospective evaluation of HITL; (4) extension of the evaluation framework to other SNOMED CT hierarchies; and (5) empirical testing of the ICDI optimization strategies.

Acknowledgments. This work was supported by *Sahabat Tanoto Pascasarjana 2025* program and the *Pasca Doctoral* - Community Funding Grant (No. 3567/UN1/KU.1.3/PP/PT.01.01/2026) from Faculty of Medicine, Public Health, and Nursing, Universitas Gadjah Mada.

Competing Interests. The authors declare no competing interests.

Ethics Approval. This study was performed in line with the principles of the Declaration of Helsinki. Ethical clearance was granted by the Ethics Committee of the Faculty of Medicine, Public Health and Nursing, Universitas Gadjah Mada (Protocol No. KE/FK/1895/EC/2025). Informed consent was waived as all data were fully and irreversibly anonymized prior to any LLM processing through removal of all direct patient identifiers without retention of a re-identification key.

References

1. SNOMED International: SNOMED CT Implementation Guide for Healthcare Providers. SNOMED International, London (2025)
2. Donnelly, K.: SNOMED CT: the advanced terminology and coding system for eHealth. Stud. Health Technol. Inform. **121**, 279–290 (2006)
3. Kementerian Kesehatan, R.I.: Peraturan Menteri Kesehatan No. 24 tahun 2022 tentang rekam medis elektronik. Kemenkes RI, Jakarta (2022)
4. Hossain, E., Rana, R., Higgins, N., Soar, J., Barua, P.D., Pisani, A.R., et al.: Natural language processing in electronic health records in relation to healthcare decision-making: a systematic review. Comput. Biol. Med. **155**, 106649 (2023). https://doi.org/10.1016/j.compbiomed.2023.106649
5. Assale, M., Dui, L.G., Cina, A., Seveso, A., Cabitza, F.: The revival of the notes field: leveraging the unstructured content in electronic health records. Front. Med. **6**, 66 (2019). https://doi.org/10.3389/fmed.2019.00066
6. Ayuningtyas, N.W., Mufidah, P.K., Yusuf, S., Sitompul, T., Susanti, M.I.: Bringing interoperability into action: adoption of health terminology standard in Indonesia. Stud. Health Technol. Inform. **329**, 83–87 (2025). https://doi.org/10.3233/SHTI250806
7. Nori, H., King, N., McKinney, S.M., Carignan, D., Horvitz, E.: Capabilities of GPT-4 on medical challenge problems. arXiv. (2023). https://doi.org/10.48550/arXiv.2303.13375
8. Singhal, K., Azizi, S., Tu, T., et al.: Large language models encode clinical knowledge. Nature. **620**(7972), 172–180 (2023). https://doi.org/10.1038/s41586-023-06291-2
9. Vanzin, V.J.D.B., Moreira, D.D.A., Marcacini, R.M.: LLM-based approaches for automated vocabulary mapping between SIGTAP and OMOP CDM concepts. Artif. Intell. Med. **168**, 103204 (2025)
10. Chang, E.K., Sung, C.: Use of SNOMED CT in large language models: scoping review. JMIR Med. Inform. **12**, e52052 (2024). https://doi.org/10.2196/52052
11. Gaudet-Blavignac, C., Foufi, V., Bjelogrlic, M., Lovis, C.: Use of the systematized nomenclature of medicine clinical terms (SNOMED CT) for processing free text in health care: systematic scoping review. J. Med. Internet Res. **23**(1), e24594 (2021). https://doi.org/10.2196/24594
12. Reddy, S.: Ethical considerations in AI-based clinical coding. NPJ Digit. Med. **6**(1), 78 (2023). https://doi.org/10.1038/s41746-023-00823-8
13. Kurniasih, S.L.D., Sukawan, A.: Penerapan SNOMED CT, LOINC dalam rekam medis elektronik di RSUD Prof. Dr. Margono Soekarjo Purwokerto. Media Informasi. **19**(2), 2023–2086 (2023)
14. Haug, P.J., Christensen, L., Gundersen, M.: A natural language parsing system for encoding admitting diagnoses. In: Proc AMIA Symp, pp. 814–818 (1997)

15. Uzuner, O., Solti, I., Cadag, E.: Extracting medication information from clinical text. J. Am. Med. Inform. Assoc. **17**(5), 514–518 (2010)
16. Alsentzer, E., Murphy, J., Boag, W., et al.: Publicly available clinical BERT embeddings. In: Proceedings of the 2nd Clinical NLP Workshop, pp. 72–78 (2019)
17. Gallego, F., Krallinger, M., Armengol-Estapé, J., et al.: ClinLinker: medical entity linking of clinical concept mentions in Spanish. arXiv. (2024). https://doi.org/10.48550/arXiv.2404.06367
18. Kulyabin, M., Zhuravlev, A., Rubashevskii, A., et al.: SNOBERT: a benchmark for clinical notes entity linking in the SNOMED CT clinical terminology. arXiv. (2025). https://doi.org/10.48550/arXiv.2405.16115
19. Chen, Y., Hu, D., Li, M., Duan, H., Lu, X.: Automatic SNOMED CT coding of Chinese clinical terms via attention-based semantic matching. Int. J. Med. Inform. **159**, 104676 (2022). https://doi.org/10.1016/j.ijmedinf.2021.104676
20. Matentzoglu, N., Caufield, J.H., Hegde, H., et al.: MapperGPT: large language models for linking and mapping entities. arXiv. (2023). https://doi.org/10.48550/arXiv.2310.03666
21. Seo, J., Park, S., Kim, D., et al.: Clinical negation detection errors and their impact on downstream NLP tasks. J. Biomed. Inform. **156**, 104682 (2024)
22. Névéol, A., Grouin, C., Lentes, J., et al.: Clinical NLP: a review with focus on non-English languages. J. Am. Med. Inform. Assoc. **25**(3), 301–312 (2018)
23. Campillos-Llanos, L., Valverde-Mateos, A., Capllonch-Carrión, A., et al.: A clinical NLP framework for semantic annotation of clinical findings. J. Biomed. Inform. **158**, 104723 (2025)
24. Dorca Josa, A., Bleda Bejar, F.: Privacy implications of cloud-based LLM APIs in clinical settings. J. Med. Internet Res. **26**, e52234 (2024)
25. Liu, J., Koopman, B., Brown, N.J., Chu, K., Nguyen, A.: Generating synthetic clinical text with local large language models to identify misdiagnosed limb fractures in radiology reports. Artif. Intell. Med. **159**, 103027 (2025)
26. Labrak, Y., Bazoge, A., Morin, E., et al.: BioMistral: a collection of open-source pretrained large language models for medical domains. arXiv. (2024). https://doi.org/10.48550/arXiv.2402.10373
27. Landis, J.R., Koch, G.G.: The measurement of observer agreement for categorical data. Biometrics. **33**(1), 159–174 (1977)
28. Dagli, M.M., Ghenbot, Y., Ahmad, H., et al.: Development and validation of a novel AI framework using NLP with LLM integration for automated chart review. Sci. Rep. **14**(1), 26312 (2024). https://doi.org/10.1038/s41598-024-77535-y
29. Mustafa, A., Naseem, U., Rahimi Azghadi, M.: Large language models vs human for classifying clinical documents. Int. J. Med. Inform. **195**, 105800 (2025)
30. Heryawan, L. et al.: Fast healthcare interoperability resources (FHIR)–based interoperability design in Indonesia: content analysis of developer hub's social networking service. JMIR Formative Res. **9**, e51270 (2025). https://doi.org/10.2196/51270
31. Cahyawijaya, S. et al.: Cendol: Open instruction-tuned generative large language models for Indonesian languages. In: Proceedings of the 62nd Annual Meeting of the Association for Computational Linguistics (Volume 1: Long Papers), pp. 14899–14914. Association for Computational Linguistics (2024)

Bridging the Digital Divide for Nutrition Systems in Pakistan: Evidence from Nutrition Professionals and a Capacity Development Framework

Afifa Tanweer[1,2](✉), Rida Khan[3], and Samra Imran[4]

[1] Department of Medicine, Internal Medicine, Lausanne University Hospital, Lausanne, Switzerland
afifa.tanweer@umt.edu.pk
[2] Department of Nutrition and Dietetics, School of Health Sciences, University of Management and Technology, Lahore, Pakistan
[3] Faculty of Communication, Culture and Society, Institute of Communication and Public Policy, Università della Svizzera italiana, Lugano, Switzerland
[4] Government Graduate College for Women, Gulberg, Lahore, Pakistan

Abstract. Artificial intelligence (AI) has a potential of transforming nutrition research and practice through innovations in automated dietary assessment, predictive analytics, and personalized interventions. However, in low- and middle-income countries (LMICs) such as Pakistan, the transition toward digital nutrition systems remains constrained. While existing efforts have primarily focused on digitalization of tools, limited attention has been paid to the readiness of nutrition professionals who are central to the application of such tools, data generation and interpretation. This study presents qualitative evidence from semi-structured interviews with nutrition professionals ($n = 10$) to explore challenges in current dietary assessment practices and perceptions of digitalization in Pakistan. Findings reveal limitations including recall and reporting bias, cultural variability, resource-intensive data collection, and challenges in data interpretation. While the digital dietary assessment tools are perceived as feasible and beneficial, concerns remain regarding the loss of contextual understanding, reduced human interaction, and inequitable access. Building on these findings, it is suggested that there is not only a digital divide, but a capacity divide too. Therefore, a Nutrition AI Capacity Framework is proposed, emphasizing human-centred capacity development across individual, practice, and system levels. This work contributes to the emerging field of digitalization in nutrition by providing empirically grounded insights and a model for workforce readiness in resource-constrained settings.

Keywords: Artificial Intelligence (AI) · Digital nutrition systems · Nutrition workforce capacity · Dietary Assessment

N. Mohd Sharef et al. (Eds.): CAMP 2026, CCIS 3094, pp. 32–38, 2027.
https://doi.org/10.1007/978-981-92-4305-1_3

1 Introduction

Dietary assessment plays a central role in nutritional epidemiology, clinical decision-making, and public health policy. However, traditional methods such as 24-h dietary recalls (24HR), food frequency questionnaires (FFQs), and food diaries are subject to significant limitations, including recall bias, reporting bias, and inaccuracies in portion size estimation [1]. These limitations compromise data validity and subsequently, may lead to suboptimal interventions. There are now digital versions of tools for the field of nutrition such as electronic spreadsheets, nutrient analysis software, mobile messaging applications, online dietary recall forms, and electronic medical records for dietary assessment and patient follow-up.

Artificial intelligence (AI) is increasingly being integrated into healthcare and nutrition systems through automated dietary assessment, image-based food recognition, and personalized nutrition recommendations [2]. These innovations have the potential to enhance the accuracy, efficiency, and scalability of dietary data collection and analysis.

In LMICs such as Pakistan, these challenges are compounded by structural and contextual factors, including fragmented health systems, cultural diversity in dietary practices, and limited integration of digital tools. Although digital dietary assessment tools have been proposed as solutions to improve data quality and efficiency, their implementation introduces new challenges related to usability, accessibility, and interpretation [3, 4]

The adoption of AI in nutrition systems can be conceptualized across three interrelated dimensions including data quality layer, system layer and human capacity layer (Fig. 1). The transition toward AI-enabled systems requires not only technological infrastructure but also a workforce capable of interacting with and interpreting digital outputs. Emerging evidence suggests that human capacity is a critical determinant of successful digital health implementation in LMICs [5] World Health Organization 2021.

This study focuses on the human capacity layer and examines the perspectives of nutrition professionals in Pakistan on dietary assessment and digitalization. It aims to:

a. Explore perceptions regarding digital dietary assessment tools
b. Derive potential implications for AI-enabled nutrition systems

2 Methods

A qualitative research design was employed using semi-structured interviews to explore the experiences and perceptions of nutrition professionals. Ten nutrition professionals, including clinical dietitians, academics, and researchers, were recruited through purposive sampling to ensure diversity in professional roles and experience. Participants were identified through professional networks and institutional contacts and invited via email or WhatsApp messages. Data saturation was considered as an ending point for further recruitment. Interviews were conducted online and lasted approximately 30–45 min. A semi-structured guide was used to explore dietary assessment practices, perceived gaps, and views on digitalization. Questions explored current dietary assessment practices, challenges encountered, perceptions of digital tools and anticipated competencies required for future AI-enabled systems. Data were transcribed verbatim and

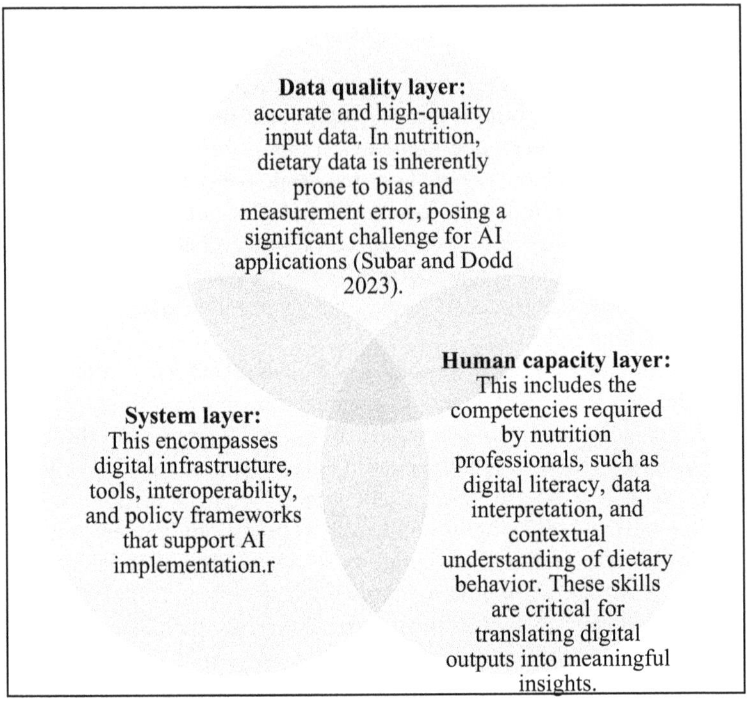

Fig. 1. Conceptual Framework: AI Capability in Nutrition Systems

analysed using thematic analysis. Codes were developed iteratively and grouped into themes reflecting challenges and opportunities in dietary assessment. The results were used to develop the framework for capacity building need for AI enabled nutrition systems. Framework development occurred in two stages. First, themes describing barriers and enabling factors were identified through thematic analysis. Second, these themes were synthesized into three interrelated domains representing capacities required for AI-enabled nutrition systems: individual competencies, practice-level integration, and system-level support structures.

3 Results and Discussion

3.1 Limitations of Traditional Dietary Assessment

Participants identified several limitations in commonly used dietary assessment tools. These included recall bias, reporting bias, and inaccuracies in portion size estimation.

One participant highlighted the lack of detail in reported intake:

> "If I tell you that I just had a bowl of salad and I don't tell you that it has a dressing of cream or fat… telling a bowl of salad does not tell what exactly I have consumed." (Q-R3)

Participants also emphasized that respondents often omit or misreport information, either unintentionally or due to social desirability:

> "People say we take good enough water… but they are taking less. Fruit juice is confused with milkshake… things need to be probed." (Q-R10)

These findings indicate that dietary data collected through conventional methods is frequently incomplete and subject to systematic bias.

Dietary assessment was also described as resource-intensive, requiring significant time and effort for both practitioners and respondents.

A substantial portion of consultation time was reported to be dedicated to dietary history-taking.

> "Most of the time… 45-minute session… 20–30 minutes is spent on history taking." (Q-R2)

This burden limits scalability, particularly in settings involving large patient loads or population-level research.

Dietary assessment was described as highly context-dependent, with cultural diversity presenting significant challenges. Participants reported difficulties in capturing accurate dietary data across different regions and linguistic groups.

> "People come from far-flung areas… language barrier… their foods are different, cooking methods are different, portion sizes are different." (Q-R7)

Such variability requires extensive probing and adaptation by the nutrition professional, increasing both complexity and time requirements. Participants also reported difficulties in interpreting dietary data, particularly when translating qualitative intake into standardized quantitative measures.

> "The methods are easy to understand… but when we sit as a researcher… how am I going to interpret this information?" (Q-R9)

This reflects a gap in analytical and statistical capacity, which may further limit the effective use of digital or AI-based systems.

Participants emphasized the importance of human interaction in dietary assessment, particularly for probing, clarification, and interpretation of responses. Dietary intake was viewed as a complex behaviour influenced by cultural, psychological, and social factors.

A key theme across all interviews was the importance of human judgment in dietary assessment. Participants emphasized that effective assessment goes beyond quantification and requires interpretation of behavioural and emotional factors.

> "Food is not just for the sake of filling appetite… it may be stress management, habit loops… human element must be added." (Q-R3)

Similarly, interpersonal interaction was considered essential for accurate data collection:

"I want to talk to the patient… I judge personality, compliance, attitudes… I can never be satisfied by making the patient fill a written form." (Q-R2)

These findings highlight that dietary assessment is not purely a technical process but a relational and interpretive one.

3.2 Opportunities and Concerns for Digitalization

Digital tools were perceived as beneficial for automating calculations, reducing workload, and improving data management

Participants recognized several advantages of digital tools, particularly in improving efficiency and reducing workload.

"If something is made… where calculations are done automatically… it will save effort." (Q-R4)

Digital tools were also viewed as helpful for record-keeping and monitoring:

"When I have a patient's data maintained in a system… I remember that more." (Q-R4)

These findings suggest that digitalization is perceived as both feasible and beneficial in the local context.

These findings align with previous studies highlighting the efficiency and scalability of digital dietary assessment tools [6, 7].

Despite their benefits, participants expressed concerns regarding loss of human interaction, inability to capture behavioural context, risk of excluding certain populations. These concerns reflect broader challenges in digital health implementation, including equity and usability [2].

Participants expressed concerns regarding over-reliance on digital systems. A major concern was the potential loss of human interaction and contextual understanding.

"I don't think if a patient fills a computer form… it will be effective… the patient needs interaction." (Q-R2)

Additionally, excessive quantification was seen as problematic:

"If we only record numbers… some important determinants may be missed… qualitative data is very important." (Q-R5)

There were also concerns regarding accessibility and equity:

"We are already addressing a limited population… if we make it digital, the population may become even more limited." (Q-R1)

3.3 Hybrid Approaches as Emerging Solution

Participants implicitly supported a hybrid model combining digital tools with human expertise. Digital systems were viewed as useful for initial data collection and automation, while human interaction remained essential for interpretation and counselling. While AI systems offer efficiency and scalability, they lack the contextual understanding necessary for interpreting dietary behaviours. Human expertise remains essential for ensuring accuracy and relevance [8]. A hybrid model combining digital tools with human expertise is likely to be most effective in the Pakistani context. Such approaches can leverage the strengths of both systems while mitigating their limitations.

3.4 Potential Implications of AI Systems

The findings highlight those existing limitations in dietary data may compromise the performance of AI systems. At the data level, issues such as recall bias, cultural variability, and loss of nuance during quantification affect the reliability of dietary information. At the human level, the process remains highly dependent on professional judgment, communication skills, and interpretive capacity. Digital literacy and data interpretation emerged inductively from participants' accounts of difficulties in translating dietary information into standardized measures, interpreting outputs, and adapting assessments to culturally diverse contexts. These competencies were prioritized because they were consistently discussed across interviews and were considered essential for meaningful use of digital and AI-enabled nutrition tools. At the system level, while digital tools offer efficiency and scalability, they also introduce concerns related to equity, usability, and the potential loss of human interaction. Collectively, these themes illustrate the emergence of an AI capability divide, where the effective use of digital and AI-enabled systems is constrained by human and contextual factors rather than access to technology alone.

Based on our findings we propose a three-level framework for nutrition AI capacity building (Table 1).

Table 1. Nutrition AI Capacity Framework for resource-constrained settings

Levels	Components	Key gaps identified	Recommended actions
Individual	Nutrition professionals	Limited digital literacy, weak data interpretation skills	Training in digital tools, data interpretation skills
Practice	Clinical/research workflows	Over-reliance on manual methods, lack of integration	Hybrid models combining digital tools with human expertise
System	Policy and education	Absence of AI-focused curriculum and guidelines	Curriculum reform, policy development, investment in digital health training

Implementation of the proposed framework requires coordinated action from multiple stakeholders. Universities could integrate digital nutrition, data science, and AI-related competencies into undergraduate and postgraduate curricula. Professional associations may support continuing professional development programmes focused on digital literacy and responsible use of AI. Health facilities can promote hybrid workflows that combine digital tools with professional expertise, while government agencies should establish policies, standards, and investments to support digital nutrition infrastructure. A phased roadmap beginning with foundational digital literacy training, followed by advanced competencies in data analytics and AI interpretation, may facilitate scalable workforce development in resource-constrained settings.

4 Conclusion

This study demonstrates that the transition toward AI-enabled nutrition systems in Pakistan is constrained by gaps in human capacity rather than technology alone. By introducing the concept of the AI capability divide and proposing a capacity development framework, this work provides a foundation for future research and policy initiatives aimed at strengthening nutrition systems in LMICs.

References

1. Tanweer, A., Khan, S., Mustafa, F.N., Imran, S., Humayun, A.: Improving dietary data collection tools for better nutritional assessment–a systematic review. Comput. Methods Programs Biomed. Update. **2**, 100067 (2022)
2. Cvrkel, T.: The ethics of mHealth: moving forward. J. Dent. **74**, S15–S20 (2018)
3. Das, S.K., Miki, A.J., Blanchard, C.M., Sazonov, E., Gilhooly, C.H., Dey, S., et al.: Perspective: opportunities and challenges of technology tools in dietary and activity assessment: bridging stakeholder viewpoints. Adv. Nutr. **13**(1), 1–15 (2022)
4. Illner, A., Freisling, H., Boeing, H., Huybrechts, I., Crispim, S., Slimani, N.: Review and evaluation of innovative technologies for measuring diet in nutritional epidemiology. Int. J. Epidemiol. **41**(4), 1187–1203 (2012)
5. Labrique, A.B., Wadhwani, C., Williams, K.A., Lamptey, P., Hesp, C., Luk, R., et al.: Best practices in scaling digital health in low and middle income countries. Glob. Health. **14**(1), 103 (2018)
6. Boushey, C., Spoden, M., Zhu, F., Delp, E., Kerr, D.: New mobile methods for dietary assessment: review of image-assisted and image-based dietary assessment methods. Proc. Nutr. Soc. **76**(3), 283–294 (2017)
7. Gemming, L., Utter, J., Mhurchu, C.N.: Image-assisted dietary assessment: a systematic review of the evidence. J. Acad. Nutr. Diet. **115**(1), 64–77 (2015)
8. Hebden, L., Cook, A., van der Ploeg, H.P., Allman-Farinelli, M.: Development of smartphone applications for nutrition and physical activity behavior change. JMIR Res. Protoc. **1**(2), e2205 (2012)

Implementation of Hybrid Model: Auto-Encoder and Isolation Forest for Anomaly Detection in BPJS Healthcare Billing Data

Clara Shinta Dea Ananda Rosalie, Lukman Heryawan(✉), and Yohana Butar Butar

Department of Computer Science and Electronics, Universitas Gadjah Mada, Yogyakarta, Indonesia
lukmanh@ugm.ac.id

Abstract. The integrity of healthcare billing data is paramount for the financial sustainability of national health insurance systems. This research presents a comprehensive hybrid framework combining the deep learning based Auto-Encoder (AE) with the tree based Isolation Forest (IF) to detect anomalies in BPJS Kesehatan billing records. The study details the end-to-end process including data collection, preprocessing, model design with mathematical formulations, and implementation. Experimental results demonstrate that the hybrid AE-IF model outperforms standalone models, effectively identifying suspicious patterns that represent potential fraud or administrative irregularities.

Keywords: Hybrid Model · Anomaly · Auto-Encoder · Isolation Forest · Healthcare Billing Data · BPJS Kesehatan

1 Introduction

The National Health Insurance (JKN) system, administered by BPJS Kesehatan, constitutes one of the largest healthcare programs globally. The massive scale of daily billing transactions presents a formidable challenge for data integrity and financial oversight. Conventional auditing methodologies, predominantly reliant on rule systems and manual expert verification, exhibit significant limitations in terms of scalability and adaptability to evolving irregularity patterns [9].

The scarcity of labelled fraudulent cases, unsupervised machine learning offers a robust alternative by modelling the latent distribution of normal billing behaviours [2, 8].

This research proposes a hybrid framework integrating Deep Auto-Encoders (AE) for non-linear dimensionality reduction and Isolation Forest (IF) for efficient outlier isolation in latent space [6, 15], aiming to enhance the detection precision in national healthcare insurance data.

N. Mohd Sharef et al. (Eds.): CAMP 2026, CCIS 3094, pp. 39–47, 2027.
https://doi.org/10.1007/978-981-92-4305-1_4

2 Data Collection and Preparation

This section details the systematic approach to handle the healthcare billing data, beginning with the acquisition of raw datasets from the BPJS Kesehatan database and followed by the essential preprocessing and normalization required to prepare the features for deep learning analysis.

2.1 Data Acquisition

The primary dataset was sourced from the BPJS Kesehatan billing database, obtained through the official BPJS Kesehatan platform which provide data for research purposes. The dataset includes claim records sampled from across Indonesia during 2023 and is derived BPJS Kesehatan sample data intended to represent national healthcare utilization. Each record represents a single FKRTL healthcare claim submitted to BPJS Kesehatan under the JKN reimbursement system.

The initial dataset consisted of 1,488,080 records and 55 variables, while the final dataset retained the same records with 49 variables after removing redundant fields and applying feature engineering. The main variable groups used in this study are summarized in Table 1.

The dataset comprises several critical healthcare metrics used as input features for the model. These include Clinical Case Definition Fraud (CCDF) indicators, Case Based Groups (CBG) classifications, and Length of Stay (LOS) data [10, 14].

These variables provide the necessary clinical context for the Auto-Encoder to learn the underlying patterns of legitimate versus anomalous claims [12].

2.2 Preprocessing and Feature Engineering

To prepare the high dimension raw data for neural network input, several critical cleaning and engineering steps were performed:

- ***Data Cleaning and Filtering:*** Records with missing essential field, such as primary diagnosis codes or zero verified billing amount were removed to ensure data integrity. Inconsistent records such as negative claimed costs or impossible admission to discharge dates were also excluded to prevent the model from learning erroneous pattern.
- ***Derived Feature Engineering:*** Empirical median were calculated from the full dataset to establish expected benchmarks. Specially, feature such as cbg_expected_cost and icd_expected_LOS were derived to serve as a baseline. Gap features, quantifying the deviation of a specific claim from these medians were added to directly expose outliers to the model [8].
- ***Categorical Encoding:*** High cardinality administrative codes, including hospital class (FKL09) and casemix main group (FKL20) were numerically encoded to be compatible with deep learning layers.

Table 1. Summary of variable groups and their roles in the BPJS Kesehatan dataset

Variable group	Examples	Role in model
Dataset metadata	2023, nationwide Indonesia, 1,488,080 records, 55 to 49 variables	Describe scope and dataset size
Clinical variables	ICD-10 diagnosis codes, INA-CBG classifications, LOS, CCDF indicators	Capture diagnosis structure and clinical complexity
Financial variables	Billing amount, verified fee, log-transformed billing amount, log-transformed verified fee, log-transformed weight	Capture monetary irregularities and cost patterns
Administrative variables	Hospital ID, province, district/city, hospital ownership, hospital type, hospital class, service dates, referral facility characteristics	Capture facility and service context
Engineered variables	Length of stay derived from admission and discharge dates, admission month, discharge month, log-transformed financial variables	Improve model input and expose deviations
Identifier fields removed	Participant ID, family ID, claim ID, referral visit ID	Prevent data leakage and preserve privacy

2.3 Normalization

To ensure numerical stability and optimal convergence during the neural network training phase, feature scaling is applied. Numerical attributes, such as total claim costs and patient age, are transformed into a normalized range.

While the initial approach utilized Min-Max Scaling to a [0, 1] range, the final implementation favoured the StandardScaler method, which rescales features to have a zero mean and unit variance. This procedure is critical for healthcare datasets because it mitigates the risk of features with larger magnitudes such as extreme surgical billing costs disproportionately influencing the model's objective function, thereby allowing the Auto-Encoder to weigh clinical and financial signals equally [2, 4].

3 Model Design and Architecture

3.1 Hybrid Framework

The proposed architecture operates in a sequential of two stage pipeline. The first stage utilizes a Deep Auto-Encoder (AE) as a non-linear feature extractor to learn a compressed, latent representation of the healthcare billing data. By forcing the data through a bottleneck layer, the model filters out statistical noise and redundancy [3].

The second stage involves the Isolation Forest (IF) algorithm, which consumes the learned latent features and reconstruction errors to score anomalies based on their isolation depth in the latent space [6, 15].

3.2 Auto Encoder Reconstruction Loss (MSE)

The Deep Auto-Encoder is composed of an Encoder, which maps the input vector x to a lower dimensional latent representation z, and a Decoder, which attempts to reconstruct the original input as $\hat{x}$.

The model is optimized by minimizing the Mean Squared Error (MSE) between the original input vector and its reconstruction:

$$\mathrm{MSE} = \frac{1}{n}\sum_{\{\mathrm{i}=1\}}^{\{\mathrm{n}\}} (\mathrm{x_i} - \hat{x}_\mathrm{i})^2$$

Each claim's individual reconstruction error will be calculated as the squared Euclidean distance [2, 8]:

$$Err_i = \|x_i - \hat{x}_\mathrm{i}\|_2^2$$

When trained on predominantly legitimate billing records, the Auto-Encoder learns to reconstruct normal patterns with high precision. Consequently, anomalous claims characterized by unusual clinical or financial combinations will produce significantly higher reconstruction errors, serving as a robust primary indicator of atypical behaviour [3].

3.3 Isolation Forest Scoring

To refine the detection process, the Isolation Forest isolates observations by recursively and randomly partitioning the feature space into an ensemble of isolation trees. Unlike distance-based methods, IF focuses on the principle that anomalies are few or different and thus easier to isolate with fewer splits [2, 6].

The anomaly score $s(x, n)$ is derived from the average path length $E(h(x))$ across the forest [11, 15]:

$$s(x, n) = 2^{-\frac{E[h(x)]}{c(n)}}$$

Where $E(h(x))$ is the average path length for point x, $c(n)$ is a normalization constant for the average path length of an unsuccessful search in a Binary Search Tree, and n is the total number of records.

3.4 Cumulative Distribution Function (CCDF)

CCDF measures the probability that a claim has an anomaly score greater than a chosen threshold and will show how the model behaves in the score distribution.

In this study, let S_i denote the anomaly score of claim i, with higher values indicating more anomalous behaviour. The empirical complementary cumulative distribution function (CCDF) of the anomaly scores is defined as [8]:

$$CCDF_s(s) = \frac{1}{n}\sum_{\{i=1\}}^{\{n\}}(S_i \geq \mathrm{s})$$

To ensure the model captures structural irregularities rather than stochastic noise, Spearman Rank Correlation is used to evaluate the consistency of anomaly rankings across different random seeds [5, 13].

Furthermore, Jaccard Similarity is applied to measure the consistency of the top-percentile (P99) flagged cases between repeated model runs [7].

4 Implementation

4.1 Experimental Environment

The implementation and evaluation of the hybrid model were conducted on a high performance system capable of processing large scale healthcare datasets. The software environment utilized TensorFlow and Keras for constructing and optimizing the Deep Auto-Encoder, while Scikit Learn was employed for the Isolation Forest implementation and unsupervised evaluation metrics [11, 13].

4.2 Code Implementation and Detailed Explanation

The implementation follows a modular approach where the Auto-Encoder acts as a feature extractor [3, 4]. Below is the core logic for building and training the hybrid model:

```
def build_autoencoder(input_dim: int, latent_dim: int = 16):
inp = keras.Input(shape=(input_dim,), name="input")
# encoder
x = layers.Dense(64, activation="relu")(inp)
x = layers.Dense(32, activation="relu")(x)
z = layers.Dense(latent_dim, activation="linear",
name="latent")(x)
# decoder
y = layers.Dense(32, activation="relu")(z)
y = layers.Dense(64, activation="relu")(y)
out = layers.Dense(input_dim, activation="linear",
name="recon")(y)
ae = keras.Model(inp, out, name="autoencoder")
ae.compile(optimizer=keras.optimizers.Adam(1e-3), loss="mse")
encoder = keras.Model(inp, z, name="encoder")
return ae, encoder
```

Implementation Details and Logical Flow:

- ***Progressive Dimensionality Reduction:*** The encoder follows a 64-32-16 structure. This bottleneck architecture is critical for forcing the network to prioritize the most significant clinical and financial features, effectively filtering out noise inherent in BPJS administrative claims [3, 15].
- ***Linear Bottleneck Layer:*** For non-linear mapping, the latent (bottleneck) and output layers utilize Linear activation. This preserves the continuous range of the latent features and ensures that the reconstruction error directly reflects the magnitude of deviation in the normalized input space [4].
- ***Model Optimization***: The network was trained for 20 epochs with a batch size of 1024. The Adam optimizer with a learning rate of 10^{-3} 10 was chosen to ensure stable convergence across the 1.48 million claim records.
- ***Hybrid Model Integration***: Post training, the encoder extracts a noise free data representation. This latent representation is concatenated with the calculated reconstruction error to form a new feature set. This combined input allows the Isolation Forest to isolate anomalies by considering both the structural rarity (latent space) and the degree of clinical mismatch (reconstruction error) [6, 11].
- ***Anomaly Thresholding:*** For final flagging, a P99 threshold was applied to the Isolation Forest scores. Claims falling into the top 1% of the anomaly score distribution are identified as the most suspicious records, requiring priority auditing.

5 Results and Discussion

5.1 Performance Visualization and Analysis

The model's performance is evaluated by analysing how the hybrid architecture responds to the complex clinical and operational features inherent in the BPJS Kesehatan dataset.

As demonstrated in Table 2, the model shows a distinct sensitivity to medical complexity and resource utilization.

The data reveals a proportional relationship between reconstruction error and CBG severity. For routine or mild clinical cases, the Auto-Encoder exhibits high reconstruction accuracy, suggesting that the model has successfully internalized the baseline signature of standard healthcare billing.

Conversely, severe cases or unusually prolonged hospitalizations ($\geq$ P99 Length of Stay) result in significantly higher errors. This behaviour confirms that the model naturally identifies complex or rare resource intensive claims as deviations from the learned normal distribution [3, 4].

Furthermore, the Isolation Forest scoring distribution confirms the hybrid model's calibration. By assigning high anomaly scores primarily to records with both high reconstruction errors and sparse latent representations, the model effectively isolates rare patterns as deviations from the learned distribution, consistent with established hybrid AE-IF anomaly detection behaviour [6, 11, 15].

Critically, the model does not flag all complex cases — only those whose feature combinations are statistically inconsistent with similar claims. This distinction is important: a severe (Severity 3) claim with expected billing patterns will score low, while a

Table 2. Summary model reconstruction error and behavioural insights

Feature category	Data group	Median reconstruction error	Behaviour insight
CBG Severity	Severity 0 (Mild)	0.000053	High familiarity with standard clinical patterns.
CBG Severity	Severity 3 (Severe)	0.002057	Increased complexity leads to higher reconstruction error.
Length of Stay	High ($\geq$ P99)	2.24	Model reconstruction error is highest, reflecting feature combinations that deviate from learned patterns
Length of Stay	Low ($\leq$ P1)	0.63	Lowest reconstruction error, consistent with dominance in training data

mild (Severity 0) claim with unusual cost or procedure combinations will score high. This selectivity is what makes the model useful as a prioritisation tool rather than a simple complexity filter

5.2 Quantitative Comparison

The integration of a Deep Auto-Encoder as a latent feature extractor provides several advantages over traditional single-model approaches:

- ***Feature Refinement:*** By compressing the high dimensional input space, the model filters out statistical noise. This allows the subsequent isolation stage to focus on structural irregularities rather than random variations in the data [4, 13].
- ***Ranking Stability:*** Stability testing using Spearman Rank Correlation yielded a coefficient of $\rho > 0.98$ across multiple random seeds, indicating that the anomaly rankings are highly consistent and not artifacts of stochastic initialization [5].
- ***Computational Efficiency:*** The hybrid architecture maintains stable convergence and fast inference times, which is a critical requirement for processing the millions of daily billing transactions generated by the national health insurance system [1, 11].

5.3 Research Limitations

Despite the promising results, this study identifies several limitations:

- ***Feature Constraints:*** The current analysis primarily relies on numerical administrative data. Future work could incorporate Natural Language Processing (NLP) to analyse unstructured text in ICD-10 diagnoses for deeper fraud detection [12, 14].
- ***Temporal Sensitivity:*** The model was trained on a dataset from a specific period. Changes in healthcare regulations or disease patterns may require periodic model retraining to maintain detection accuracy.

- ***Resource Allocation:*** While the model is effective at *detecting* anomalies, it serves as a triaging tool rather than a full diagnostic framework. It identifies *what* is suspicious, but human expert review remains necessary to determine *why* a claim was flagged as fraudulent or merely a rare legitimate case [9].
- ***Policy Relevance:*** The current analysis is limited to structured claim-level administrative variables. As a result, it partially captures policy-relevant fraud signals through demographic plausibility and clinical-context features, but it does not model longitudinal patient history or cross-facility utilization because patient identifiers were removed during preprocessing. These missing elements remain important for future work on comprehensive fraud investigation.

6 Conclusion

This study successfully presents a hybrid machine learning framework combining a Deep Auto-Encoder and Isolation Forest to detect potential fraud and anomalies in BPJS Kesehatan insurance claims. The architecture utilizes a 64–32-16 encoder structure with linear activation to compress high-dimensional healthcare data into a refined latent space, effectively filtering out statistical noise and administrative redundancies [3, 15].

Experimental results confirm that the model is highly sensitive to clinical complexities. Specifically, reconstruction errors were found to increase proportionally with CBG severity and unusually prolonged Length of Stay (LOS), identifying these patterns as significant structural irregularities. Stability tests, including Spearman Rank Correlation and Jaccard Similarity, further validate that the model produces consistent and reproducible anomaly rankings across different random initializations.

In conclusion, the proposed hybrid approach offers a precise and scalable solution for automated healthcare auditing by isolating atypical billing patterns more effectively than traditional single-model or rule-based methods. While the current framework is focused on numerical administrative features, its strong convergence and stability provide a robust foundation for next-generation fraud detection systems in the national healthcare insurance landscape. However, its policy value is constrained by the absence of longitudinal and cross-facility features, which limits its ability to support comprehensive fraud investigation. Future work may further enhance this system by integrating unstructured clinical text and real-time auditing capabilities [1, 6].

Acknowledgments. This work was supported by the Department of Computer Science and Electronics, Universitas Gadjah Mada under the Publication Funding Year 2026.

References

1. Ahmed, M., Naser Mahmood, A., Hu, J.: A survey of network anomaly detection techniques. J. Netw. Comput. Appl. **60**, 19–31 (2016). https://doi.org/10.1016/j.jnca.2015.11.016
2. Chandola, V., Banerjee, A., Kumar, V.: Anomaly detection: a survey. ACM Comput. Surv. **41**(3), 1–58 (2009). https://doi.org/10.1145/1541880.1541882

3. Chen, J., Sathe, S., Aggarwal, C., Turaga, D.: Outlier detection with autoencoder ensembles. In: Wang, W. (ed.) Proceedings of the 2017 SIAM International Conference on Data Mining, pp. 90–98. Society for Industrial and Applied Mathematics (2017). https://doi.org/10.1137/1.9781611974973
4. Erfani, S.M., Rajasegarar, S., Karunasekera, S., Leckie, C.: High-dimensional and large-scale anomaly detection using a linear one-class SVM with deep learning. Pattern Recogn. **58**, 121–134 (2016). https://doi.org/10.1016/j.patcog.2016.03.028
5. Fayzrakhmanov, R., Kulikov, A., Repp, P.: The difference between precision-recall and roc curves for evaluating the performance of credit card fraud detection models. In: Proc. of the 6th International Conference on Applied Innovations in IT, (ICAIIT) (2018)
6. Maheswari, G.N., Vinith, A., Sathyanarayanan, A.S., Sowmi Saltonya, M., Sambath, M.: An ensemble framework for network anomaly detection using isolation forest and autoencoders. In: 2024 International Conference on Advances in Data Engineering and Intelligent Computing Systems (ADICS), pp. 1–6 (2024). https://doi.org/10.1109/ADICS58448.2024.10533499
7. He, H., Garcia, E.A.: Learning from imbalanced data. IEEE Trans. Knowl. Data Eng. **21**(9), 1263–1284 (2009). https://doi.org/10.1109/TKDE.2008.239
8. Hodge, V., Austin, J.: A survey of outlier detection methodologies. Artif. Intell. Rev. **22**(2), 85–126 (2004). https://doi.org/10.1023/B:AIRE.0000045502.10941.a9
9. Inggi, P.G., Achadi, A.: Evaluation of fraud prevention policies in the National Health Insurance System in Indonesia: narrative literature review. Media Publikasi Promosi Kesehatan Indonesia (MPPKI). **7**(10), 2449–2457 (2024). https://doi.org/10.56338/mppki.v7i10.6164
10. Kaur, H., Wasan, S.K.: Empirical study on applications of data mining techniques in healthcare. J. Comput. Sci. **2**(2), 194–200 (2006). https://doi.org/10.3844/jcssp.2006.194.200
11. Kim, K.: An effective anomaly detection approach based on hybrid unsupervised learning technologies in NIDS. KSII Trans. Internet Inf. Syst. **18**(2), 515–532 (2024). https://doi.org/10.3837/tiis.2024.02.012
12. Kourou, K., Exarchos, T.P., Exarchos, K.P., Karamouzis, M.V., Fotiadis, D.I.: Machine learning applications in cancer prognosis and prediction. Comput. Struct. Biotechnol. J. **13**, 8–17 (2015). https://doi.org/10.1016/j.csbj.2014.11.005
13. Lee, K., Lee, C., Lee, J.: Semi-supervised anomaly detection algorithm using probabilistic labeling (SAD-PL). IEEE Access. **9**, 1 (2021). https://doi.org/10.1109/ACCESS.2021.3120710
14. Obermeyer, Z., Emanuel, E.J.: Predicting the future—big data, machine learning, and clinical medicine. N. Engl. J. Med. **375**(13), 1216–1219 (2016). https://doi.org/10.1056/NEJMp1606181
15. Toufigh, V., Ranjbar, I.: Unsupervised deep learning framework for ultrasonic-based distributed damage detection in concrete: integration of a deep auto-encoder and isolation Forest for anomaly detection. Struct. Health Monit. **23**(3), 1313–1333 (2024). https://doi.org/10.1177/14759217231183143

Integrating Local Stroke Registry Data into a Predictive Digital Twin Platform: A Proof-of-Concept for Acute Stroke Monitoring

Fatih Nurul Izzah[1], Annisa Ristya Rahmanti[2,3(✉)], Lukman Heryawan[4], Farida Niken Astari Nugroho Hati[5], Lutfan Lazuardi[2], Minh Cong Tran[6,7], Mai Duy Ton[8,9,10], and Huan X. Nguyen[3]

[1] Master Program in Biomedical Engineering, Graduate School, Universitas Gadjah Mada, Yogyakarta, Indonesia
[2] Department of Health Policy and Management, Universitas Gadjah Mada, Yogyakarta, Indonesia
annisaristya@ugm.ac.id
[3] London Digital Twin Research Center, Middlesex University, London, UK
[4] Department of Computer Science and Electronics, Faculty of Mathematics and Natural Sciences, Universitas Gadjah Mada, Yogyakarta, Indonesia
[5] Universitas Gadjah Mada Academic Hospital, Yogyakarta, Indonesia
[6] Nuffield Department of Medicine, University of Oxford, Oxford, UK
[7] School of Computing and Mathematical Sciences, University of Leicester, Leicester, UK
[8] Stroke Center, Bach Mai Hospital, Hanoi, Viet Nam
[9] Hanoi Medical University, Hanoi, Viet Nam
[10] VNU-University of Medicine and Pharmacy, Hanoi, Viet Nam

Abstract. Stroke remains one of the leading healthcare challenges in Indonesia, yet the use of digital twin technology for stroke monitoring is still relatively uncommon. This study describes a proof-of-concept of an AI-enabled digital twin platform for predicting discharge modified Ranking Scale (mRS) of acute stroke patients using local stroke registry data from Universitas Gadjah Mada Academic Hospital, Indonesia. The architecture comprises three layers: the Clinical Data Layer, the Data Integration and Intelligence Layer, and the Digital Twin Representation and Monitoring Layer. The prediction model incorporated clinical variables including demographics, medical history, stroke type, neurological status (NIHSS, GCS), functional status, imaging results, and acute clinical parameters. System evaluation focused on integration testing, variable harmonization, predictive validity, digital twin representation, and expert review. The platform achieved an area under the ROC curve (AUROC) of 0.81 (95% CI: 0.76–0.85) for admission prediction against ground-truth discharge mRS outcomes (n = 347). In addition, the system successfully synchronized clinical state updates, generated updated outcome predictions, and visualized patient progression through Human Twin and trajectory monitoring interfaces. Overall, the study demonstrates the feasibility of building a predictive digital twin platform from local stroke registry data as a foundation for data-driven stroke monitoring and future clinical decision support.

Keywords: Stroke Digital Twin · Stroke Registry · Functional Outcome Prediction

N. Mohd Sharef et al. (Eds.): CAMP 2026, CCIS 3094, pp. 48–58, 2027.
https://doi.org/10.1007/978-981-92-4305-1_5

1 Introduction

The increasing trend of stroke in Indonesia not only impacts patients' quality of life but also leads to a high burden of health care costs. In 2023, stroke was recorded as the third-highest costly disease, costing approximately USD 297 million (Muhawarman 2024; Yaputra 2024). To mitigate this impact, effective management that accommodates limited access to rehabilitation and assessment standards in Low- and Middle-Income Countries (LMICs) is required (Venketasubramanian et al. 2022; Feigin et al. 2024).

In recent years, artificial intelligence (AI) has demonstrated potential in predicting stroke risk and outcomes through large-scale clinical data analysis (Heo et al. 2019, Chen et al. 2024). Machine learning models are able to detect complex patterns that are difficult to recognize manually, thereby allowing more informed clinical decision-making (Vargas et al. 2024). On the other hand, the concept of digital twins (DT) in healthcare offers a more integrative approach as virtual representations built from clinical and physiological parameters to dynamically model patient conditions (Allen et al. 2021). By integrating AI models as enabling technology, digital twins can perform better disease prediction and modeling tasks (Roopa and Venugopal 2025).

However, development of clinical digital twins in developing countries still faces major challenges related to limited data, predictive models, and the integrative framework that can integrate them (Li et al. 2025; Upadrista et al. 2025). The majority of predictive models developed in developed countries are often irrelevant to the local context, due to differences in data representation and clinical procedures (Rahmanti et al. 2026). This presents a challenge in the integration of existing predictive models into local registry-based monitoring systems. On the other hand, the discharge modified Rankin Scale (mRS) is gradually acknowledged as a simpler yet still effective parameter for predicting long-term stroke recovery results (Liu et al. 2023).

Therefore, this research focuses on designing a digital twin prototype based on structured clinical data capable of visualizing the dynamics of a patient's neurological condition as a proof-of-concept. This research emphasizes an integrative architecture that accommodates the output of existing functional outcome prediction models (discharge mRS) into a single web-based digital twin platform. By utilizing data from the Stroke Registry of Gadjah Mada University Hospital (RSA UGM), this research is expected to provide an initial design for a virtual representation of acute stroke patients, thereby assisting clinicians in interpreting the progression of the patient's condition in a more structured, data-driven manner.

2 Related Works

Digital twin development in stroke care has followed three main directions: predictive modeling, physiological simulation, and data integration architectures.

In the predictive modeling domain, Allen et al. (2021) developed a Variational Autoencoder-based digital twin to model temporal disease trajectories in stroke patients. Similarly, Hussain et al. (2021) explored several machine learning models for stroke classification using EEG data and reported strong predictive performance. While these studies demonstrate the potential of AI-enabled digital twins for prediction, they primarily focus on model development and provide limited discussion of integrated clinical monitoring systems.

In the physiological and rehabilitation domain, digital twins have been applied to stroke therapy and motor rehabilitation. Chen et al. (2023) developed a digital twin for robotic upper-limb rehabilitation, while Lauer-Schmaltz et al. (2024) proposed human digital twin approaches combining dashboards, exoskeletons, and serious games. These studies contribute to personalized rehabilitation but remain focused on biomechanical simulation rather than longitudinal monitoring of clinical variables during hospitalization.

Recent studies have also explored data integration architectures for stroke digital twins. Upadrista et al. (2025) integrated blockchain technology with EHR-based stroke prediction, while Guan (2025) developed an AI-driven clinical digital twin combining clinical data, imaging, and vital signs for outcome prediction. Although these systems demonstrated strong predictive performance, limited attention was given to explicit patient-state visualization and longitudinal representation of clinical evolution. Herrgårdh et al. (2022) further proposed a hybrid framework combining physiological simulation and machine learning, but its application to routine clinical monitoring remains limited.

Overall, existing studies have largely focused on predictive models, rehabilitation-oriented simulations, or data integration frameworks. However, the integration of structured registry-based clinical data, longitudinal patient-state updates, outcome prediction, and visual patient representation within a single monitoring platform remains limited. Therefore, this study develops a proof-of-concept digital twin platform that integrates local stroke registry data, predictive modeling, and patient-state visualization to support monitoring of acute stroke patients..

3 Methodology

3.1 System Design

This research is a systems engineering research focused on the development of a website-based digital twin prototype to support functional outcome prediction for stroke patients. The development process was conducted using the Systems Development Life Cycle (SDLC) approach because this approach provides a systematic and structured flow, encompassing the stages of needs analysis, design, implementation, and system evaluation (Presman 2001).

In the context of this research, the development of the digital twin prototype is targeted to achieve Level 3 (Predictive). This is based on the integration of clinical variables from the stroke registry with existing functional outcome prediction models.

The system not only displays the patient's clinical status but also provides an estimated hospital discharge time as a basis for clinical decision-making. Two-way communication at this level is realized through a clinician-mediated feedback mechanism, where predictive information from the system is received by a neurologist and translated into clinical decisions for the patient. This approach is consistent with the characteristics of retrospective data-driven clinical DT, which does not yet have real-time monitoring integration, where the feedback loop runs through a user interface (dashboard) as a meeting point between virtual representations and physical decisions (Mohamed et al. 2023; Metcalfe et al. 2023). To meet these needs, the system was designed following the flowchart (Fig. 1).

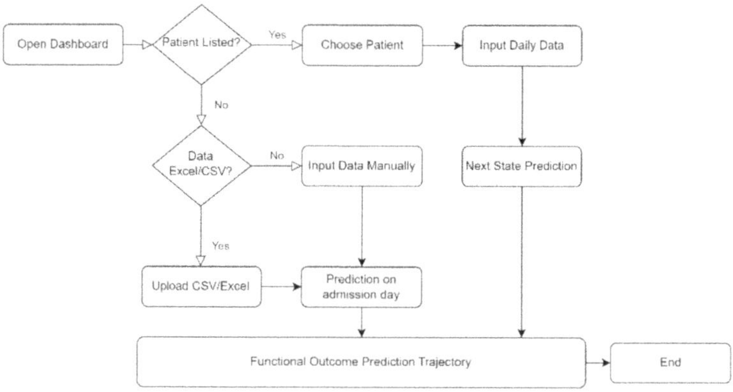

Fig. 1 Flowchart design

3.2 Data Source and Variable Selection

This study used secondary data from the Universitas Gadjah Mada Academic Hospital's Stroke Registry, which documented patients' clinical variables from the acute phase to hospital discharge. The available registry dataset covers the period from January 2020 to December 2025, with approximately 302 clinical variables and 412 stroke cases, both ischemic and non-ischemic.

The clinical variables used in the system refer to the 31 most influential variables in the prediction model trained using ResQ Vietnam data and validated using Universitas Gadjah Mada Academic Hospital's stroke registry data (Rahmanti et al. 2026). The variables then divided into two groups. The first group is a static baseline variable that includes demographic data (age, gender), pre-stroke condition (prestroke mRS), admission parameters (mode of arrival, onset-to-door, door-to-imaging), stroke type and etiology, imaging results, history of atrial fibrillation, history of anticoagulant use and previous stroke risk factors. The second group is a dynamic clinical variable that can be updated in each additional state. This group including neurological status (NIHSS, GCS), acute condition (fever, ventilator use), functional status (walking ability, dysphagia), laboratory parameters (cholesterol), interventions (thrombectomy), post-acute management, and post-stroke complications.

The stroke registry data is episode-based, so each row represents a single episode of patient care from admission to discharge. The snapshot dataset is used to initialize the initial state of the digital twin, while the system architecture is designed to accept clinical data updates temporally, allowing the patient representation to evolve into a digital representation.

3.3 System Architecture

To support this, the digital twin stroke system was built based on the architecture (Fig. 2). It shows that this digital twin architecture consists of three main layers to support monitoring of stroke patients.

Clinical Data Layer is responsible for acquiring and storing patient data. Data are obtained through CSV/Excel uploads or manual input. It includes baseline patient data such as demographics and initial clinical parameters as well as dynamic clinical state data such as NIHSS score, GCS, ventilator status and mobility, which can be updated periodically as the patient's condition evolves. All clinical data are stored and managed in a database to support monitoring.

The Data Integration and Intelligence Layer handles data harmonization and predictive inference. This layer standardizes clinical variables and adjusts data formats before processing the data with an XGBoost-based machine learning model (Rahmanti et al. 2026) to generate patient outcome probabilities for each clinical state.

The Digital Twin Representation and Monitoring Layer presents a virtual representation of the patient and predictive results through an interactive dashboard. This layer includes a body map visualization that anatomically represents the patient's clinical condition, a trajectory chart that displays the evolution of outcome probabilities over time, and a summary of the patient's condition to support clinical monitoring.

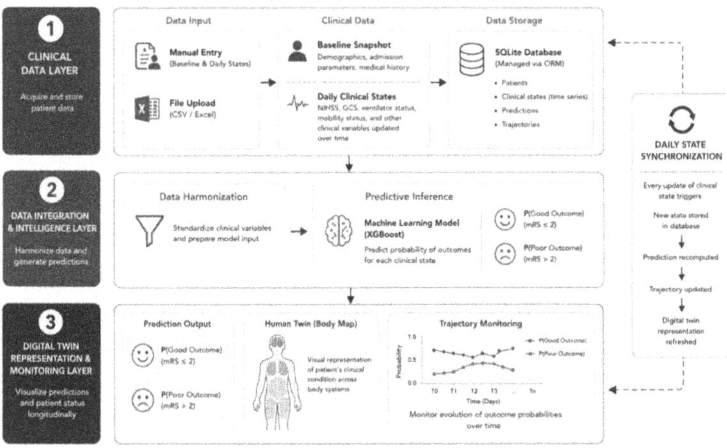

Fig. 2. Digital twin architecture

All three layers are connected through a Daily State Synchronization mechanism, where each update to the patient's clinical condition automatically updates the prediction results, trajectory and virtual representation of the patient, making sure that the digital twin continuously reflects the actual condition of the patient.

3.4 Evaluation Criteria

System evaluation followed criteria mapped to the three architectural layers and the study objectives. For the Clinical Data Layer, the criterion was data acquisition and storage integrity: registry records and clinical state updates are imported and persisted without loss or error. For the Data Integration and Intelligence Layer, two criteria applied: variable harmonization, ensuring that local RSA variables are correctly mapped to the model input schema; and predictive validity, assessing whether the resulting predictions are clinically meaningful. Predictive validity was evaluated by comparing admission predictions generated through the system endpoint against ground-truth discharge mRS using the area under the ROC curve (AUROC), with 95% confidence intervals obtained through bootstrap resampling. For the Digital Twin Representation and Monitoring Layer, the criterion was representation fidelity: clinical state updates correctly propagate to the prediction, trajectory, and visual representation, including explicit handling of missing data. Cross-cutting these layers, component connectivity was verified through integration testing of the REST API. An expert review was conducted involving a neurologist with 10 years of experience and a representative from the hospital information technology (IT) department, both from Universitas Gadjah Mada Academic Hospital. The evaluation aimed to assess the clinical plausibility of the prediction workflow, the interpretability of the digital twin interface, and the practical feasibility of the system within a hospital setting.

4 Result

As summarized in Table 1, all endpoints returned a status code of 200 with acceptable response times. The POST /upload endpoint processed 350 patient records in 1513 ms, while the prediction endpoints, POST /patients and POST /patients/{id}/states, generated responses within 155 ms and 159 ms, respectively. The shorter response time of the GET/patients/{id}/trajectory endpoint (13 ms) reflects its lower computational load, as it retrieves stored prediction data rather than executing the machine learning inference pipeline.

Table 1. Integration testing results

Request	Status code	Response time	Result
POST/patients	200	155 ms	Predictions Success
POST/patients/{id}/states	200	159 ms	Prediction value changes
GET/patients/{id}/trajectory	200	13 ms	Trajectory prediction
POST/upload	200	1513 ms	Historical prediction values obtained

To support deployment of the prediction model, local RSA stroke registry variables were harmonized with the model input schema as shown in → Table 2. Of the 28 required predictor variables, 16 were mapped directly, 6 required renaming, 3 required value transformation, 1 required both renaming and transformation, and 1 was derived from existing registry fields. Two variables were unavailable in the local registry and were handled as missing inputs during inference.

Table 2. Mapping variables

Mapping method	Count
Direct	16
Renamed	6
Transformation	3
Renamed + Transformation	1
Derived Variable	1
Not Available	2
Total	28

Predictive validity was evaluated using admission predictions generated through the deployed system pipeline and compared against discharge functional outcomes (mRS) from the local stroke registry. Of the 350 admission predictions generated by the system, 347 were successfully matched to complete discharge outcome records and included in the analysis. The deployed pipeline achieved an AUROC of 0.81 (95% CI: 0.76–0.85), indicating good discrimination between favorable and unfavorable discharge outcomes. These findings suggest that the prediction model retained clinically meaningful performance after integration into the digital twin platform (Table 3).

Table 3. Predictive validity evaluation results

Metric	Value
Patients evaluated	347
Poor outcome prevalence	46.4%
AUROC	0.81
95% CI	0.76–0.85

Figure 3 presents the Human Twin Dashboard used for individual patient monitoring. The interface integrates baseline patient information, current clinical status, predicted functional outcome, and an anatomical body-map representation within a single view. The Human Twin visualizes neurological and systemic conditions using color-coded indicators, where green indicates favorable status, yellow indicates moderate concern, red indicates clinically unfavorable status, and grey indicates unavailable data. The system simultaneously displays predicted probabilities and clinically relevant factors associated with the current prediction.

Figure 4 illustrates the disease progression monitoring interface. The trajectory visualization displays changes in predicted outcome probabilities across multiple clinical states, enabling longitudinal monitoring of patient condition over time. The accompanying Clinical State Evolution table summarizes the corresponding clinical variables and prediction results for each recorded state, providing a transparent representation of how patient status changes throughout hospitalization.

The prototype was reviewed by a neurologist to obtain feedback regarding the clinical relevance of the proposed digital twin interface. The review confirmed that the integrated monitoring workflow was clinically interpretable and identified several opportunities to improve information presentation. Based on this feedback, the system was enhanced with a Patient Overview dashboard to provide a concise summary of the patient's current condition and prediction status, as well as a Clinical Factors panel to display variables associated with the current prediction result. The review also highlighted a potential limitation related to the inclusion of variables that may introduce information leakage into the prediction model, informing future model refinement efforts.

Overall, the test results demonstrated that all system components can be integrated and run as a single, integrated system. This validates the proposed architecture's feasibility as a proof-of-concept for a digital twin stroke platform based on functional outcome prediction.

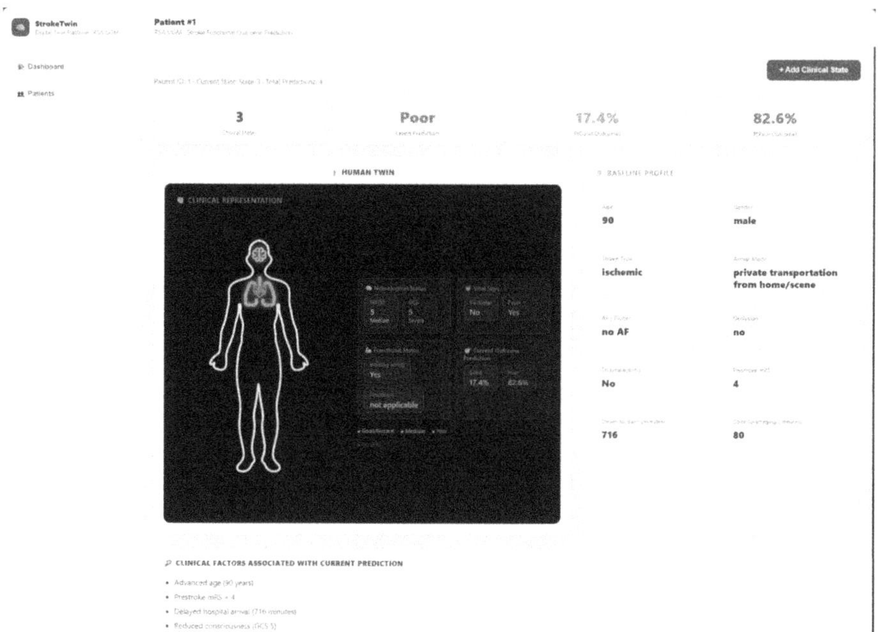

Fig. 3 Human twin dashboard and patient monitoring interface

5 Conclusion

This study presents a proof-of-concept digital twin platform for monitoring stroke patients using local registry data. The platform integrates three architectural layers: the Clinical Data Layer, the Data Integration and Intelligence Layer, and the Digital

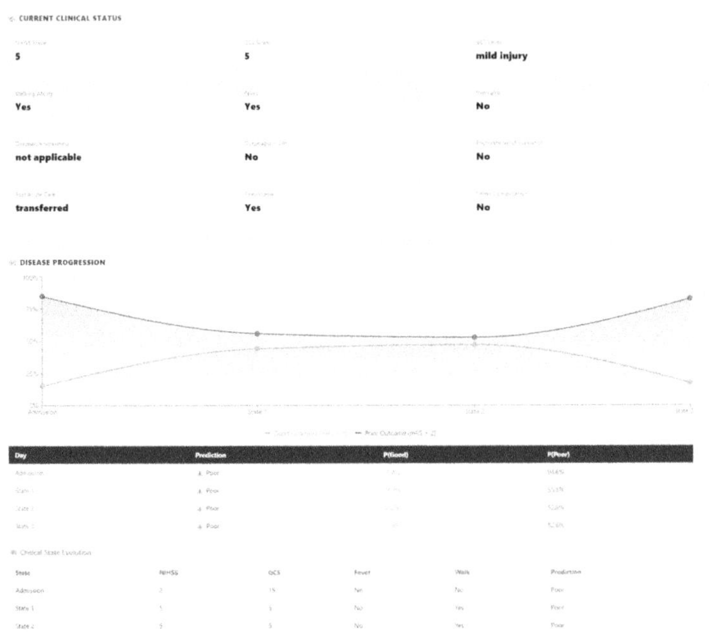

Fig. 4 Disease progression monitoring and prediction trajectory

Twin Representation and Monitoring Layer to support functional outcome prediction and longitudinal patient monitoring.

The evaluation results demonstrated the feasibility of the proposed architecture. Integration testing confirmed successful communication between system components, while variable harmonization enabled local stroke registry data to be aligned with the prediction model input schema. Predictive validity evaluation showed that the deployed pipeline achieved an AUROC of 0.81 (95% CI: 0.76–0.85) for admission prediction against discharge mRS outcomes. The platform was also able to synchronize clinical state updates, generate updated predictions, and visualize patient progression through Human Twin and trajectory monitoring interfaces.

The main contribution of this study is the integration of clinical data acquisition, variable harmonization, predictive model inference, clinical state synchronization, and interactive digital twin visualization within a single platform. Unlike conventional stroke prediction systems that provide static predictions, the proposed approach supports evolving patient representations through longitudinal monitoring of clinical states and predicted outcomes.

As a proof-of-concept, several areas remain for future development. The current platform is limited to predictive inference and does not yet support dynamic simulation or what-if scenario analysis. Integration with hospital Electronic Health Records (EHR) would enable automated real-time data acquisition. To further address questions of clinical validity and system adoptability, future studies should incorporate a formal usability evaluation using a validated instrument such as the System Usability Scale (SUS) and prospective clinical validation to assess prediction performance in real-world settings. Model refinement to mitigate the information leakage limitation identified during expert evaluation is also warranted.

Acknowledgments. This work was supported by The British Council ISPF Research Collaborations Programme, Grant ID #1203701580; the Academy of Medical Sciences' ISPF Networking Grant, Grant ID: NGR1\1961; and the Pasca Doctoral - Community Funding Grant (3567/UN1/KU.1.3/PP/PT.01.01/2026).

Competing Interests The authors declare that they have no competing interests.

Ethics Approval The dataset used in this study was derived from the UGM Academic Hospital Stroke Registry and approved by the Medical and Health Research Ethics Committee (MHREC), Faculty of Medicine, Public Health and Nursing, Universitas Gadjah Mada - Dr. Sardjito General Hospital under approval number KE/FK/1408/EC/2025.

References

Allen, A., Siefkas, A., Pellegrini, E., et al.: A digital twins machine learning model for forecasting disease progression in stroke patients. Appl. Sci. (Switz.). **11** (2021). https://doi.org/10.3390/app11125576

Chen, M., Qian, D., Wang, Y., et al.: Systematic review of machine learning applied to the secondary prevention of ischemic stroke. J. Med. Syst. **48**, 8 (2024). https://doi.org/10.1007/s10916-023-02020-4

Chen, Y., Wang, W., Diao, J., et al.: Digital-twin-based patient evaluation during stroke rehabilitation∗. In: ICCPS 2023 - Proceedings of the 2023 ACM/IEEE 14th International Conference on Cyber-Physical Systems with CPS-IoT Week 2023, pp. 22–33. Association for Computing Machinery, Inc. (2023)

Feigin, V.L., Abate, M.D., Abate, Y.H., et al.: Global, regional, and national burden of stroke and its risk factors, 1990–2021: a systematic analysis for the global burden of disease study 2021. Lancet Neurol. **23**, 973–1003 (2024). https://doi.org/10.1016/S1474-4422(24)00369-7

Guan, W.: Construction of digital twin model for acute ischemic stroke patients and its application in disease management. Front. Med. Sci. Res. **7** (2025). https://doi.org/10.25236/fmsr.2025.070210

Heo, J., Yoon, J.G., Park, H., et al.: Machine Learning–Based Model for Prediction of Outcomes in Acute Stroke. Stroke. **50**, 1263–1265 (2019). https://doi.org/10.1161/STROKEAHA.118.024293

Herrgårdh, T., Hunter, E., Tunedal, K., et al.: Digital twins and hybrid modelling for simulation of physiological variables and stroke risk. bioRxiv. (2022). https://doi.org/10.1101/2022.03.25.485803

Hussain, I., MdA, H., Park, S.-J.: A healthcare digital twin for diagnosis of stroke. In: 2021 IEEE International Conference on Biomedical Engineering, Computer and Information Technology for Health (BECITHCON), pp. 18–21. IEEE (2021)

Lauer-Schmaltz, M.W., Cash, P., Paulin Hansen, J., Das, N.: Human digital twins in rehabilitation: a case study on exoskeleton and serious-game-based stroke rehabilitation using the ETHICA methodology. IEEE Access. **12**, 180968–180991 (2024). https://doi.org/10.1109/ACCESS.2024.3508029

Li, X., Loscalzo, J., Mahmud, A.K.M.F., et al.: Digital twins as global learning health and disease models for preventive and personalized medicine. Genome Med. **17**, 11 (2025). https://doi.org/10.1186/s13073-025-01435-7

Liu, Y., Yu, Y., Ouyang, J., et al.: Functional outcome prediction in acute ischemic stroke using a fused imaging and clinical deep learning model. Stroke. **54**, 2316–2327 (2023). https://doi.org/10.1161/STROKEAHA.123.044072

Metcalfe, B., Boshuizen, H.C., Bulens, J., Koehorst, J.J.: Digital twin maturity levels: a theoretical framework for defining capabilities and goals in the life and environmental sciences. F1000Res. **12**, 961 (2023). https://doi.org/10.12688/f1000research.137262.1

Mohamed, N., Al-Jaroodi, J., Jawhar, I., Kesserwan, N.: Leveraging digital twins for healthcare systems engineering. IEEE Access. **11**, 69841–69853 (2023). https://doi.org/10.1109/ACCESS.2023.3292119

Muhawarman A.: Cegah Stroke dengan Aktivitas Fisik (2024). https://kemkes.go.id/eng/cegah-stroke-dengan-aktivitas-fisik

Presman, R.S.: Software Engineering, 5th edn. McGraw Hill, New York (2001)

Rahmanti, A.R., et al.: Prediction of early functional outcome after acute ischemic stroke using real-world clinical data in Vietnam and Indonesia: retrospective cohort study. Stud. Health Technol. Inform. **330**, 218–225 (2026). https://doi.org/10.3233/SHTI260140

Roopa, M.S., Venugopal, K.R.: Digital twins for cyber-physical healthcare systems: architecture, requirements, systematic analysis, and future prospects. IEEE Access. **13**, 44963–44996 (2025). https://doi.org/10.1109/ACCESS.2025.3547991

Upadrista, V., Nazir, S., Tianfield, H.: Blockchain-enabled digital twin system for brain stroke prediction. Brain Inform. **12** (2025). https://doi.org/10.1186/s40708-024-00247-6

Vargas, P., Maldonado-Diaz, M., Gutiérrez-Panchana, T.: Early prediction of functional mobility severity after stroke: two key milestones. J. Neurol. Sci. **466** (2024). https://doi.org/10.1016/j.jns.2024.123278

Venketasubramanian, N., Yudiarto, F.L., Tugasworo, D.: Stroke burden and stroke services in Indonesia. Cerebrovasc Dis Extra. **12**, 53–57 (2022). https://doi.org/10.1159/000524161

Yaputra, H.: Kemenkes: Tren Kasus Stroke Alami Peningkatan dan Jadi Penyebab Kematian Tertinggi (2024)

Longitudinal Modelling of Cognitive Impairment Progression among Malaysian Older Adults Using Multimodal Assessments

Afiqah Anum Yusof[1], Nurfadhlina Mohd Sharef[1,2,3](✉), Fakhrul Zaman Rokhani[2,4], and Mohamad Fazdillah Bagat[2]

[1] Faculty of Computer Science and Information Technology, Universiti Putra Malaysia, Serdang, Malaysia
nurfadhlina@upm.edu.my
[2] Institute of Ageing Research (MyAgeing™), Universiti Putra Malaysia, Serdang, Malaysia
[3] Institute of Mathematical Research, Universiti Putra Malaysia, Serdang, Malaysia
[4] Faculty of Engineering, Universiti Putra Malaysia, Serdang, Malaysia

Abstract. Early detection of cognitive decline is essential to facilitate effective early interventions and to foster healthy ageing. Although the current assessment tools to screen cognitive impairment (CI) are numerous, their direct predictive value is limited. To advance the prediction of cognitive impairment progression in aged individuals, this article studies using different machine learning models built upon demographic, psychosocial and neuropsychiatric evaluations that were obtained from the Malaysian LRGS-TUA longitudinal cohort. The model integrates variables including age, education level, GDS-15, RAVLT, Digit Span, Digit Symbol, MoCA and MMSE to model the progression of MMSE over the course of 18 months. We examined three machine learning classifiers i.e., Multinomial Logistic Regression (MLR), Random Forest (RF) and Support Vector Machine (SVM) in various settings of resampling, feature selection and hyperparameter optimisation. The SMOTE+ENN performance addressed class imbalance effectively, while the RF performed the best in terms of predicting both the best accuracy without feature selection (96.16%) and accuracy with a set of 17 fewer features (94.13%). The results show that cognitive impairment progression can be well-modelled using non-invasive, accessible tests without the use of neuroimaging, genetic information, or biomarker data. These methods may facilitate the surveillance and decision making in ageing populations for cognitive decline.

Keywords: Cognitive Impairment Modeling · Cognitive Impairment Progression · Longitudinal Ageing Study · Machine Learning

1 Introduction

Alzheimer's Disease (AD) represents a significant public health challenge, particularly as the global population ages. The prevalence of AD is increasing at an alarming rate, leading to substantial economic and social burdens. According to recent statistics, the

N. Mohd Sharef et al. (Eds.): CAMP 2026, CCIS 3094, pp. 59–71, 2027.
https://doi.org/10.1007/978-981-92-4305-1_6

number of individuals affected by AD is projected to rise dramatically, with estimates suggesting that by 2050, nearly 14 million people in the United States alone will be living with the disease (Alzheimer's Association, 2023). This surge underscores the urgent need for effective diagnostic and therapeutic strategies to manage and mitigate the impact of AD on individuals and healthcare systems.

Early diagnosis of AD is crucial for several reasons. It allows for timely intervention, which can slow disease progression, improve quality of life, and reduce the overall burden on caregivers and healthcare providers. Early detection also facilitates participation in clinical trials, which are essential for developing new treatments (Rasmussen & Langerman, 2019). Despite the availability of various diagnostic tools, their standalone predictive capabilities are often limited, necessitating the exploration of combined assessment approaches. This, in turn, enhances the likelihood of successful cognitive aging (Muhammed Niyas & Thiyagarajan, 2023).

Several screening instruments including MMSE, MoCA, and neuropsychological assessments (Vanoh et al., 2016a) are widely used for cognitive impairment (CI) screening. However, MMSE alone has limited predictive capability for progression from CI to dementia, motivating the need for multimodal prediction approaches (Arevalo-Rodriguez et al., 2021; Devos et al., 2022).

These challenges highlight the need for alternative approaches, such as machine learning-based CI progression prediction. Our hypothesis is that, combining demographic factors (e.g., education level and age), psychosocial assessments (e.g., GDS15), and neuropsychiatric evaluations (e.g., RAVLT and digit span) from both initial and subsequent assessments, machine learning models can enhance early detection and monitoring of CI. This approach offers the potential to not only overcome the constraints of neuroimaging but also serve as a scalable and accessible screening approach for monitoring cognitive impairment progression in older adults.

While MMSE remains one of the most widely used cognitive screening instruments, evidence suggests that MMSE alone is insufficient for predicting cognitive impairment progression. There remains a need to investigate how complementary assessments such as MoCA, RAVLT, Digit Span, Digit Symbol and GDS-15 can collectively improve progression modelling.

Hence, this paper presents the investigation of various computational techniques used to predict the CI progression of the elderly using machine learning based models that utilized MMSE as the screening tool. While the MMSE serves as a commonly used instrument for assessing global cognitive function, this study acknowledges the established understanding that its standalone predictive capabilities for cognitive decline, particularly the progression from MCI to dementia, are limited.

This study makes four contributions. First, it studies cognitive impairment progression prediction based on longitudinal data of the Malaysian LRGS-TUA cohort, one of the few ageing datasets representing Malaysian older adults which has previously been used to investigate cognitive frailty among Malaysian older adults (Rivan et al., 2019). Second, it combines demographic, psychosocial and neuropsychiatric assessments within a single machine learning framework. Third, it assesses the effects of class imbalance handling, feature selection and hyperparameter optimisation on predictive performance. Finally, it demonstrates that cognitive impairment progression can be modelled effectively using accessible non-invasive assessments without relying on neuroimaging, genetic information or biomarker data.

This paper is constructed over five parts. The first section provides the introduction on a Malaysian older adult cognitive frailty assessment study and gives a background on our proposed work. The second section covers a review of related works in CI prediction. The proposed work predictive models are elaborated in section three, followed by results and discussion. The final section of the paper provides the conclusion and future work.

2 Related Work

CI is commonly assessed using screening instruments such as the Mini-Mental State Examination (MMSE), Montreal Cognitive Assessment (MoCA), and other neuropsychological tests. MMSE remains one of the most widely used tools due to its simplicity and low administration cost. However, previous studies have reported that MMSE alone has limited sensitivity for identifying mild CI (MCI) and predicting future cognitive decline. Consequently, researchers have increasingly explored the integration of multiple assessment modalities, including demographic, psychosocial, neuropsychological, biomarker, and neuroimaging data, to improve prediction performance and support early intervention.

Various approaches to data-driven analysis have been introduced. Researchers have also combined structured data like clinical information and unstructured data like MRI scans, PET scans from diverse sources to explore MCI and its progression to Alzheimer's disease (Kang et al., 2019; Muhammed Niyas & Thiyagarajan, 2023). There were also works that combined MRI data with neuro-psychological metrics such as MMSE and clinical demographics while another study employs a comprehensive mix of MRI, PET, CSF, demographics, and neuro-psychological data as well as genetic details. Socio-demographics, vascular risk factors, and structural neuroimaging markers for early diagnosis of CI in a multi-ethnic Asian population has been recently introduced for early diagnosis of CI to facilitate timely interventions to mitigate progression (Tan et al., 2023).

Besides the above issues, the number of existing studies on predicting the progression of MCI using the MMSE is still limited. While some studies have used longitudinal data to predict MCI, not all of them have specifically focused on disease progression, which is an important aspect of MCI research for identifying those at greater risk for developing dementia and who may benefit from early interventions. Although the asymptomatic behavior in the early stages makes the identification of the MCI onset particularly challenging, there are several actively ongoing works are in various countries. The scope of works addresses the early identification of MCI (Campbell et al., 2018; Kang et al., 2019; Michaud et al., 2017), the design of personalized interventions and the promotion of patients empowerment towards healthy ageing. Existing works can be further divided into several categories such as risk factors of conversion from normal cognition to MCI, and progression prediction (Cremers et al., 2020; Li et al., 2017; Li & Luo, 2019; Williams et al., 2021) using statistics, mathematical model and machine learning. Multimodal parameters are used covering biochemical, to imaging, pathological, and personal information (Albright, 2019; Bandosz, 2017; Hamid et al., 2021; Id et al.,

2019; Nguyen et al., 2020; Parnetti et al., 2019; Tang et al., 2015; L. Wang et al., 2019; T. Wang et al., 2018; Yu et al., 2020).

Understanding and predicting the progression of MCI remains a critical challenge in cognitive health research, particularly in identifying individuals at higher risk of developing dementia. While considerable advancements have been made in leveraging multimodal parameters ranging from biochemical markers to imaging and personal data imbalanced data within these studies poses a significant barrier to achieving accurate and equitable predictions. Disparities in the representation of subgroups or stages of cognitive decline can skew machine learning models, limiting their ability to generalize effectively and increasing the risk of biased outcomes. Addressing these imbalances is essential for developing fair and reliable predictive tools that can support early interventions and improve patient outcomes.

Existing studies can generally be categorised into imaging-based approaches, biomarker-driven approaches, and assessment-based approaches, each with different trade-offs in terms of accessibility, cost, and predictive capability. Although numerous studies have explored cognitive impairment prediction using machine learning, several challenges remain. First, many recent approaches rely heavily on neuroimaging, biomarker, or genetic information to achieve high predictive performance, limiting their practicality for large-scale community screening and resource-constrained settings (Devos et al., 2022; Tan et al., 2023). Second, despite increasing interest in machine learning for cognitive impairment research, relatively fewer studies focus on modelling cognitive impairment progression using longitudinal assessments collected from older adults, with many investigations concentrating on cross-sectional diagnosis instead (Muhammed Niyas & Thiyagarajan, 2023). Third, while multimodal prediction frameworks are gaining attention, the combined contribution of demographic, psychosocial, and neuropsychiatric assessments remains underexplored, particularly within Malaysian ageing populations. These gaps motivate the need for an accessible and non-invasive prediction framework capable of modelling cognitive impairment progression using routinely collected assessment data from older adults.

3 Cognitive Impairment Progression Modeling with Descriptive and Predictive Analytics

To address the challenges of predicting CI progression, we employed a comprehensive methodological framework that integrates demographic, psychosocial, and neuropsychiatric data, as detailed in this section. This section presents the data, experiments and output from the analysis.

3.1 Data Source

In this study, the LRGS TUA dataset (secondary data) provided by the Malaysian Research Institute on Ageing (MyAgeing) was utilized. The dataset contains 1,844 longitudinal observations collected from 922 individuals aged 60 to 90 years old across two assessment periods (T0 and T18). The target variable is MMSE_Range_T18. This dataset is genuine and was gathered based on the interactions with elderly individuals in four different states in Malaysia, namely Kelantan, Selangor, Perak, and Johor conducted over a span of three years, 466 males (50.5%) and 456 females (49.45%) were asked to complete a series of assessments to determine their cognitive status. In the dataset, 960 individuals are categorized as having a normal MMSE score, 575 individuals are categorized as having MCI, 284 individuals are categorized as moderate, and 25 individuals are categorized as severe.

3.2 Ethical Consideration

The respondents were provided with an information sheet and the study was explained to them verbally, providing an opportunity for them to discuss any concerns prior to providing written consent. The study protocol was approved by the Medical Research and Ethics Committee, Ministry of Health Malaysia (NMRR-13-1023-14660 IIR). All methods were performed in accordance with the relevant guidelines and regulations (Declaration of Helsinki). Informed consent was obtained from all subjects involved in the study.

3.3 Dataset Content Description

The dataset is a combination of two instruments which are psychosocial and neuropsychiatric as shown in Table 1. The scope of the study revolves around utilising MMSE as the CI test. While the MMSE can be a valuable screening instrument, it is not a definitive diagnostic test for dementia or MCI. It should be used in conjunction with other tests and evaluations by a healthcare expert. The dataset contains a total of 3667 attributes but only 25 attributes from the neuropsychiatric instruments are used. All the selected features and their descriptions are as shown in Tables 2 and 3.

Table 1. Dataset description

Category	Variables	Description	Score interpretation
Demographic	Gender, Age_T0, Age_T18, Education_Level, Race, Marital_Status, Living, Smoking, Current_Job, UrbanRural, States	Participant demographic and lifestyle information	N/A
MMSE	MMSE_T0, MMSE_T18	Global cognitive assessment	Normal (25–30), MCI (21–24), Moderate (13–20), Severe (<12)
MoCA	MoCA_T0, MoCA_T18	Cognitive screening assessment	Normal (25–30), MCI (18–24), Severe (<17)
Digit Span	DigitSpan_T0, DigitSpan_T18	Working memory assessment	Continuous score

(*continued*)

Table 1. (*continued*)

Category	Variables	Description	Score interpretation
Digit Symbol	DigitSymbol_T0, DigitSymbol_T18	Processing speed and attention	Continuous score
RAVLT Immediate	RAVLTImmediate_T0, RAVLTImmediate_T18	Verbal learning and memory	Continuous score
RAVLT Forgetting	RAVLTPercentForgetting_T0, RAVLTPercentForgetting_T18	Memory retention performance	Continuous score
GDS-15	GDS15_T0, GDS15_T18	Depression screening	Normal (0–4), Mild (5–8), Moderate (9–11), Severe (12–15)

3.4 MMSE Progression Prediction

We leverage the LRGS TUA dataset to explore the potential of machine learning models to predict the trajectory of cognitive level based on MMSE assessments (target prediction is MMSE level at month 18, T18), offering a more dynamic and personalized approach to understanding the progression of CI. By incorporating various features and factors, these models aim to provide valuable insights into the future cognitive health of individuals, contributing to earlier detection and tailored interventions for those at risk of cognitive decline.

Several predictor variables were obtained during the T18 follow-up assessment, during the same assessment period as the target variable (MMSE_Range_T18). These variables were included as they stem from independent demographic, psychosocial, and neuropsychiatric instruments that examine complementary aspects of cognitive health and were not determined via MMSE scores. As a result, this study aims to explore whether the sensitivity of cognitive impairment severity can be modelled at T18 follow-up using multimodal assessment. However, the concurrent follow-up assessments might create information overlap and induce optimistic performance estimates as compared to a purely prospective prediction scenario. Future studies should evaluate models using only baseline assessments (T0) to establish true prospective prediction performance.

The flow and setting of the MMSE progression prediction is captured in Fig. 1 which depicts that the process begins with data understanding, followed by data pre-preparation and data pre-processing, before the model development. Model optimization is also implemented before the best model is selected for deployment. Further details are provided below.

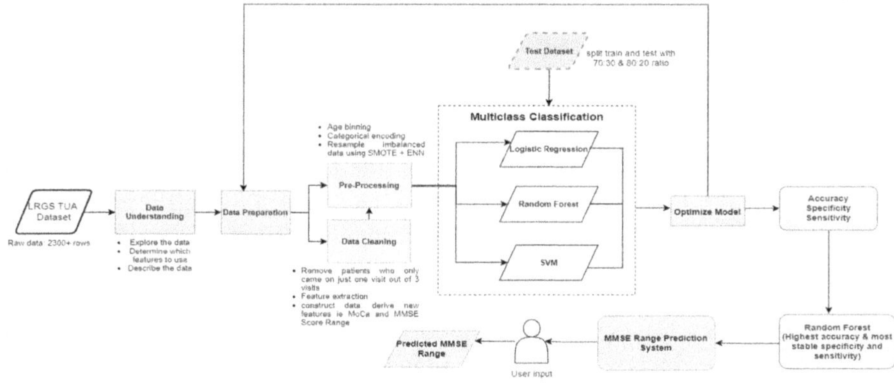

Fig. 1. MMSE progression prediction processes

3.4.1 Data Pre-processing

To address the issue of class imbalance in the MMSE case distribution, we employed SMOTE (Synthetic Minority Over-sampling Technique) to generate synthetic data points for the minority class. Additionally, we used ENN (Edited Nearest Neighbors) to balance the dataset by removing misclassified data points from the majority class. ENN works by identifying and eliminating instances in the majority class that are incorrectly classified by their nearest neighbors. This helps reduce the dominance of the majority class and can improve the classifier's performance on imbalanced datasets.

When combined, SMOTE and ENN (often referred to as SMOTE-ENN) effectively balance the dataset by oversampling the minority class and cleaning the majority class. SMOTE-ENN removes additional noisy or overlapping instances from the dataset, further refining the data. Specifically, ENN evaluates the nearest neighbors of each majority class instance and removes it if those neighbors misclassify it. This process, when applied alongside SMOTE's synthetic data generation, ensures a more balanced and cleaner dataset, enhancing the classifier's performance.

Feature selection employs an integrated approach that combines statistical ranking and domain knowledge. Chi-square ranking was first applied to find out 10 most informative variables associated with MMSE_Range_T18. Nonetheless, statistically rank feature selection will not consider clinically significant features whose statistical importance is less within that specific dataset. As such, seven more variables were retained, based on evidence of cognitive ageing literature and expert knowledge, being Gender, Education_Level, RAVLTPercentForgetting_T0, RAVLTPercentForgetting_T18, GDS15_Range_T0, GDS15_Range_T18 and AGE_binned_T18. These constructs were included in the study because in previous studies, the age, the education, the depressive symptoms, the performance of memory retention were found to be robust predictors of cognitive impairment and cognitive reserve.

The age group is one of the most potent known risk factors for cognitive impairment, and it is well established that the education level is an appropriate proxy to estimate the cognitive reserve (Campbell et al., 2013; Devos et al., 2022; Vanoh et al., 2016;

Williams et al., 2021). Depressive symptoms assessed by means of the GDS-15 score have been linked with an elevated risk of cognitive decline, and RAVLT forgetting scores capture deficits in memory retention that are frequently seen in those already experiencing cognitive impairment. If these clinically relevant variables are retained, the hybrid feature selection strategy presents us with balanced statistical evidence and domain know-how, thus reducing model complexity while keeping predictors as such that have previous value to predicting the progression of cognitive impairment.

We also added seven additional features to improve the predictive model: Gender, Education_Level, RAVLTPercentForgetting_T0, RAVLTPercentForgetting_T18, GDS15_Range_T0, GDS15_Range_T18, and AGE_binned_t18. To evaluate the model's performance, we used RepeatedStratifiedKFold, which splits the dataset into 10 stratified folds and repeats the process 3 times. This ensures that the performance results are robust and account for variability in the data. Finally, we applied grid search to optimize the model's hyperparameters, ensuring the best possible performance for our predictive model.

3.4.2 Model Development

The number of attributes was decreased after the pre-processing in the data preparation stage, and the dataset is now more refined for creating data mining models. For quantitatively predicting the occurrence of a target variable, a variety of data mining techniques can be employed. In this study, classification is used to make predictions. Classification, one of the most well-known data mining supervised learning techniques, extracts valuable information from enormous datasets and can be applied to forecast unknown classes.

In this study, our goal is to develop a model that can predict the MMSE_Range_T18, MMSE_Range_T18 represents the level of cognitive function (classified as Normal, Mild, Moderate, or Severe) measured 18 months after the initial assessment. To achieve this, we use three different machine learning techniques: Multinomial Logistic Regression (MLR), Random Forest (RF), and Support Vector Machine (SVM). These methods help us find patterns in the data that can map the input features to the cognitive function categories.

RF is a method that builds multiple "decision trees" (like flowcharts) and combines their results to make a final prediction. It is known for being accurate, reliable, and able to handle large or complex datasets, even when there are many features or some noise in the data. This makes it a powerful tool for our analysis. MLR is a statistical approach used when the outcome we want to predict has more than two categories (in this case, four levels of cognitive function). It calculates the probability of each category based on the input features. MLR is straightforward to interpret and works well for multi-class problems like ours. SVM is another technique that tries to find the best dividing line (or boundary) between different categories of data. It focuses on maximizing the gap between this boundary and the closest data points, which helps it perform well even with complex or high-dimensional data. SVM is also robust to noise and can handle both simple and intricate patterns in the data.

4 Experiments and Results

Combining demography, psychosocial and neuropsychiatric features, three experiments are devised as the empirical identification of the best approach for developing the MCI progression prediction (target is MMSE_Range_T18), by focusing on (i) training ratios and resampling, (ii) cross-validation without feature selection, and (iii) cross-validation with feature selection. The details of how this is done are explained below.

4.1 Experiment 1: Training Ratios and Resampling

Experiment 1 evaluated the effect of resampling strategies on prediction performance. Table 2 shows the performance of the classifiers across multiple resampling strategies. SMOTE+ENN showed above average accuracy, precision, and recall, across all classifiers, in a significant way in the context of training against the original imbalanced dataset and Random Oversampling. The fact that the same precision in the classifier increased with recall suggests that the performance improvements which SMOTE+ENN achieved are indicative of true class separability, rather than the optimization based on metrics. The highest performance of SVM among classifiers evaluated was obtained after SMOTE+ENN, where it is accurate 91.13%, precision 0.88 and recall 0.90. The same trends apply to MLR and RF. This suggests how the class imbalance is a major problem in the modelling stages of the cognitive impairment progression and the oversampling and cleansing mechanism of SMOTE+ENN, together it is the best representation of minority cognitive states.

Table 2. Accuracy, precision and recall under different resampling methods

	Classifier	Train (%)	Test (%)	Before resample	SMOTE + ENN	Random over sampler
Accuracy	MLR	70	30	65.22	87.77	69.57
		80	20	68.48	88.13	71.60
	RF	70	30	71.01	87.77	72.53
		80	20	*71.20	86.69	73.83
	SVM	70	30	65.94	*91.13	78.95
		80	20	67.39	89.93	*79.51
Precision	MLR	70	30	0.44	0.85	0.85
		80	20	0.46	0.86	0.86
	RF	70	30	*0.48	0.84	0.84
		80	20	0.45	0.83	0.83
	SVM	70	30	0.46	*0.88	*0.88
		80	20	0.47	0.86	0.86
Recall	MLR	70	30	0.43	0.85	0.69

(continued)

Table 2. (*continued*)

Classifier	Train (%)	Test (%)	Before resample	SMOTE + ENN	Random over sampler
	80	20	0.45	0.87	0.71
RF	70	30	*0.51	0.86	0.72
	80	20	0.47	0.86	0.73
SVM	70	30	0.45	*0.90	0.79
	80	20	0.45	0.89	*0.80

4.2 Experiment 2 and 3: Feature Selection with Hyperparameter Tuning

Table 3 presents the predictive performance of the classifiers with and without feature selection. Experiment 2, with all 25 feature parameters retained, show RF resulted in the highest accuracy of 96.16%, followed by SVM (92.81%) and MLR (88.13%). RF may be particularly suitable for this problem because it can capture complex non-linear interactions among demographic, psychosocial, and neuropsychiatric factors while remaining robust to noise and feature correlations.

For Experiment 3, the feature selection resulted in a reduction in the number of variables from 25 to 17 using chi-square order. Even with this decrease, prediction performance still seemed strong for all classifiers. RF outperformed the others in accuracy with 94.13%, while SVM and MLR with accuracies of 92.67% and 84.35%, respectively. The smaller decrease in accuracy implies that the feature subset chosen captured nearly all the predictive features needed to model MMSE progression. RF exhibited the greatest robustness when feature reduction was used to compare both the results of two separate experiments using different groups of models with only a small degradation in accuracy from 96.16% to 94.13%.

This suggests that cognitive impairment progression can be well modelled with a shorter set of demographic, psychosocial, and neuropsychiatric variables, enabling the reduction of model complexity while maintaining strong predictive accuracy.

Table 3. Accuracy for experiment 2 and 3

Classifier	Train (%)	Test (%)	Experiment 2 (No feature selection)	Experiment 3 (with feature selection)
MLR	70	30	87.05	84.35
	80	20	88.13	83.52
RF	70	30	96.16	94.13
	80	20	94.96	93.41
SVM	70	30	92.57	92.67
	80	20	92.81	90.84

4.3 Interpretation of Predictive Factors

The selected predictors span multiple dimensions of cognitive health, including global cognition (MMSE, MoCA), memory (RAVLT), working memory (Digit Span), processing speed (Digit Symbol), psychological well-being (GDS-15), and demographic characteristics such as age, gender, and education level. The retention of these variables after feature selection suggests that cognitive impairment progression is influenced by multiple interacting domains rather than a single cognitive assessment.

The inclusion of memory-related measures such as RAVLT and Digit Span aligns with existing evidence that memory deterioration is a key characteristic of cognitive decline. Similarly, GDS-15 highlights the potential influence of psychological well-being on cognitive outcomes, while age and education level have long been recognised as important determinants of cognitive reserve. The simultaneous retention of baseline and follow-up cognitive assessments further suggests that cognitive impairment progression is best understood as a longitudinal process involving multiple interacting cognitive domains rather than a single point-in-time measurement. These findings support the use of multimodal non-invasive assessments for modelling cognitive impairment progression among older adults.

4.4 Comparative Analysis Across Experiments

Best predictive performances of each classifier over three experiments are presented in Table 4. In Experiment 1, which concentrated on the method of correcting class imbalance with resampling, SVM performed best with an accuracy of 91.13%, which suggests that it gained more from the SMOTE+ENN method. RF and MLR obtained similar results with an 87.77% and 88.13% accuracy, respectively. In Experiment 2, RF achieved highest overall accuracy of 96.16%, outperforming both SVM (92.81%) and MLR (88.13%).

This result indicates that when the complete group of features was used, the RF was particularly good at capturing the intricate interaction between demographical, psychosocial and neuropsychiatric variables. Experiment 3 tested the reduction of features using a mixed feature selection strategy. The number of variables was dropped to 17 from 25. Nevertheless, RF still achieved the best accuracy at 94.13%, followed by SVM (92.67%) and MLR (84.35%). RF accuracy was only decreased from 96.16% to 94.13%, suggesting a modest reduction in prediction power, since overall predictive information should be retained even in spite of the reduced model complexity. In general, SVM showed the highest sensitivity to class imbalance while RF proved to be the most robust classifier in the whole modeling algorithm.

These results also imply that cognitive impairment progression could be effectively modelled with a much smaller set of demographic, psychosocial, and neuropsychiatric measures, as the predictive performance of predicting cognitive impairment stays strong.

Table 4. Summary of best predictive performance across experiments

Classifier	Experiment 1 (Best resampling configuration)	Experiment 2 (25 features)	Experiment 3 (17 features)
MLR	88.13 (SMOTE + ENN, 80:20)	88.13	84.35
RF	87.77 (SMOTE + ENN, 70:30)	96.16	94.13
SVM	91.13 (SMOTE + ENN, 70:30)	92.81	92.67

5 Conclusion and Future Work

This study proposes a predictive approach for MMSE progression by integrating demographic factors (e.g., education level, age), psychosocial indicators (e.g., GDS-15), and neuropsychiatric evaluations (e.g., RAVLT and Digit Span) collected during initial and follow-up assessments. By combining multiple assessment tools and demographic factors, this study enhances the predictive accuracy of CI progression, offering a robust framework for early intervention and personalized care strategies.

The findings align with existing literature, which highlights the limitations of using MMSE as a standalone predictor of cognitive decline and underscores the need for a more comprehensive approach (Arevalo-Rodriguez et al., 2021; Devos et al., 2022). RF achieved the highest performance, reaching 96.16% accuracy without feature selection and 94.13% accuracy after feature reduction. The relatively small performance reduction after feature selection suggests that a compact subset of demographic, psychosocial and neuropsychiatric variables is sufficient for modelling cognitive impairment progression.

While this study offers valuable contributions to cognitive aging research, it is important to acknowledge its limitations. The socio-cultural context, education levels, healthcare access, and prevalence of certain health conditions in Malaysia may differ significantly in other populations. These variations could influence the generalizability of the predictive models. Future research should focus on validating these models in more diverse cohorts, including different ethnic groups, nationalities, and healthcare settings, to ensure broader applicability in predicting cognitive decline.

References

Albright, J.: Forecasting the progression of Alzheimer's disease using neural networks and a novel preprocessing algorithm. Alzheimer's & Dementia Transl. Res. Clin. Interventions. **5**, 483–491 (2019). https://doi.org/10.1016/j.trci.2019.07.001

Alzheimer's Association: 2023 Alzheimer's disease facts and figures. Alzheimer's Dementia. **19**(4), 1598–1695 (2023). https://doi.org/10.1002/alz.13016

Arevalo-Rodriguez, I. et al.: Mini-mental state examination (MMSE) for the early detection of dementia in people with mild cognitive impairment (MCI). Cochrane Database Syst. Rev. **2021**(7) (2021). https://doi.org/10.1002/14651858.CD010783.pub3

Campbell, N.L., Unverzagt, F., Lamantia, M.A., Khan, B.A., Boustani, M.A.: Risk factors for the progression of mild cognitive impairment to dementia. Clin. Geriatr. Med. **29**(4), 873–893 (2013). https://doi.org/10.1016/j.cger.2013.07.009

Cremers, L.G.M. et al.: Predicting global cognitive decline in the general population using the disease state index. Front. Aging Neurosci. **11**, 379 (2020). https://doi.org/10.3389/fnagi.2019.00379

Devos, H. et al.: Predicting progression to dementia with comprehensive visual rating scale and machine learning algorithms. Front. Neurol. **13**, 906257 (2022). https://doi.org/10.3389/fneur.2022.906257

Hassler, A.P., Menasalvas, E., García-García, F.J., Rodríguez-Mañas, L., Holzinger, A., Martin-Sánchez, F.: Importance of medical data preprocessing in predictive modeling and risk factor discovery for the frailty syndrome. BMC Med. Inform. Decis. Mak. **19**(1), 33 (2019). https://doi.org/10.1186/s12911-019-0747-6

Kang, M.J. et al.: Prediction of cognitive impairment via deep learning trained with multi-center neuropsychological test data. BMC Med. Inform. Decis. Mak. **19**(231), 1–9 (2019). https://doi.org/10.1186/s12911-019-0952-3

Li, K., Chan, W., Doody, R.S., Quinn, J., Luo, S.: Prediction of conversion to Alzheimer's disease with longitudinal measures and time-to-event data. J. Alzheimer's Dis. **58**(2), 361–371 (2017). https://doi.org/10.3233/JAD-161201

Li, K., Luo, S.: Dynamic prediction of Alzheimer's disease progression using features of multiple longitudinal outcomes and time-to-event data. Stat. Med. **38**(24), 4804–4818 (2019). https://doi.org/10.1002/sim.8334

Michaud, T.L., Su, D., Siahpush, M., Murman, D.L.: The risk of incident mild cognitive impairment and progression to dementia considering mild cognitive impairment subtypes. Dementia Geriatric Cognit. Disord.Extra. **7**(1), 15–29 (2017). https://doi.org/10.1159/000452486

Muhammed Niyas, K.P., Thiyagarajan, P.: A systematic review on early prediction of mild cognitive impairment to Alzheimer's using machine learning algorithms. Int. J. Intell. Netw. **4**, 74–88 (2023). https://doi.org/10.1016/j.ijin.2023.03.004

Rasmussen, J., Langerman, H.: Alzheimer's disease – why we need early diagnosis. Degenerative Neurol. Neuromuscular Dis. **9**, 123–130 (2019). https://doi.org/10.2147/DNND.S228939

Rivan, N.F.M. et al.: Cognitive frailty among Malaysian older adults: baseline findings from the LRGS TUA cohort study. Clin. Interv. Aging. **14**, 1343–1352 (2019). https://doi.org/10.2147/CIA.S211027

Tan, W.Y., Chen, C.H.C., Hilal, S.: A machine learning approach for early diagnosis of cognitive impairment using population-based data. J. Alzheimer's Dis. **91**(1), 449–461 (2023)

Tang, E.Y.H. et al.: Current developments in dementia risk prediction modelling: an updated systematic review. PLoS One. **10**(9), e0136181 (2015). https://doi.org/10.1371/journal.pone.0136181

Vanoh, D., Shahar, S., Rosdinom, R., Din, N.C., Yahya, H.M., Omar, A.: Development of TUA-WELLNESS screening tool for screening risk of mild cognitive impairment among community-dwelling older adults. Clin. Interv. Aging. **11**, 579–587 (2016). https://doi.org/10.2147/CIA.S102925

Williams, M.E. et al.: Twelve-year prediction of mild cognitive impairment aided by Alzheimer's brain signatures at mean age 56. Brain Commun. **3**(3), fcab167 (2021). https://doi.org/10.1093/braincomms/fcab167

Social Media-Based Classification of Mental Health Using Comparative Machine Learning and Deep Learning Techniques

Nurina Humaira Batrisyia Mohd Badli[1], Mohd Shahrul Nizam Mohd Danuri[1(✉)], Rohizah Abd Rahman[2], Che Wan Shamsul Bahri Che Wan Ahmad[3], and Khirulnizam Abd Rahman[3]

[1] Universiti Malaya, Kuala Lumpur, Malaysia
s2197968@siswa.um.edu.my, msnizam@um.edu.my
[2] Universiti Kebangsaan Malaysia, Bangi, Malaysia
rohizah@ukm.edu.my
[3] Universiti Islam Selangor, Kajang, Malaysia
{cwshamsul,khirulnizam}@uis.edu.my

Abstract. Mental health detection through social media platforms offers a scalable and anonymous method to identify early signs of psychological distress. This project investigates the application of both traditional machine learning (ML) and deep learning (DL) techniques for multi-class classification of mental health conditions, specifically stress, anxiety, and depression, using X text data. A structured pipeline was developed that included multilingual text pre-processing, TF-IDF feature extraction, and data balancing using SMOTE and undersampling. Models such as Support Vector Machine (SVM), Random Forest, Naive Bayes, and a Deep Neural Network (DNN) were trained and evaluated on classification performance metrics. Among these, the Random Forest model yielded the highest overall performance across categories, offering both accuracy and interpretability, with macro-F1 scores of 0.83, 0.81, and 0.78 for stress, anxiety, and depression classification, respectively. The system also includes visualization components, such as word clouds and confusion matrices, to improve interpretability. This project provides a reproducible, practical framework for mental health screening on social media, potentially aiding healthcare providers in decision support, subject to clinical validation and support. The findings demonstrate the feasibility of classifying mental health-related expressions in X posts and may support future research on non-clinical screening and human-supervised decision-support systems.

Keywords: Mental health classification · Stress · Anxiety · Depression · Machine Learning · Deep Learning

1 Introduction

Mental health disorders such as depression, anxiety, and stress have become major global public health concerns, contributing significantly to disability, reduced quality of life, and premature mortality worldwide. According to the World Health Organization

N. Mohd Sharef et al. (Eds.): CAMP 2026, CCIS 3094, pp. 72–87, 2027.
https://doi.org/10.1007/978-981-92-4305-1_7

(2022), approximately one in eight people globally live with a mental disorder, with anxiety and depressive disorders being among the most prevalent conditions. Mental health challenges are particularly critical among adolescents and young adults, where mental disorders remain one of the leading causes of disability and impaired functioning. Furthermore, the WHO emphasizes that mental health is not merely the absence of illness, but an essential component of overall well-being and a fundamental human right. Despite increasing awareness, many individuals experiencing mental health problems continue to face stigma, discrimination, limited access to professional care, and reluctance to seek help. In recent years, the rapid growth of social media platforms has created opportunities for researchers to explore digital footprints, such as online posts and textual expressions, as alternative indicators of emotional states and psychological well-being. Consequently, social media data provide an opportunity to examine linguistic patterns associated with self-reported emotional distress and mental health-related experiences (World Health Organization, 2022).

The X, formerly known as Twitter, is a platform where individuals publicly and spontaneously convey their thoughts, and, as such, it provides an opportunity to examine linguistic markers of distress in the wild. Previous research has shown that, compared with non-depressed individuals, those who experience depressive symptoms use not only more first person's pronouns, but also negative emotion words and words that convey a sense of loneliness and helplessness (Guntuku et al., 2017; Al-Mosaiwi & Johnstone, 2018). These patterns can be detected and analyzed in scale using natural language processing (NLP) and machine learning (ML).

Early research contributions to Sentiment Analysis were based on lexicogrammatical and statistical approaches and now seem very limiting because they cannot account for context and ambiguity in human language (Calvo et al., 2017). Since social media can also be multilingual, informal, and noisy, recent studies highlight the use of multilingual NLP models and sophisticated preprocessing to make it more inclusive and accurate (Islam et al., 2018; Shen et al., 2017).

To mitigate the challenge, this project aims to construct a multi-classification system to classify the severity of stress, anxiety and depression as Normal, Mild, Moderate and Severe, using social media data from X. The system incorporates preprocessing stages (translating, cleaning, lemmatizing), as well as, TF-IDF feature extraction and class re-balancing with SMOTE. The long-term vision is to develop a reliable and scalable framework to enable automated classification or preliminary screening, triage and partial intervention for mental health problems emerging from social media behavior.

2 Literature Review

Yazdavar et al. (2017) further developed into the multimodal domain by fusing multimodal information like user profile, behavioral and network interaction to textual modulator information. They proposed a deep learning model to derive emotion tokens from short texts to predict depressive symptoms more accurately. They focused on context-aware models and integration of heterogeneous features in their methodology. A challenge across these studies is the existence of noisy, informal, and multilingual text that may decrease modal performance. Accordingly, translation aids and complicated preprocessing methods are used more often. Moreover, while early research was primarily concerned with binary classification (depressed/not depressed), more recent models seek to predict severity levels and delineate between comorbid disorders such as anxiety and stress, which is congruent with the research directions of the current project.

The ascendance of deep learning (DL) along models such as LSTM and transformers, has enhanced the detection of subtle signals of emotion (Orabi et al., 2018; Trotzek et al., 2018). However, the deep learning models are typically more resource-demanding and especially can need big annotations data. Consequently, many research works, such as this one, use a mixed approach where traditional ML classifiers (e.g., SVM, Random Forest) and DNNs are compared to conclude which is the most suitable for industry applications.

Mental health conditions such as depression, anxiety and stress are increasing internationally. However, stigma, poor clinical access and delayed intervention is leaving many undiagnosed or untreated. Social media, such as X, has been suggested as a candidate source of health-related early warning signals, providing real-time, unfiltered information on user well-being (Guntuku et al., 2017; Islam et al., 2018). Noisy and multilingual text data, class imbalance, and the requirement of explainable models have made it difficult to realize robust, scalable detection models. They are usually very specialized for binary classification or a specific disease only and do not have enough capacity for complex multi-class prediction problems. Accordingly, an inclusive, accurate, and interpretable model is required for the early screening and identification of high risk patients in the real-life mental health care setting.

This study demonstrates the increasing feasibility of using NLP and machine learning for predicting mental health from social media. Yet there is still a gap in developing DL models which are not only interpretable but also computationally efficient enough for real-world deployment.

3 Methodology

In this section, data science approaches concerning the proposed study to develop machine learning and deep learning systems for multi-level classification in mental detection will be explored. The methodology is composed of Data Sourcing, Data Pre-processing, Data Representation, Data Augmentation and Model Training.

3.1 Data Sourcing

The dataset was constructed using X posts sharing personal thoughts or experiences about mental health. We obtained tweets from 19,744 tweets, with each tweet being labeled for three psychological conditions, stress, anxiety and depression each on a four point scale such as Normal (0), Mild (1), Moderate (2) and Severe (3). Data collection was initiated by extracting a targeted dataset from X (formerly Twitter) utilizing a comprehensive search syntax that paired platform-specific identifiers such as "online social network", "social media", "microblog" and "Twitter" with clinically validated mental health terminology. To ensure diagnostic alignment, these clinical search terms were derived from the United Kingdom National Institute for Health and Care Excellence (NICE) guidelines, incorporating descriptors such as "mental health", "mental illness", "mental disorder", "psychological stress", "depression", "anxiety", "suicide" and "post-traumatic stress disorder (PTSD)". Following initial data extraction, the corpus underwent rigorous manual annotation and labeling, which three certified psychologists independently evaluated to establish ground-truth annotations for subsequent analysis. The summary of the whole process is illustrated in Fig. 1.

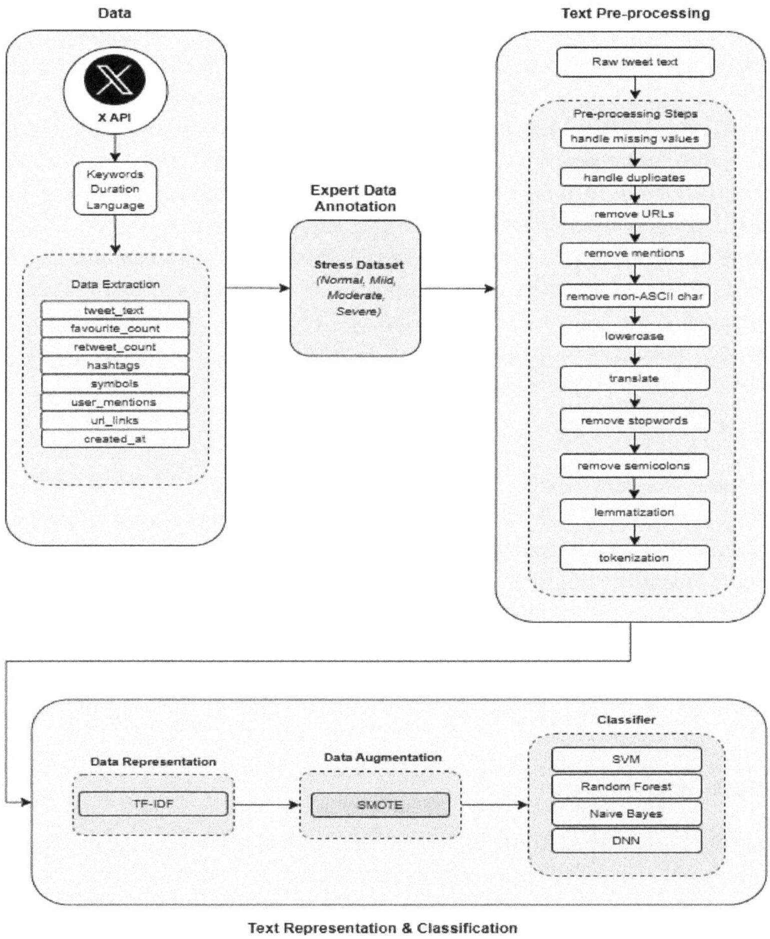

Fig. 1. Data science methodology

3.2 Data Pre-processing

To prepare the X dataset for model training, a comprehensive preprocessing pipeline was implemented to enhance data quality and maintain textual consistency throughout the analysis. Initially, missing or null tweets were removed to ensure that only valid textual data were retained. Duplicate entries were also eliminated to prevent redundancy and reduce potential bias in the dataset. Subsequently, all text was converted to lowercase to standardize word representation and improve uniformity during feature extraction.

In addition, URLs and X mentions such as usernames (@usernames) were removed because they do not contribute significantly to sentiment or mental health classification. Non-ASCII characters, including emojis and special symbols, as well as punctuation marks and semicolons, were also discarded to further clean the textual content. Since the dataset contained multilingual tweets, all tweets were translated into English using the Google Translate API to ensure language consistency across the dataset.

Furthermore, stop words were removed using the predefined stop word list provided by the Natural Language Toolkit (NLTK) library to reduce noise and improve model efficiency. Lemmatization was then applied to transform words into their root or base forms, allowing semantically similar words to be treated uniformly. Finally, tokenization was performed to split the text into individual word tokens, enabling the data to be processed effectively by both machine learning and deep learning models. This preprocessing pipeline ensures that the raw X data is cleaned, standardized and linguistically normalized. By removing noise and transforming text into a consistent format, the data becomes suitable for feature extraction and model training, ultimately improving the accuracy and robustness of mental health classification.

3.3 Data Representation

In this project, we represented the data with the Term Frequency-Inverse Document Frequency (TF-IDF). TF-IDF transforms text into a word importance numerical vector to the data distribution. Unigrams (single words) and bigrams (two-word n-grams) were considered to capture simple and more complex words or phrases that might suggest mental health issues. To reduce dimensionality and computation cost, the 3000 most relevant features were selected, thus preserving related information. Normalization was done on the TF-IDF vectors to have a uniform scaling of features. This allow the models to learn from meaningful contextual patterns with reducing interference of "stop words".

3.4 Data Augmentation

Regarding the data augmentation, in order to deal with consistent class differences in the number of instances across classes, with a preponderance of "normal" over stress, anxiety and depression classes, the Synthetic Minority Oversampling Technique (SMOTE) was used. SMOTE aims to generate artificial samples for the minority class in the feature space and it does not simply duplicate observations. This serves to scale the model out and avoid biases to the majority class. By equalizing the number of samples between classes, SMOTE improves the model's sensitivity to less prevalent but crucial instances of emotional distress. The SMOTE result will be displayed in the next section.

3.5 Model Training

Both traditional machine learning and deep learning applied to train the model to classify mental health states from the tweet text. The machine learning models utilized were Support Vector Machine (SVM), Random Forest and Naive Bayes that were trained over TF-IDF features. Furthermore, a DNN with 2 hidden layers (512 and 256 neurons) with dropout layers to avoid overfitting was also built. All of them were trained with an 80:20 train-test split and the deep learning model was also optimized with early stopping to prevent overfitting and unnecessary training epochs. The performance was measured in terms of classification measures, including accuracy, precision, recall and F1-score.

4 Result and Findings

4.1 Exploratory Data Analysis

Exploratory data analysis started with the creation of word clouds to visualize the language of the tweet associated with stress, anxiety and depression. After preprocessing, text cloud visualizations were produced, using lemmatization and tokenization. These visualizations illustrate commonly used words and emotional indicators under each of the mental health categories.

Fig. 2. Word cloud for stress

As an illustration, in Fig. 2 we also display the World Cloud for Stress, displaying the most dominant terms such as "stress", "sad", "depression", "anxiety", indicating the emotional and psychological depth present on stressful posts. Terms like "insane", "need" and "life" also imply issues of excessive expectations, desperation and existential crisis. These findings parallel symptoms frequently described in general stress literature.

Another visual representation Word Cloud for Anxiety in Fig. 3 shows some recurrent words such as "stress", "sad" and "depression" in the anxiety category, and emotionally charged words like "insane", "need" and "still". Time, help and people communicate how urgent and frightening in social terms this condition is. The overlap with the wording of stress also suggests mental health could converge.

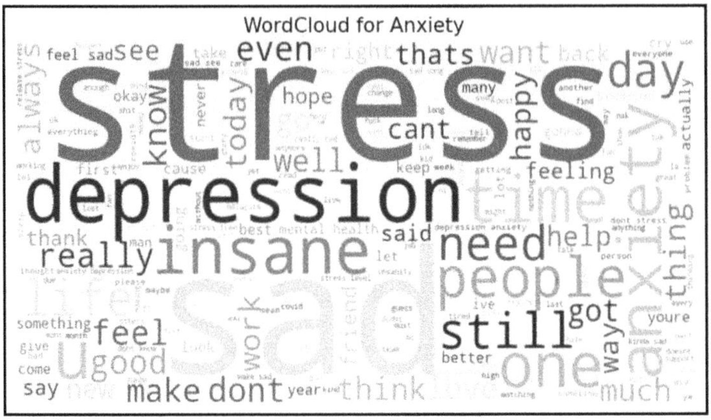

Fig. 3. Word cloud for anxiety

Fig. 4. Word cloud for depression

Figure 4, the word cloud for depression, features "sad", "stress", "depression" and "insane" significant words. The reinforcing words such as "alone", "cry", "problem" and "nothing" also enforce a sense of emotional emptiness, hopelessness and loneliness that are typical of depressive tweets. These patterns of speech are consistent with clinical language for diagnosing depression and for self-reporting on social networks. The Word Clouds and bar charts about the most used words for each of the target class and level.

For Fig. 5, the bigram display indicates typical two-word phrases such as 'feel sad', 'mental health', 'don't stress' and 'covid 19' and thus real emotional states and real-world events that were contributing to distress. These kinds of sentences bear more precise signals for classification models as they embed semantic cues that are directly related to psychological conditions. This also highlights the role of contextual patterns in mental health research.

Fig. 5. Bigram word cloud

4.2 Data Balancing

Exploratory analysis was carried out to investigate the distribution of classes across the target variables such as Stress, Anxiety and Depression prior to classification models training.

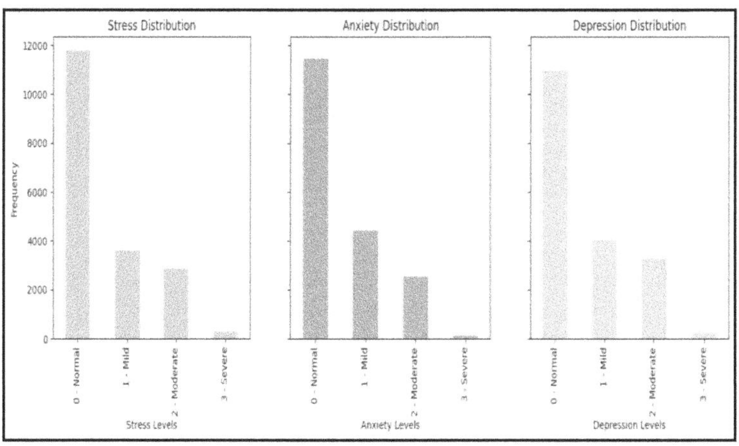

Fig. 6. Bar graph showing class distribution before SMOTE for stress, anxiety and depression

For each bar graph:

- X-axis indicates severity levels: Normal (0), Mild (1), Moderate (2) and Severe (3).
- Y-axis represents frequency of data points for each severity level.

Figure 6, the distribution of instances in each mental health condition in the testing data set. On the other hand, the "Mild", "Moderate" and "Severe" classes were highly underrepresented, most especially the Severe class, which had less than 2,000 samples in both the targets. This distribution imbalance may lead to biased model predictions and poor generalization performance for minority classes. To alleviate this problem, the Synthetic Minority Over-sampling Technique (SMOTE) was applied (Chawla et al., 2002). SMOTE synthesizes minority samples by interpolating between the existing samples in the feature space, instead of copying. SMOTE application produced a fairly balanced distribution among all severity levels, all of which containing over 10,000 observations, as depicted in Fig. 7, 8, 9.

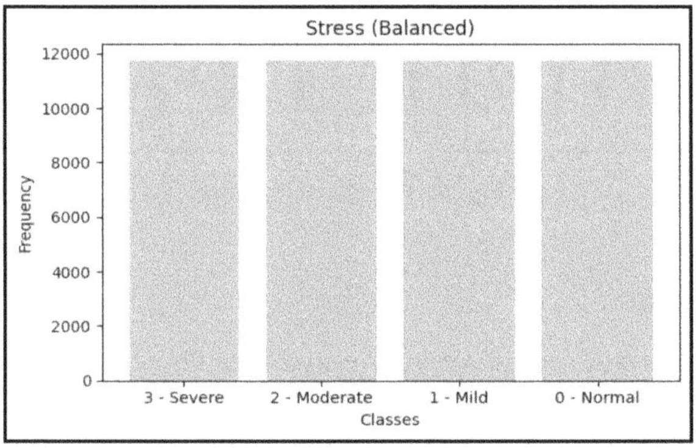

Fig. 7. Bar graph showing class distribution after SMOTE for stress

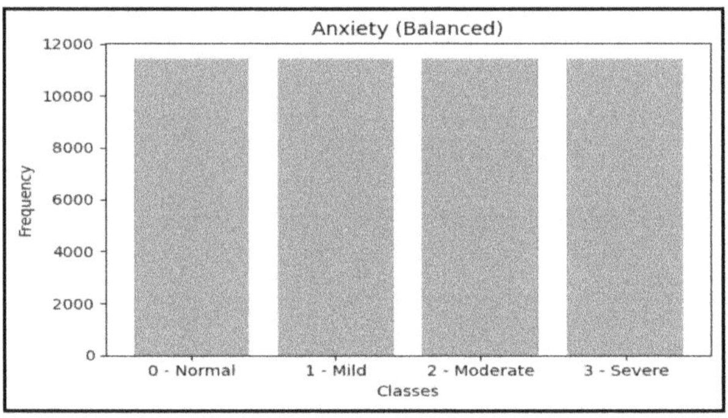

Fig. 8. Bar graph showing class distribution after SMOTE for anxiety

This balancing scheme improves the fairness and performance of the classification models by avoiding that the majority class becomes dominant during learning. Therefore, the learnt models will be more sensitive in detecting subtle signs of Mild, Moderate and Severe mental health.

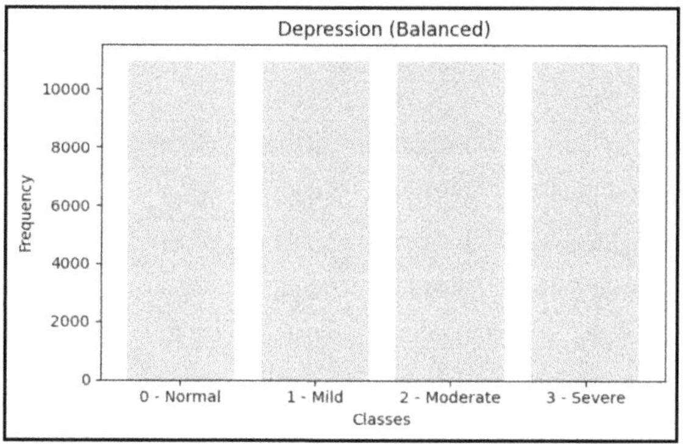

Fig. 9. Bar graph showing class distribution after SMOTE for depression

4.3 Model Performance

The dataset was vectorized with TF-IDF and four different classifiers were tested: (i) Support Vector Machine (SVM), Random Forest, Naive Bayes and a Deep Neural Network (DNN), as reported in Data Representation and Model Training. For all the models, each Mental Health class (i.e. Stress, Anxiety and Depression) were trained separately. Initially, all three targets were trained and tested using TF-IDF vectorized features with three machine learning models, SVM, Random Forest, and Naive Bayes. Classification reports of each model and target were noted and are shown in the tables below. (Table 1, 2 and 3).

Table 1. Classification report for SVM, Random Forest and Naive Bayes on Stress

Stress						
Model	Class	Precision	Recall	F1-Score	Support	Accuracy
SVM	Normal (0)	0.69	0.77	0.73	2298	0.7
	Mild (1)	0.57	0.67	0.62	2369	
	Moderate (2)	0.66	0.43	0.52	2371	
	Severe (3)	0.87	0.91	0.89	2367	
	Macro Avg	0.7	0.7	0.69	9405	
	Weighted Avg	0.7	0.7	0.69	9405	
Random forest	Normal (0)	0.79	0.82	0.8	2298	0.81
	Mild (1)	0.75	0.75	0.75	2369	
	Moderate (2)	0.81	0.71	0.76	2371	
	Severe (3)	0.9	0.97	0.93	2367	

(continued)

Table 1. (*continued*)

Stress						
Model	Class	Precision	Recall	F1-Score	Support	Accuracy
	Macro Avg	0.81	0.81	0.81	9405	
	Weighted Avg	0.81	0.81	0.81	9405	
Naive Bayes	Normal (0)	0.58	0.68	0.63	2298	0.65
	Mild (1)	0.58	0.49	0.53	2369	
	Moderate (2)	0.59	0.48	0.53	2371	
	Severe (3)	0.8	0.95	0.87	2367	
	Macro Avg	0.64	0.65	0.64	9405	
	Weighted Avg	0.64	0.65	0.64	9405	

Table 2. Classification report for SVM, Random Forest and Naive Bayes on Anxiety

Anxiety						
Model	Class	Precision	Recall	F1-score	Support	Accuracy
SVM	Normal (0)	0.56	0.6	0.58	2320	0.68
	Mild (1)	0.61	0.57	0.59	2312	
	Moderate (2)	0.61	0.58	0.59	2236	
	Severe (3)	0.94	0.98	0.96	2282	
	Macro Avg	0.68	0.68	0.68	9150	
	Weighted Avg	0.68	0.68	0.68	9150	
Random forest	Normal (0)	0.68	0.76	0.72	2320	0.81
	Mild (1)	0.73	0.71	0.72	2312	
	Moderate (2)	0.86	0.77	0.81	2236	
	Severe (3)	0.98	0.99	0.98	2282	
	Macro Avg	0.81	0.81	0.81	9150	
	Weighted Avg	0.81	0.81	0.81	9150	
Naive Bayes	Normal (0)	0.52	0.48	0.5	2320	0.61
	Mild (1)	0.5	0.46	0.48	2312	
	Moderate (2)	0.52	0.55	0.53	2236	
	Severe (3)	0.86	0.98	0.92	2282	
	Macro Avg	0.6	0.61	0.61	9150	
	Weighted Avg	0.6	0.61	0.61	9150	

After evaluation of performance, Random Forest was found to be the best performing model for all targets, and with highest accuracy and precision for each stress (0.81), anxiety (0.81) and depression (0.77). This model was the best one for the multi-class classification of the imbalanced data, especially over SMOTE.

Table 3. Classification report for SVM, Random Forest and Naive Bayes on Depression

Depression						
Model	Class	Precision	Recall	F1-Score	Support	Accuracy
SVM	Normal (0)	0.59	0.62	0.6	2169	0.66
	Mild (1)	0.55	0.46	0.5	2221	
	Moderate (2)	0.56	0.61	0.58	2181	
	Severe (3)	0.93	0.98	0.95	2195	
	Macro Avg	0.66	0.67	0.66	8766	
	Weighted Avg	0.66	0.66	0.66	8766	
Random forest	Normal (0)	0.69	0.76	0.72	2169	0.77
	Mild (1)	0.72	0.62	0.67	2221	
	Moderate (2)	0.71	0.71	0.71	2181	
	Severe (3)	0.95	0.98	0.97	2195	
	Macro Avg	0.77	0.77	0.77	8766	
	Weighted Avg	0.77	0.77	0.77	8766	
Naive Bayes	Normal (0)	0.51	0.51	0.51	2169	0.60
	Mild (1)	0.5	0.43	0.46	2221	
	Moderate (2)	0.52	0.5	0.51	2181	
	Severe (3)	0.82	0.95	0.88	2195	
	Macro Avg	0.59	0.6	0.59	8766	
	Weighted Avg	0.59	0.6	0.59	8766	

Then, a DNN model was also trained on the same TF-IDF features for further comparison with machine learning models. However, the performance of the DNN was not superior to the Random Forest in any of the three targets, demonstrating the Random Forest to be the most appropriate classifier in this research as stated in (Table 4).

To get an insight of classification accuracy the confusion matrix of the best model (Random Forest) for all targets is shown in Fig. 10, 11, 12. These matrices confirm the model's robustness to correctly predict all severity cases.

Table 4. Classification report for DNN - Random Forest on all three targets

Deep neural network – random forest						
Disorder	Class	Precision	Recall	F1-Score	Support	Accuracy
Stress	Normal (0)	0.8	0.83	0.82	2298	0.83
	Mild (1)	0.79	0.72	0.76	2369	
	Moderate (2)	0.79	0.76	0.78	2371	
	Severe (3)	0.9	0.99	0.95	2367	
	Macro Avg	0.82	0.83	0.82	9405	
	Weighted Avg	0.82	0.83	0.82	9405	
Anxiety	Normal (0)	0.69	0.77	0.72	2310	0.81
	Mild (1)	0.74	0.69	0.72	2287	
	Moderate (2)	0.85	0.8	0.82	2255	
	Severe (3)	0.98	0.98	0.98	2300	
	Macro Avg	0.81	0.81	0.81	9152	
	Weighted Avg	0.81	0.81	0.81	9152	
Depression	Normal (0)	0.7	0.77	0.74	2180	0.78
	Mild (1)	0.75	0.62	0.68	2243	
	Moderate (2)	0.71	0.76	0.73	2185	
	Severe (3)	0.96	0.98	0.97	2160	
	Macro Avg	0.78	0.78	0.78	8768	
	Weighted Avg	0.78	0.78	0.78	8768	

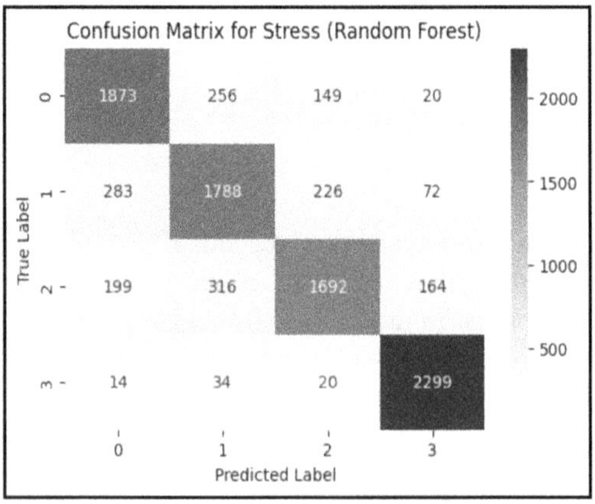

Fig. 10. Confusion Matrix - Random Forest (Stress)

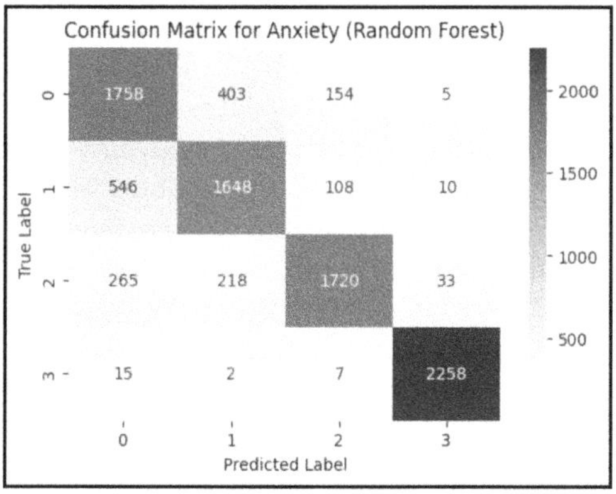

Fig. 11. Confusion Matrix - Random Forest (Anxiety)

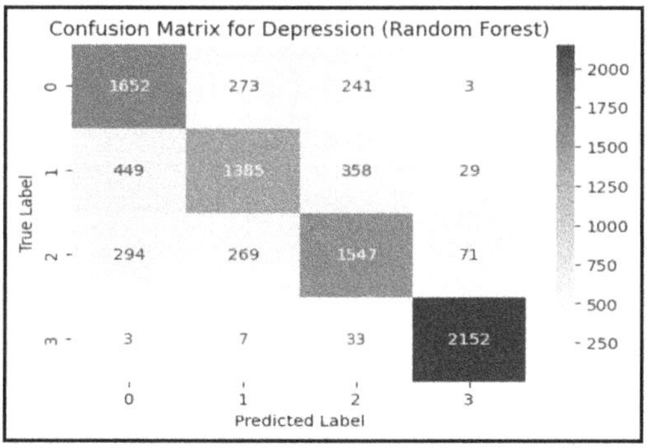

Fig. 12. Confusion matrix - Random Forest (Depression)

5 Discussion

This study demonstrates the feasibility of using text-based machine learning and deep learning models to classify pre-labeled X posts into Normal, Mild, Moderate, and Severe categories for stress, anxiety, and depression. The findings should be interpreted as post-level classification of linguistic patterns rather than clinical diagnosis, prospective prediction, or early identification of at-risk individuals. Among the evaluated models, Random Forest produced the most consistent results, achieving F1-scores of 0.83, 0.81, and 0.78 for stress, anxiety, and depression, respectively, while also requiring less computational complexity than the DNN. Nevertheless, overlapping vocabulary among the three conditions and adjacent severity categories indicates that short social media posts may not provide sufficient contextual information for reliable psychological assessment. The validity of the results also depends on the transparency of the data collection and annotation processes, including the keywords or hashtags used, the collection period, the inclusion and exclusion criteria, the annotator expertise, the labeling guidelines, and the level of inter-annotator agreement. Translation of multilingual posts, removal of emojis, TF-IDF representation, and the use of SMOTE may additionally affect the original meaning and distribution of the data. In practical settings, the proposed approach could support population-level monitoring, voluntary screening, or prioritization of posts for human review; however, its outputs should be used only as non-clinical decision support under qualified human supervision. Privacy protection, informed consent, fairness assessment, secure data handling, and procedures for managing false-positive and false-negative predictions are essential before deployment. Furthermore, the study is limited by its retrospective, single-platform dataset, post-level analysis, and absence of external or clinical validation. Future research should evaluate the models using independently collected datasets, longitudinal user data obtained with appropriate consent, validated psychological instruments, multilingual contextual models, and fairness and explainability analyses before claims concerning early screening or intervention can be established.

6 Conclusion

In summary, the project built a text classification system that can classify posts into labeled severity categories of stress, anxiety, and depression, based on X posts. NLP pipeline which consists of pre-processing, TF-IDF feature engineering, and SMOTE data augmentation, the system trained a series of machine learning models for testing. The best trade-off between accuracy and intelligibility was obtained by Random Forest among them and were taken up for deployment on web-app. The end result, a web application written using Streamlit and Visual Studio Code (VS Code) in which users can enter free-text and receive predicted level of severity scores for all of the three mental health conditions. For further models, considering more complex models such as LSTM or BERT and extend to multi-label classification to enhance the performance and practicality.

References

Al-Mosaiwi, M., Johnstone, T.: In an absolute state: elevated use of absolutist words is a marker specific to anxiety, depression, and suicidal ideation. Clin. Psychol. Sci. 6(4), 529–542 (2018). PMID: 30886766; PMCID: PMC6376956. https://doi.org/10.1177/2167702617747074

Calvo, R.A., Milne, D.N., Hussain, M.S., Christensen, H.: Natural language processing in mental health applications using non-clinical texts. Nat. Lang. Eng. **23**(5), 649–685 (2017). https://doi.org/10.1017/S1351324916000383

Chawla, N.V., Bowyer, K.W., Hall, L.O., Kegelmeyer, W.P.: SMOTE: synthetic minority oversampling technique. J. Artif. Intell. Res. **16**, 321–357 (2002). https://doi.org/10.1613/jair.953

Guntuku, S. C., Yaden, D. B., Kern, M. L., Ungar, L. H., & Eichstaedt, J. C. (2017). Detecting depression and mental illness on social media: an integrative review. Curr. Opin. Behav. Sci., 18, 43–49. https://doi.org/10.1016/j.cobeha.2017.07.005

Islam, M.R., Kabir, M.A., Ahmed, A., Kamal, A.R.M., Wang, H., Ulhaq, A.: Depression detection from social network data using machine learning techniques. Health Inf. Sci. Syst. **6**(1), 8 (2018). https://doi.org/10.1007/s13755-018-0059-1

Orabi, A.H., Buddhitha, P., Orabi, M.H., Inkpen, D.: Deep learning for depression detection of Twitter users. In: Proceedings of the Fifth Workshop on Computational Linguistics and Clinical Psychology: From Keyboard to Clinic, pp. 88–97 (2018) https://aclanthology.org/W18-0609/

Shen, G. et al.. (2017). Depression detection via harvesting social media: a multimodal dictionary learning solution. Proceedings of the Twenty-Sixth International Joint Conference on Artificial Intelligence (IJCAI-17), 3838–3844. https://doi.org/10.24963/ijcai.2017/537

Trotzek, M., Koitka, S., Friedrich, C.M.: Utilizing neural networks and linguistic metadata for early depression detection in social media. J. Artif. Intell. Res. **62**, 435–461 (2018). https://doi.org/10.1613/jair.1.11302

World Health Organization: World Mental Health Report: Transforming Mental Health for All. World Health Organization (2022)

Yazdavar, A. H., Al-Olimat, H. S., Ebrahimi, M., Bajaj, G., Banerjee, T., Thirunarayan, K.,Sheth, A.: Semi-supervised approach to monitoring clinical depressive symptoms in social media (2017). https://doi.org/10.1145/3110025.3123028

AI-Resilient Computer Science Education: Assessment Innovation, Curriculum Redesign, and Learning Analytics in the Era of Generative AI

Beyond Vibe Coding: Human–AI Teaming for Generative AI-Enhanced Learning in Undergraduate Data Mining Education

Nurfadhlina Mohd Sharef(✉)

Faculty of Computer Science and Information Technology, Universiti Putra Malaysia, Serdang, Malaysia
nurfadhlina@upm.edu.my

Abstract. Existing studies on artificial intelligence (AI) in higher education mainly concentrate on the technology adoption while scarce attention is paid to how numerous AI modalities can be orchestrated and woven into a human-centred learning ecosystem. In this paper, a multimodal human-AI teaming learning design integrating cognitive, computational, knowledge, social and governance dimensions is also proposed. It was introduced in an undergraduate Data Mining course, incorporating conversational AI, AI-assisted coding environments, document intelligence tools, mentoring, collaborative learning and self-directed learning activities. Using a survey instrument, a post-course evaluation was conducted comprises of learning support, engagement, learning outcomes, responsible AI behaviour, mentoring, and pedagogical design. The findings reflected positive perceptions across various dimensions, such as learning outcomes, AI learning support, and responsible human-AI teaming behaviour. The qualitative research also showed that students validated the AI outputs using a manual process of reviewing them; discuss information amongst peers and feedback from lecturers. The results indicate that the most effective AI-enabled technical education approach would not only need the AI tools, but also structured human oversight and instructional design.

Keywords: Human-AI Teaming · Generative AI in Teaching and Learning · Data Mining Education · Responsible AI · Higher Education · Vibe Coding

1 Introduction

Artificial intelligence (AI), such as generative AI and large language model (LLM) systems, is taking over higher education to change the way students learn, tackle problems, and engage with learning materials. Many examples include conversational AI assistants, code generation platforms, and document intelligence systems that aid in explanation, programming, debugging, summarisation, and independent experimentation, so are particularly relevant in technical subjects such as computer science, data analytics, or data mining (Kasneci et al., 2023; Lo, 2024). These tools present significant opportunities to break down learning barriers, expedite experiments, and facilitate learning self-directed by undergraduate learners.

N. Mohd Sharef et al. (Eds.): CAMP 2026, CCIS 3094, pp. 91–107, 2027.
https://doi.org/10.1007/978-981-92-4305-1_8

Yet, the educational benefit of AI integration is not based solely on having tools to access. Unstructured AI usage could lead to shallow processing, over reliance on produced output, lack of critical thinking and mindless acceptance of wrong or hallucinated responses (Cotton et al., 2024; Tlili et al., 2023). The extent of this concern is particularly notable in the context of computational and analytical courses, where understanding concepts, debugging and approach reasoning are critical aspects of learning outcomes. AI should never replace student reasoning, only aid in deeper engagement in problem-solving in these contexts. Studies of AI in education thus far tend to focus on adoption readiness, acceptance, academic integrity issue, and usage of a chatbot (Ghazvini et al., 2024; Sharef et al., 2025a,b; Zawacki-Richter et al., 2019)

While these studies offer valuable foundational insights, they often treat AI as an isolated educational tool, rather than as a component in a larger education ecosystem that comprises collaboration. Students are increasingly interfacing with a range of other AI modalities in real, interactive learning environments, such as conversational reasoning assistants, notebook-based coding assistance, document intelligence apps, and online knowledge resources. However, very little research has explored how these modalities can be intentionally structured across curricula to maintain human decision making, critical thinking, collaboration and responsible practice.

This challenge can be approached in the context of Human-AI Teaming (HAIT), which refers to a collaborative relationship in which humans and AI systems work together toward shared goals by leveraging complementary strengths while maintaining meaningful human oversight and decision-making authority (Dellermann et al., 2019; Endsley, 2023; Ghazvini & Sharef, 2026). Existing HMT research has primarily focused on domains such as healthcare, military operations, autonomous systems, and organisational decision making, where trust, transparency, coordination, and oversight are critical (O'Neill et al., 2022). In contrast, applications of HMT within higher education remain relatively limited. Current educational studies predominantly examine AI adoption, acceptance, learning performance, or perceptions of generative AI tools (Zawacki-Richter et al., 2019; Kasneci et al., 2023; Lo, 2024), often treating AI as a learning aid rather than a collaborative teammate.

The gap is particularly evident in computer science and data science education, where generative AI is increasingly used for coding, debugging, algorithm exploration, and software development. Existing studies largely focus on the benefits, risks, or acceptance of AI tools (Kasneci et al., 2023; Tlili et al., 2023; Cotton et al., 2024), while comparatively fewer studies investigate how students, educators, and AI systems can interact through structured roles, feedback loops, validation mechanisms, and governance practices that support both technical learning and responsible AI use.

In recent years, advances in generative AI have popularised the notion of "vibe coding" where users use AI programs and intuitive language to incrementally develop software, and through interaction with these systems solve technical problems in natural language. Such techniques can expedite experimentation and implementation through lowering the programming barrier and reducing the need to memorise syntax. But technical education is more than just making working code. Students should also possess understanding of ideas, reason logically, validate facts and responsibly think about AI practices. So the challenge is not so much how to use AI to deliver solutions but how to actually embed AI in a learning environment that will continue to allow for human judgement, critical thinking, collaboration and meaningful lessons. From this vantage point, our study goes beyond vibe coding to consider AI not only as an assistive coding but as a team player within the greater learning community. A HAIT approach explicitly describes the roles in which students, educators, and AI systems are complementary, with mentoring, feedback, validation, reflection, and governance embedded throughout the learning process.

Therefore, we investigate one undergraduate data mining course in which learning with different AI-enabled modalities (conversational AI for conceptual reasoning, AI-assisted coding assistance, document intelligence, peer collaboration, mentoring, self-paced learning) were seamlessly embedded into an intentionally designed instructional course. Drawing from this implementation, the study proposes a multimodal HAIT learning design that theorises how cognitive, computational, knowledge, and social collaboration can be orchestrated for the delivery of AI-augmented technical education. Iit is crucial that the development of such a model can not only support technical learning, but also hold the student interest, and secure responsible AI use behaviours. For this reason this study assesses how students perceive the learning design following its introduction in an undergraduate Data Mining course. This study contributes:

(a) A Practical HAIT Learning Ecosystem that Explicitly Defines the Complementary Roles of Lecturers, Students, and AI Tools across the Data Mining Project Lifecycle.
(b) An Operational Model Showing how AI Supports Concept Understanding, Coding, Experimentation, Validation, and Communication while Preserving Human Oversight and Critical Judgement.
(c) Empirical Evidence Demonstrating that AI-Enhanced Learning Shall Coexist with Mentoring, Collaboration, Validation, and Responsible AI Use.

2 Related Work

2.1 Artificial Intelligence in Higher Education

The use of artificial intelligence has increasingly penetrated higher education teaching and learning by providing the ability to personalise learning, automate support tasks, and enhance student access to knowledge resources. Past research in AI in education centered on algorithms for intelligent tutoring systems, adaptive learning platforms, and automated assessment mechanisms (Zawacki-Richter et al., 2019).

In the more modern era, the advent of generative AI and LLMs plays a pivotal role in expanding the educational realm, providing conversational interaction, code generation, summarisation, explanation, and the retrieval of contextual knowledge (Kasneci et al., 2023). Indeed, in technical education these capabilities are promising, for students can use AI for programming support, debugging, concept clarification, and exploratory problem solving. Previous studies have also shown positive perceptions of generative AI that enhance efficiency, accessibility, and support in learning (Lo, 2024; Tlili et al., 2023). Students typically regard AI assistants as convenient companions to study independently, especially during times when immediate human support is unavailable.

But with these advantages come significant potential risks. The hazards such as hallucinated outputs, factual inaccuracies, reduced academic integrity, superficial learning, and overreliance on AI-generated responses have been identified (Cotton et al., 2024; Kasneci et al., 2023). In undergraduate contexts, however, these concerns might be compounded as learners are still forming domain knowledge, exercising critical reasoning, and developing an ability to evaluate technical outputs. Thus, integration of AI into education will take more than a mere adoption of technology; it entails pedagogical architectures providing a context for its meaningful and responsible use.

2.2 AI in Technical and Computational Learning

Different educational requirements emerge in the technical disciplines such as computer science, programming, analytics, and data science. This is not common knowledge-based learning either; students should have learning requirements not just for picking up information like general programming (a skill in a technical area in general) in knowledge domain but also for developing their procedural logic, troubleshooting ability, algorithmic ability and how to work within the system method. AI tools in turn, help students understand a lot more about their language and can help generate code, teach syntax, advise on debugging and accelerate experimentation (Sharef et al., 2020).

New studies say generative AI can contribute to programming education by lowering the barriers to entry, and boost students' confidence on the job when solving problems (Li et al., 2025). Yet, there are concerns over whether too much of the AI code used in our task is reducing the understanding of abstraction, independent debugging, solution acceptance with no verification or understanding. As such, there is a pedagogical tension between using AI as a tool for productivity and maintaining real computational learning.

Modern student learning increasingly involves interaction with multiple AI-enabled modalities rather than a single standalone assistant. A learner may use conversational AI for concept clarification, notebook environments for code experimentation, document intelligence systems for content synthesis, and online learning resources for self-paced exploration. At the same time, human interaction through peer collaboration, mentoring, and instructor guidance remains essential.

This creates a multimodal learning ecosystem in which knowledge construction emerges through interaction across human and machine actors. However, current educational literature offers limited conceptual models explaining how such multimodal AI interactions should be pedagogically orchestrated. Existing literature mostly focuses on isolated and technical tasks, specifically on AI coding support or chatbot-based tutor education, and does not explicitly address how different AI modalities might coexist with others, in a system learning ecology context. But real student learning now takes place in real time and may include interacting with many tools combined (conversational LLMs, notebook-based code environments, document intelligence systems, and external online resources). The learning implications of such multimodal AI interaction are still largely unknown.

2.3 HAIT as a Learning Paradigm

HAIT extends beyond traditional human-computer interaction by emphasising collaborative relationships between humans and AI systems working toward shared objectives (Dellermann et al., 2019). Effective teaming requires complementary capabilities, continuous interaction, feedback exchange, and meaningful human oversight (Endsley, 2023; O'Neill et al., 2022).

In some areas like healthcare, defence and decision support systems, HAIT is emerging as a path towards performance improvement without compromising accountability and human control. Fundamental values of this framework encompass role complementarity, transparency, human oversight, calibration of trust, and an iterative model for the interaction between humans and AI actors. These principles are particularly applicable to education, and to technical education. Learning involves so much more than getting the task done. It calls for reasoning, reflection, error correction and conceptual growth. AI should thus not act as a substitute decision-making tool, but as a learning partner whose results are still accountable to human evaluation and contextual analysis.

However, HAIT still shows limited impact in educational settings, particularly in undergraduate classrooms, where various AI tools are under active use. AI-in-education research has centered on adoption of tools as opposed to designs for collaborative teamwork. Therefore, a structured HAIT perspective may offer a stronger theoretical basis in AI-integrated pedagogy.

3 Multimodal Human–AI Teaming Learning Ecosystem

Instead of treating AI as either prohibited technology or unrestricted convenience, an instructional design is proposed which positioned AI as a collaborative learning partner operating under pedagogical guidance and human oversight. This setting provided an authentic environment to investigate HAIT in undergraduate technical education. The proposed design was organised around five complementary teaming dimensions that integrated multiple human and AI learning modalities in which AI augmented human learning rather than replacing it, as elaborated further in this section.

This study was conducted in an undergraduate Data Mining course within a computer science programme at a Malaysian public university. The course introduces students to key concepts and practical techniques in data mining, including data preprocessing, classification, clustering, association rule mining, model evaluation, and data-driven analytical problem solving. Beyond theoretical understanding, students are expected to develop practical computational competencies such as coding, experimentation, debugging, result interpretation, and critical reasoning.

3.1 HAIT Operationalisation

To operationalise the proposed HAIT learning ecosystem, the roles of lecturers, students, and AI tools were explicitly defined across different stages of the data mining learning process. Unlike conventional AI-assisted learning approaches that focus primarily on tool usage, the operationalisation emphasises complementary responsibilities, continuous interaction, and feedback loops among human and AI actors. Table 1 summarises the roles performed by each actor throughout the learning lifecycle.

In contrast to traditional AI-assisted learning paradigms which largely discuss the use of AI tools, Table 1 demonstrates that it is intentional that learning responsibilities are distributed across lecturers, students and AI systems. Lecturers provide instruction,

Table 1 Operationalisation of the Human–AI teaming learning ecosystem

Learning stage	Lecturer role	Student role	AI role/tools
Learning design	Create authentic challenges and define project scope	Understand objectives and expectations	Gemini, Canva support content generation and learning preparation
Problem investigation	Facilitate inquiry and guide exploration	Understand datasets, identify problems, formulate questions	Gemini supports concept clarification and exploratory questioning
Feature engineering	Guide discussion and challenge assumptions	Analyse datasets, identify relevant features, compare alternatives	Gemini assists feature exploration and analytical reasoning
Model development and experimentation	Provide feedback and challenge modelling decisions	Build models, test alternatives, tune parameters	Gemini and Colab support coding, debugging, implementation, and experimentation
Validation and verification	Curate knowledge and encourage critical evaluation	Compare evidence, review outputs, validate findings	NotebookLM supports retrieval, summarisation, and evidence comparison
Communication and reflection	Review outputs and facilitate reflection	Present findings, explain decisions, reflect on learning	Docs, Sheets, Slides, Gmail, and Drive support collaboration and communication

mentoring and governance; students are responsible for problem formulation, experimentation, interpretation, and decision making; and AI tools offer cognitive, computational and knowledge support. This operationalisation creates the HAIT environment in which AI supports learning but never replaces human judgment. In addition, the progression across learning stages creates continuous feedback, validation, reflection, and improvement cycles that are essential characteristics of effective HAIT.

3.2 Multi-Modal HAIT Dimensions

3.2.1 HAIT Dimension: Cognitive Teaming

In the first dimension, students used conversational generative AI tools, particularly Gemini, Copilot and ChatGPT, to support conceptual understanding, clarify technical concepts, brainstorm analytical approaches, and explore alternative interpretations (Table 2). In this role, AI functioned as a reasoning support mechanism rather than an authoritative answer provider. This learning approach has empowered students to leap their competency in data mining model development using coding skills although some of them are novice. Although some sessions is still spent on coding syntax and skeleton, the total hours spent is much lesser compared to the traditional approach used. Students have also been referred to the past projects in the previous semesters.

Table 2. Operational approach for cognitive teaming

Component	Description
Objective	Support conceptual understanding and analytical reasoning in data mining
AI role	Provides explanations, examples, concept clarification, alternative analytical approaches, and reasoning support
Human role	Critically evaluates responses, asks follow-up questions, relates explanations to course concepts, and validates understanding
AI tools	Gemini
Learning activities	Concept exploration, lecture reinforcement, self-directed inquiry, quiz preparation
Expected outcomes	Improved conceptual understanding, analytical reasoning, and critical thinking

3.2.2 HAIT Dimension: Computational Teaming

AI-assisted computational support was integrated through notebook-based environments such as Google Colab that is enhanced with Gemini as the AI assistance. Students used these tools for coding support, debugging, syntax explanation, implementation refinement, and experimentation with data mining workflows. Several sessions in the class are conducted to collaboratively explore and justify data mining design alternatives. This enforces that human validation remained necessary to ensure correctness and conceptual understanding (Table 3). Students performed group discussions to complete the assessments.

Table 3. Operational approach for computational teaming

Component	Description
Objective	Support implementation, experimentation, and debugging of data mining workflows
AI role	Assists with code generation, debugging, syntax explanation, workflow implementation, and error diagnosis
Human role	Validates generated code, modifies solutions, interprets outputs, and performs experimentation
AI tools	Gemini, Google Colab

(*continued*)

Table 3. *(continued)*

Component	Description
Learning activities	Hands-on laboratories, coding exercises, implementation of classification, clustering, and association analysis tasks
Expected outcomes	Improved coding confidence, debugging skills, computational thinking, and technical competency

3.2.3 HAIT Dimension: Knowledge Teaming

In knowledge teaming (Table 4), students leveraged document intelligence tools such as NotebookLM, online references such as YouTube, Kaggle, IEEE data port, and certification platforms such as Putra Experiential Learning, LinkedIn Learning and Coursera to support information retrieval, synthesis, comparison of methods, and independent knowledge exploration. This dimension emphasised knowledge augmentation rather than passive information consumption. The students also get to be exposed to a diverse range of knowledge content and solutions.

Table 4. Operational approach for knowledge teaming

Component	Description
Objective	Enhance information retrieval, knowledge synthesis, and independent learning
AI role	Retrieves, summarises, synthesises, and compares information from multiple sources
Human role	Verifies information, contextualises findings, and integrates knowledge from diverse resources
AI tools	NotebookLM, Gemini
Learning activities	Literature exploration, certification activities, independent study, supplementary learning
Expected outcomes	Enhanced information literacy, knowledge synthesis capability, and self-directed learning

3.2.4 HAIT Dimension: Social Teaming

Here, human interaction remained a core component of the learning ecosystem (Table 5). Peer discussions, collaborative projects, mentoring sessions, and presentation activities supported shared learning, discussion, and reflective articulation of analytical reasoning. Students spent time going over each data mining pipeline together and debate on possible choices and approaches for completing their assignments.

Table 5. Operational approach for social teaming

Component	Description
Objective	Promote collaborative learning, communication, and peer-supported problem solving
AI role	Supports idea generation, discussion preparation, and presentation refinement
Human role	Engages in teamwork, mentoring interactions, peer discussions, and collaborative decision making
AI tools	Gemini, NotebookLM
Learning activities	Group projects, mentoring sessions, peer discussions, presentation and video pitching activities
Expected outcomes	Improved communication, collaboration, teamwork, and reflective learning

3.2.5 HAIT Dimension: Governance and Instructional Oversight

The lecturer served as the instructional orchestrator, establishing expectations for responsible AI use, encouraging verification of AI-generated outputs, promoting critical evaluation, and designing activities that preserved meaningful human cognitive engagement (Table 6). There are slots provided for group and personal mentoring, to understand issues and provide personalised learning intervention. Challenge based learning approach is adopted to encourage students to explore their interests. The emphasis is the students' ability to defend and justify their choices. These activities allow the students to express themselves and make them be familiarised with collective intelligence in completing AI experiments.

Table 6. Operational approach for social teaming

Component	Description
Objective	Promote responsible, ethical, and effective use of AI in learning
AI role	Supports learning activities within defined instructional boundaries
Human role	Establishes usage guidelines, provides mentoring, promotes verification practices, and calibrates trust in AI outputs
AI tools	All AI tools used in the course
Learning activities	Responsible AI discussions, mentoring sessions, feedback activities, reflective exercises, assessment design
Expected outcomes	Responsible AI behaviour, critical evaluation, trust calibration, and effective Human–AI collaboration

4 Results and Discussion

An evaluation was conducted to examine whether the proposed multimodal HAIT learning design achieved its intended objectives within an undergraduate Data Mining course. The instrument (downloadable http://putrarepo.upm.edu.my/direct/UPM-NL-266) was organised into several sections covering AI usage patterns, perceived impact of AI tools, learning activities, learning experiences, self-perceived learning outcomes, mentoring experiences, and pedagogical design, to evaluate the proposed Multimodal HAIT dimensions.

4.1 Data Collection

The survey (design in Table 7) sought to determine whether the learning design could:

i. support technical learning through cognitive, computational, and knowledge teaming;
ii. maintain student engagement and positive learning experiences;
iii. preserve human judgement and responsible AI use through mentoring, collaboration, and instructional governance.

A total of 28 students participated in the evaluation anonymously following completion of the course. Responses were captured using a five-point Likert scale ranging from strongly disagree (1) to strongly agree (5). The instrument also included open-ended questions to capture qualitative reflections on learning benefits, challenges, and experiences with AI-supported learning.

Table 7. Components in the survey

Section	Focus	#Items	Example Item
Background and AI usage	Student learning habits, computational tools, AI tools, and AI usage patterns	10	*Which AI/GenAI tools did you use to support your learning?*
Perceived impact of AI tools	Perceived usefulness and influence of AI on learning and problem solving	6	*AI tools helped me resolve coding difficulties more efficiently.*
Effectiveness of learning activities	Evaluation of course activities and learning interventions	7	*Mentoring Sessions; Group Project; Video Pitching*
Learning experience	Student engagement, responsible AI use, critical thinking, and learning behaviours	11	*I was encouraged to think deeply rather than just complete tasks.*

(continued)

Table 7. (*continued*)

Section	Focus	#Items	Example Item
Self-perceived learning outcomes	Perceived improvement in knowledge, confidence, and capability	5	*I am more confident analysing real datasets.*
Mentoring and interactive sessions	Effectiveness of mentoring and lecturer feedback	3	*Mentoring sessions helped me clarify difficult concepts in data mining.*
Teaching strategies and course design	Evaluation of pedagogical design and instructional effectiveness	5	*The combination of activities fostered a deeper understanding of the subject.*
Confidence and reflection	Confidence level and qualitative reflections	Open-ended	*How did this course change the way you approach learning technical subjects?*

4.2 Survey Response Analysis

4.2.1 AI Usage Patterns and Learning Activities

Most students used Google Colab (92.9%), ChatGPT (85.7%), and Gemini (78.6%) throughout the course, indicating widespread adoption of AI-assisted learning tools (Fig. 1). AI was primarily used to understand concepts (96.4%), debug code (71.4%), generate code examples (71.4%), and interpret model outputs (60.7%) (Fig. 2). These findings suggest that students utilised AI not merely as an answer-generation tool, but as a learning companion supporting conceptual understanding, coding, and analytical reasoning (Fig. 3).

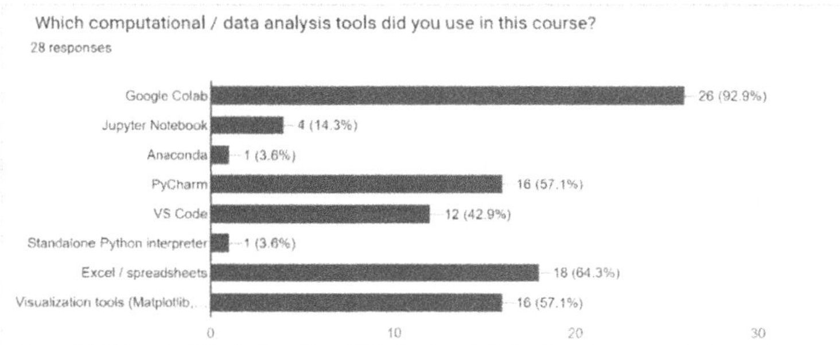

Fig. 1 Computational/data analysis tools that students used in the course

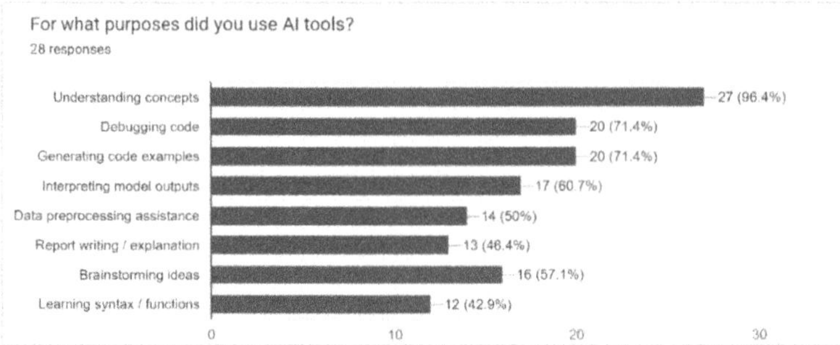

Fig. 2 Purpose of AI tools usage

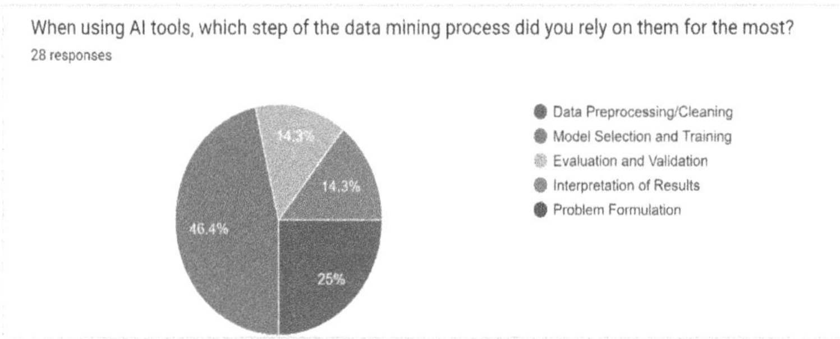

Fig. 3 AI application specifically for data mining processes

Among the learning activities, mentoring sessions, group projects, online quizzes, and external certification activities were consistently perceived as effective in supporting learning. The results indicate that AI-supported learning was complemented by substantial human interaction, collaborative work, and guided learning activities rather than relying solely on AI assistance.

4.2.2 Evaluation of the Multimodal HAIT Learning Design

All constructs demonstrated satisfactory reliability ($\alpha > 0.80$) (Table 8). Positive ratings were observed across all evaluated dimensions, with Learning Outcomes receiving the highest mean score ($M = 4.89$). The findings suggest that students perceived the learning design as effective in supporting understanding, confidence, engagement, and technical competency. The consistently high scores across AI Learning Support, Learning Engagement, and Learning Outcomes provide preliminary evidence that the cognitive, computational, and knowledge teaming dimensions contributed positively to technical learning. At the same time, positive ratings for Responsible AI Behaviour, Mentoring and Social Support, and Pedagogical Design suggest that the social and governance dimensions were also successfully integrated into the learning environment.

Table 8. Summary of the responses to the quantitative questions

Construct	α	Mean	SD
AI Learning Support	0.914	4.57	0.64
Responsible AI Behaviour	0.804	4.57	0.69
Learning Engagement	0.922	4.55	0.61
Learning Outcomes	0.861	4.89	1.33
Mentoring & Social Support	0.852	4.49	0.72
Pedagogical Design	0.978	4.39	0.89

4.2.3 Evidence of Human-AI Teaming

This section provide evidences, summaries and excerpts from qualitative responses (Figs. 4, 5, 6, 7, 8, 9, 10) which can be used as guide for similar course deliveries. Students noted that AI tools assisted them to better understand complex phenomena, differentiate between data mining methods, debug code, optimise models and lead them through the stages of the data mining pipeline. AI troubleshooting computational problems and expediting experiments were among the things some students emphasized.

But students did not rely only on AI results. Rather, they have reported validating code generated by AI as well as recommendations that they receive from AI through cross-checking with several AI tools, comparing results against lecture materials, manual code reviews, performance evaluation of the outputs, looking into AI output metrics and seeking feedback from the lecturer or from lecturers/fellow students when uncertain.

Mentoring and project learning were also found to be some of the most beneficial topics to engage students during the course. Mentoring offered clarity and direction to the project, deepening students' understanding of concepts and offering constructive feedback to students while project tasks have helped the students to be able to apply theory to practical data and practical analytic problems.

Many students say the class improved the practice of a technical discipline. Instead of memorising, they placed increased value on experiment, validation, procedure and evaluation of output. Students reported that they also learned to utilize AI as a support tool but remained responsible for verification and final decision making. These findings indicate that the proposed learning design effectively placed AI not in a place of human reasoning, but as a collaborative, friendly partner.

Project Direction and Experiment Planning: *Multiple students found the sessions helpful for clarifying project direction, planning experiments, and understanding the step-by-step flow of a data mining project, including how to take the first step and determining versioning for dataset outputs for comparison.*
Constructive Feedback and Insight: *The lecturer's idea sharing, constructive feedback, and insights provided new perspectives, helped identify shortcomings, and led to better decisions, enabling students to realize aspects they had missed and make the project better.*
Conceptual Understanding and Q&A: *Mentoring helped in understanding the overall concept of data mining tasks, explaining machine learning models, and providing clarification on why certain methods fit specific scenarios; the direct ability to ask questions and double-confirm on the spot was also highly valuable.*
Technical Guidance: *Students received assistance with specific technical tasks, such as data preprocessing, experimental design, and validation strategy.*
Encouragement: *The reassurance from the professor that they were "doing great so far" was helpful, especially when students felt confused.*

Fig. 4 Summary based on responses to question about which aspect of mentoring or discussion sessions helped students most

- *Conceptual Clarity and Explanation: Tools like ChatGPT and Gemini significantly helped in understanding complex terms, differentiating between concepts (e.g., feature selection vs. data partitioning; classification, regression, and time series), and connecting theoretical formulas (like those for Decision Trees and Naïve Bayes) to actual calculations and model predictions.*
- *Troubleshooting and Optimization: AI assistants were crucial for debugging code errors (e.g., logic errors due to incorrect algorithm selection) and optimizing model performance. For instance, Gemini helped diagnose why SVM was crashing Google Colab due to a large dataset and recommended more efficient algorithms like KRR with Nystroem application.*
- *Coding and Environment Setup: Tools assisted respondents with setting up the environment, writing Python code based on descriptions, understanding code syntax, and enabling step-by-step code execution in environments like Google Colab and Jupyter Notebook.*
- *Data Mining Pipeline and Analysis: The tools provided guidance throughout the data mining pipeline, including data extraction, train-test splitting, data cleaning, feature engineering, and data preprocessing. They also helped in interpreting model outputs, analysis results, and clarifying the specific purposes of different data mining tasks.*

Fig. 5 Summary based on responses to question about scenario on tool usefulness

- ***Cross-Validation and Comparison:*** *Correctness was frequently verified by comparing outputs from multiple sources, including cross-checking between different AI tools (ChatGPT, Gemini, Grok, Claude, DeepSeek) and contrasting results with peers, expected outputs, lecture examples, and what was taught in class.*
- ***Manual and Logical Review:*** *Students performed manual verification, checking code logic by reading it, testing code step-by-step with different inputs, ensuring reproducibility, and reviewing the structure and concepts to confirm the results made sense.*
- ***Performance Metrics and Data Validation:*** *Verification of models often involved checking multiple evaluation metrics such as accuracy, RMSE, MAE, precision, recall, and F1-score, ensuring logical consistency, and using techniques like future-date validation to avoid data leakage.*
- ***External Research and Clarification:*** *Many relied on external resources, including Google search, research, online learning platforms (like W3 Schools or Geeks for Geeks), and asking the lecturer or referring to lecture notes for clarification or deeper understanding, especially when contradictions or unusual results appeared.*

Fig. 6 Summary based on responses to question about AI output verification

Project/Group Work: *The project activity, particularly the end-to-end supervised learning project and group project, had the greatest impact on learning for many students because it required applying theoretical concepts to real-life data, involving proper planning, full implementation, and integrating multiple data mining tasks (e.g., data cleaning, classification, clustering, association rule mining).*
Mentoring Sessions: *Mentoring sessions were highly impactful because they facilitated two-way communication, helped correct incorrect understandings, and clarified why specific solutions were used.*
Data Hunt Activity: *The initial "Data Hunt" activity had a significant impact because it was the first fully individual assignment, served as the foundation for subsequent tasks, and provided extensive learning on data preprocessing.*
External Certification/Course Enrollment: *Enrolling in the external certification course was beneficial as it aided in better understanding the material and resulted in students receiving an e-certificate.*
Hands-on Application: *Activities that involved hands-on application, such as manual calculation for clustering and technical activities using tools (like the group assignment), made the application of concepts easier to grasp than simply reading slides.*

Fig. 7 Summary based on responses to question about AI impact

- ***Application and Implementation:*** *The primary challenge was applying theoretical machine learning concepts, especially in the coding and project phases (including training/testing, model selection, and finding a suitable dataset). This was difficult due to the hands-on focus of the course and the need to use Python (for those unfamiliar with it).*
- ***Group Project Complexity:*** *The group project was repeatedly cited as the most challenging element due to the volume of work, unclear initial instructions, lack of familiarity with the tasks, and the need to address unexpected requirements at the last minute.*
- ***Conceptual Understanding:*** *Other challenges included understanding the mathematical concepts behind algorithms, grasping specific models (like K-Means), determining the correct methods when "one solution does not fit all," and designing proper validation (e.g., for time-based data).*
- ***Overcoming Challenges:*** *Students overcame these difficulties mainly through* ***practice and experimentation*** *(coding repeatedly, testing different approaches),* ***collaboration*** *(discussing with team members and cross-referencing with friends), and seeking* ***external guidance*** *(attending mentoring sessions with the Professor, and utilizing AI assistants/Gemini and YouTube).*

Fig. 8 Summary based on responses to question about learning challenge

Focus on Practice and Application: *Multiple respondents shifted their focus from memorizing theory and definitions to practical implementation, application, and understanding of workflows. Learning now emphasizes practicing, testing models, making mistakes, and connecting formulas/algorithms to real-world applications.*
Critical Thinking and Methodology: *The course promoted a more systematic approach, focusing on understanding assumptions, validation methods, and critically evaluating outputs rather than accepting results at face value or focusing solely on high accuracy. This includes paying extra attention to details, consequences, and impacts, as well as prioritizing correct methodology and interpretation.*
Strategic Use of AI and Independent Learning: *Students learned to utilize AI as a support or learning tool—asking it to explain concepts and give suggestions—but not as a solution or a replacement for personal input; they emphasized the need to verify AI outputs and refer to the instructor for final decisions. This approach also highlighted the importance of independent study and self-learning.*
Data Understanding as a Foundation: *The course emphasized that understanding the dataset is the crucial first step before attempting any data mining tasks.*
Seeking Help and Collaboration: *One respondent noted the importance of asking peers for clarification when struggling with understanding certain aspects of the technical subject.*

Fig. 9 Summary based on responses to question about learning adaptation

Project-Based Learning and Practical Application: *Retain the project-based structure, group project, and hands-on project components, as they strengthen understanding, build real problem-solving skills, encourage self-learning, and connect theory to real-world problems.*
Mentoring and Discussion: *The mentoring sessions and discussion sessions should be retained, as they significantly help students complete the project and improve understanding and practical skills.*
Hands-on and Implementation: *Keep the practical hands-on implementation sessions, lab sessions, step-by-step demonstrations of model building, and integration of theory with coding practice. This includes using real-world datasets for assignments and focusing on key data mining concepts (classification, clustering, association rule mining).*
Course Structure and Additional Resources: *Retain the use of external certification courses (like Coursera and putraMOOC, though putraMOOC may need revision due to system errors), the course enrollment certificate, and the flexibility in marks allocation.*
Other Learning Elements: *Retain the use of AI tools, open-book exams, and crash courses.*

Fig. 10 Summary based on responses to question about learning approaches that need to be retained

Collectively, these findings demonstrate that students did not use AI merely to generate solutions. Instead, AI participated as a collaborative teammate supporting understanding, experimentation, validation, and reflection, while human learners and lecturers retained responsibility for evaluation, decision making, and final judgement.

5 Conclusion

AI should be treated as a learning teammate rather than a source of answers. In this paper a Multimodal HAIT Learning Design was presented for the undergraduate Data Mining education. A multimodal approach to AI-enhanced education is proposed by integrating cognitive, computational, knowledge, social, and governance teaming dimensions to support human judgement, collaboration, and responsible AI use.

The framework was operationalised in an undergraduate data mining course through the integration of conversational AI, AI-assisted coding environments, document intelligence tools, mentoring, collaborative learning activities, and self-directed learning resources. Responses from a survey reflect positive students perceptions with regard to AI learning support, learning engagement, learning outcomes, responsible AI behaviour, mentoring effectiveness, and design of pedagogy. These results suggest preliminary evidence of efficacy of AI-augmenting technical education based on AI, with regard to the intentional orchestration of human and AI strengths in a structured learning ecosystem, and not just AI tools.

Future research could compare the framework with larger groups, distinct technical disciplines, and learning situations in order to check the efficacy and generalisability of the findings.

References

Cotton, D.R.E., Cotton, P.A., Shipway, J.R.: Chatting and cheating? Ensuring academic integrity in the era of chat GPT. Innov. Educ. Teach. Int. **61**(2), 228–239 (2024). https://doi.org/10.1080/14703297.2023.2190148

Dellermann, D., Ebel, P., Söllner, M., Leimeister, J.M.: The future of human-AI collaboration: a taxonomy of design knowledge for hybrid intelligence systems. Bus. Inf. Syst. Eng. **61**(5), 637–643 (2019). https://doi.org/10.1007/s12599-019-00600-0

Ghazvini, A., Sharef, N.M.: A human-in-the-loop explainable interactive digital twin for student performance prediction and model governance. J. Soft Comput. Data Min. (2026)

Ghazvini, A., Sharef, N.M., Sidi, F.: Prediction of course grades in computer science higher education program via a combination of loss functions in LSTM model. IEEE Access. **12**, 30220–30241 (2024)

Kasneci, E. et al.: ChatGPT for good? On opportunities and challenges of large language models for education. Learn. Individ. Differ. **103**, 102274 (2023). https://doi.org/10.1016/j.lindif.2023.102274

Li, H.J., Huang, Q.R., Wen, L.P., Chen, W., Xu, Z.Z.: Generative artificial intelligence supported programming learning: learning effectiveness and core competence. Sage Open. **15** (2025)

Lo, C.K.: The use of ChatGPT in higher education: a review of the literature. Interact. Learn. Environ. **32**(8), 1–19 (2024). https://doi.org/10.1080/10494820.2023.2253851

O'Neill, T., McNeese, N., Barron, A., Schelble, B.: Human-autonomy teaming: a review and analysis of the empirical literature. Hum. Factors. **64**(5), 904–938 (2022). https://doi.org/10.1177/0018720820960865

Sharef, N.M., Mustapha, A., Azmi, M.N., Nordin, R.: Basketball players performance analytic as experiential learning approach in teaching undergraduate data science course. In: 2020 International Conference on Advancement in Data Science, E-Learning and Information Systems (ICADEIS 2020) (2020)

Sharef, N.M., Mansor, E.I., Nasharuddin, N.A., Rahman, M.S.A.: Preliminary acceptance evaluation of SINAR: an AI-powered personalized learning and decision support system for malaysian higher education. In: Proceedings of the 17th International Conference on Intelligent Human-Computer Interaction (IHCI) (2025a)

Sharef, N.M., Mansor, E.I., Nasharuddin, N.A.: Artificial intelligence literacy in Malay language through a teaching-purposed Chatbot. J. Inf. Knowl. Manag. **24**(06), 2550059 (2025b). https://doi.org/10.1142/S0219649225500595

Tlili, A. et al.: What if the devil is my guardian angel: ChatGPT as a case study of using chatbots in education. Smart Learn. Environ. **10**(1), 15 (2023). https://doi.org/10.1186/s40561-023-00237-x

Zawacki-Richter, O., Marín, V.I., Bond, M., Gouverneur, F.: Systematic review of research on artificial intelligence applications in higher education--where are the educators? Int. J. Educ. Technol. High. Educ. **16**(1), 39 (2019). https://doi.org/10.1186/s41239-019-0171-0

Ethical AI in Education and Responsible Research: Frameworks, Governance, and Trustworthy AI for Knowledge-Driven Institutions

Assessment of COVID-19 Vaccine Acceptance Among Healthcare Workers at Azadi Teaching Hospital, Kirkuk, Iraq

Wafa Mahmood Jasim[1(✉)], Anmar B. Al-Dewachi[2], Waleed Ghanim Ahmad[2], and Israa Al-Rawe[3]

[1] Renal Dialysis Techniques Department, College of Health and Medical Techniques, Northern Technical University, Kirkuk, Iraq
Wafamahmood@ntu.edu.iq

[2] Department of Family and Community Medicine, College of Medicine, University of Mosul, Mosul, Iraq
{f.medicine,wga}@uomosul.edu.iq

[3] Artificial Intelligence Research Center, Northern Technical University, Mosul, Iraq
israa.a@ntu.edu.iq

Abstract. The vaccination of corona virus acceptance plays an essential role in eradicating the virus. and the health care workers (HCWs) are the main dangerous group that should obtain the covid-19 vaccine before any others. The (COVID-19) novel corona virus disease was determined first during 2019 and regarded as a pandemic and an emergency care necessity were established by World Health Organization (WHO) during 30th / 2020. The aim of the study was to assess the acceptance of Covid 19 vaccine for health staff who working in Azadi teaching hospital in Kirkuk, Iraq. The study methods were a cross-sectional study which was done in Azadi teaching hospital by collecting (200), who were selected randomly by a special questionnaire form was distributed to them after obtaining the written consent and the data was maintained through interviewing with the study sample during the period from 1/12/2023 till end of February 2023 (3 months duration). The results shown that majority of them were received 2 doses of Pfizer vaccine (69.8%) with a period about (3 weeks) between each dose (60.2%), and their sources of information from Ministry of Health (53–53%). (82%) of them determine in their attitude that the vaccine should be still free and 34% were agree that the essential factor interrupting the agreements of the vaccine was the ministry of health while (41%) of them mentioned that family & friends' suggestions are the main influencing factor in vaccine acceptance. The concluded that the acceptance rate is high among health staff and more than 2/3 of them received Pfizer vaccine because of its safety with the important role of Ministry of Health in vaccine education The study recommended that further advanced large-scale studies can be done on big population with more updated educational training programs about the Covid-19 vaccine and its advantages.

Keywords: Vaccine · Acceptance · Health staff · Hospitals · Covid 19

N. Mohd Sharef et al. (Eds.): CAMP 2026, CCIS 3094, pp. 111–123, 2027.
https://doi.org/10.1007/978-981-92-4305-1_9

1 Introduction

COVID-19 vaccine acceptance plays a crucial role in preventing the spread of the virus among healthcare workers (HCWs), who are considered a high-risk group due to their direct and frequent contact with infected patients. Vaccination of healthcare workers is essential not only for their own protection but also for reducing transmission within healthcare settings and protecting vulnerable patients.

The World Health Organization (WHO) first documented the novel coronavirus disease (COVID-19) in 2019 and subsequently declared the outbreak a COVID-19 pandemic a Public Health Emergency of International Concern on 30 January 2020. Later, on 11 March 2020, WHO characterized COVID-19 as a global pandemic [1, 2]. Corona epidemic presented itself as a cluster of pneumonia of unknown cause in China [3]. The disease revision systematically and has reported as a very infectious form of the disease in about 20% of the diseased people with approximately 3% death mortality rate [4]. By February 2021, COVID-19 had spread worldwide, affecting more than 108 million people and resulting in approximately 2.38 million deaths globally. The rapid transmission of the virus posed a major public health challenge and placed considerable pressure on healthcare systems worldwide. In Pakistan, approximately 560,000 COVID-19 cases had been reported and confirmed through clinical and laboratory diagnosis by February 2021, with 12, 218 associated deaths recorded. These figures highlight the substantial impact of the pandemic on the country's healthcare system and population [5, 6]. Therefore, the importance of personal protective equipment and a good healthy social distancing measures are of a vital need to be encouraged for vaccinated against COVID-19 in order to protect the disease transmission among the families and communities throughout the world [7], Health care workers (HCWs) have an important role in educational process by encouraging the general public acceptance for receiving the vaccine and its indication in the future coming years This is being mandated throughout the West, especially for risky people who include the health care workers. To measure the determine HCW attitudes and perceptions towards the COVID-19 vaccine as it will lead to a future dissemination of knowledge among the public later [8, 9]. Healthcare workers are among the groups at highest risk of COVID-19 infection because of their direct and frequent exposure to patients infected with SARS-CoV-2. This increased risk not only affects healthcare workers themselves but also raises the possibility of transmitting the infection to their families, children, and the wider community. Therefore, protecting healthcare workers has become a primary objective of health organizations and public health authorities. Ensuring their safety through appropriate infection-control measures, personal protective equipment, and vaccination is essential for maintaining an effective healthcare workforce and reducing the spread of COVID-19 [10, 11].

Despite immunization against SARS COV-2 infection at the targeted level and decreasing the outcome of the pandemic mainly the developing countries confronted many challenges like lack of fair distribution, vaccine reluctance, and hesitancy. The COVID-19 vaccine acceptance was also variable and not plentiful in many regions and countries that were less than 60% [12]. This poor hesitancy and acceptance were correlated with other misinformation factors about the vaccination safety, afraid from severe side effects, individual factors, poor awareness, and socio-cultural factors. A similar study was preformed to evaluate the acceptance of Covid 19-from the USA showed a higher of healthcare professionals who working in contact directly with patients accepting the COVID-19 vaccine (including physicians 86.6% and nurses 86.3% than those others health workers working with some association with the patients), in addition to that, about 5.5% of healthcare providers have hesitancy about vaccination against COID-19, so they would refuse the SARS-CoV-2 vaccination [13, 14].

The World Health Organization defines vaccine hesitancy as the delay in acceptance or refusal of vaccination despite the availability of vaccination services. Vaccine hesitancy has been identified by WHO as one of the major threats to global health because it can reduce vaccination coverage and undermine disease prevention efforts. The term refers to a situation in which individuals are uncertain about receiving vaccines or have concerns regarding their safety, effectiveness, or necessity. Individuals who completely refuse vaccination despite its availability are often described as vaccine resistant. The success of any vaccination program depends on achieving high vaccination coverage within the population, which contributes to the development of herd immunity. Herd immunity helps reduce the transmission of infectious diseases and protects individuals who are unable to receive vaccines due to medical or other legitimate reasons. Therefore, addressing vaccine hesitancy is essential for improving vaccine uptake and ensuring the effectiveness of public health immunization programs [15]. The successful of the vaccination process depends on achieving maximum coverage and thereby attaining herd immunity.

This study aims to evaluate the level of COVID-19 vaccine acceptance and the factors influencing vaccine uptake among healthcare staff at Azadi Teaching Hospital in Kirkuk City, Iraq.

2 Related Works

Regarding the demographic features of the study sample, a similar study was conducted by Grabowski, et al. during this follow-up, 23 (3%) of 651 patients tested positive on a retest for SARS-CoV-2 by RT-qPCR assay in a routine health check. The median age of this retest-positive group was 56·0 years (range 27·0–89·0, IQR48·5–74·0), with roughly more women (12 [52%] of 23 patients) than male group (11[48%]). In this retest-positive group, 12 patients (52%) had moderate, nine (39%) severe, and two (9%) critical conditions during their previous hospitalization [16].

Seong et al. who have done a study on the COVID-19 Vaccination attitude and Perception between Healthcare Workers in Egypt in order to assess their opinions toward COVID-19 vaccines, with their knowledge of the essential health determinants regarding the factors that may elevate the vaccine acceptance by an observational web-based anonymous survey which was involved 385 Egyptian healthcare workers in many different various governorates. The questionnaire-included the needed necessary based on Vaccine Hesitancy Survey Questions adapted by World Health Organization which was answered by two languages (both Arabic and English), for easy understanding and answering and was detected for its reliability that found the51% of the participants were undecided, 28% refused, and 21% accepted vaccination [17].

They reported that the main reasons for acceptance the vaccine was mainly fear from risks of COVID-19 (93%), safety aspect (57.5%), and its effectiveness (56.25%) of the vaccine. On the other side, the necessary causes for vaccine hesitancy were the lack of the needed clinical trials (92.4%) and afraid from complications and side effects of the vaccine (91.4%).

The most important factor that can elevate the acceptance of taking the vaccine among the respondents was to provide a adequate and scientific information about the present vaccines. The participants revealed a high level of vaccine safety concern (3.8 of 5) that different significantly between the study groups (P-value.002). they suggested in their study that, only approximately 21% of Egyptian healthcare workers were agreed with the vaccination in spite of COVID-19 epidemic, furthermore the hesitancy of the vaccine present a main limitation for implementing a vaccination program in a good successful manner [17].

Recent studies published after the acute phase of the COVID-19 pandemic have shown that healthcare workers generally demonstrate higher acceptance rates of COVID-19 vaccines compared with the general population; however, vaccine hesitancy remains an important public health concern. Studies conducted during 2023 and 2024 reported that confidence in vaccine safety, perceived effectiveness, previous vaccination experience, and trust in health authorities were the strongest predictors of vaccine acceptance among healthcare workers [17].

A systematic review published in 2024 mentioned that the acceptance of the vaccine among healthcare workers quietly different a across the countries, with approximately acceptance rates ranging from moderate level to high. The review reported the major concerns about the vaccine safety with the fear of side effects, misunderstanding and information, and the deficient of trust in pharmaceutical companies which is regarded as the main prevalent factors related with vaccine hesitancy. On the other hand, healthcare workers who had a contact directly with COVID-19 patients or having a high risk of infection were more liable to accept vaccination [12, 13].

Furthermore, similar study was done by Shekhar *et al.* on the acceptance of COVID-19 vaccine between health care workers in United States to evaluate the attitude of them about the vaccine acceptance with the most detecting barriers for their obtaining by done a cross sectional study. They Found that 36% of participants were willing to receive the vaccine as soon as it became present while 56% were not sure enough or waiting to obtain a more information. Only 8% of HCWs do not decide to take the vaccine. The acceptance of the vaccine was elevated with advancing education, age, and income level. A little percentage of female (31%), Lantinx (30%), Black (19%), and rural (26%) HCWs were agreeing to take the vaccine as it became available than the other study groups [18].

A further more studies and reports were conducted the acceptability of the vaccine among healthcare workers in the Saudi Arabia kingdom to evaluate their agreements of towards the vaccine and the essential associated factors affecting their intention to accept it by done a across-sectional survey which was adapted between 8–14 December 2020. They revealed that of the total of 736 workers who started the questionnaire electronically, 673 fully completed it (with a compellation rate = 91.44%). Among the study workers, 50.52% were having the interest and willing to receive the vaccine, of which 49.71% were vaccinated as soon as it becomes present and available in the country, while 50.29% would delay their response until detecting the safety of the vaccine, then receive the it later [19, 20].

Recent studies from Middle Eastern countries have demonstrated similar findings. Healthcare workers who possessed greater knowledge regarding vaccine effectiveness and safety were significantly more likely to receive recommended vaccine doses and booster vaccinations. Educational level, occupation, previous history of vaccination, and confidence in public health recommendations were significant predictors of variables [19]. Another same study was performed by Arazoo Issa et al. to determine the intention of COVID-19 vaccination with the main related factors that affecting their decision among the whole population and they revealed that most of study sample (53.6%) were female. The agreements and willingness for COVID-19 vaccine was very low (33%) [21]. They moreover mentioned that vaccine disagreements or rejection was high which was interrelated for (32%) and (34.9%) of studied participants, respectively, and this resistance was associated with occupation., age, and education. In addition to that, many other factors were interrelated with study sample refusal like: loss of family members during the pandemic, and adverse effects of the vaccine. Khamis, et al. prepared a similar study during 2021 on the acceptability of COVID-19 vaccination among healthcare workers in Sudan, to investigate acceptability of the COVID-19 vaccine and its associated factors among HCWs by conducting a semi-structured questionnaire, showed that: a total of 576 HCWs participated in the survey. The Mean age was 35 years. Females (53.3%), medical doctors (55.4%) and being lived in Khartoum State (76.0%) each of them presents for more than half of the participants [22].

Sulaiman in [23] published a study to assess the rate of acceptance of Covid 19 vaccine and its related factors in addition to discuss both the most received and preferable type and the major causes behind the refusal or acceptance of this vaccine in Erbil city, Iraq since 2022, reported that study presents that 43.9% of the respondents accepted the vaccine. The major cause behind their refusal (53.8%) was that in not trusting for development, on the other hand, the reason behind obtaining were to prevent themselves from further infections (44.2%), most of them were completely vaccinated (76.5%) received Pfizer type of vaccine, which is regarded as the preferable one (67.4%), only 10.3% of them were not completing the suggested scheduled dose and the reasons for that was, neglecting and shortage of time (31%), and side effect afraid (31%). A statistically significant correlation obtained between vaccination practice with gender, occupation, education, socioeconomic status and the family income. Vaccinated persons were less likely to be infected with COVID19 (20.7%) than non-vaccinated ones (62.7%).

Recent Iraqi and regional studies have also demonstrated that healthcare workers with higher educational back ground and better access to scientific information showed greater acceptance of COVID-19 vaccination. On the other side, misinformation received through social media platforms was correlated with higher vaccine hesitancy levels. These results document the need for future programs depending on evidence-based awareness targeting healthcare professionals and workers.

Another study which is a cross-sectional was done among 1069 participants from various Iraqi states by using a prepared designed questionnaires conducted electronically. Though a total of 77.6% of them agreed to obtain the COVID19 vaccine when present, the most common (64.3%) said they would postpone for taking the vaccine. Approximately two-thirds willing to pay a price for the receiving the vaccine and the high percentage of them (48.6%) preferred the type of Oxford (AstraZeneca) vaccine. There was a really significant correlation between educational status, age factor, working status, having an infected family members with the taking a flu vaccine before Covid19 vaccine acceptance [24, 25].

Overall, contemporary evidence indicates that although COVID-19 vaccine acceptance has improved over time, vaccine hesitancy still continues to exist among little healthcare workers. Addressing concerns associated with vaccine effectiveness, its safety, and misinformation remains the more essential for maintaining a high coverage vaccination rate and supporting with adequate strengthening preparedness for future infectious disease outbreaks and epidemics, in addition to that, studies conducted in new years among healthcare professionals showed that vaccine acceptance increased after the accumulation of evidence regarding vaccine safety and effectiveness. Nevertheless, concerns regarding long-term adverse effects and the rapid development of vaccines remained among a proportion of healthcare workers. These concerns highlight the importance of continuous educational campaigns and transparent communication from health authorities.

3 Methods

3.1 Administrative Agreement and Ethical Considerations

Before data collection, the necessary approvals and official permissions were obtained from the Kirkuk Health Directorate. Participation in the study was voluntary, and informed written consent was obtained from all participants before data collection. Participants were informed about the study aims and objectives and were assured of the confidentiality of their information and the right to withdraw from the study at any time.

3.2 Study Design and Period

A descriptive cross-sectional study was conducted among healthcare staff working at Azadi Teaching Hospital in Kirkuk City, Iraq, to assess the acceptance of COVID-19 vaccination over a period of three months, from 1 December 2023 to the end of February 2024.

3.3 Study Sample and Sampling Technique

The study included 200 healthcare staff members working at Azadi Teaching Hospital. Participants were selected using a random sampling technique. Data was collected using a structured questionnaire specifically designed for the study and the needed information was obtained through direct face-to-face interviews with them.

3.4 Inclusion Criteria and Exclusion Criteria, the Study Included

- Healthcare staff working at Azadi Teaching Hospital.
- Individuals who had received at least one dose of a COVID-19 vaccine.
- Individuals willing to participate in the study.
- The following individuals were excluded from the study:
- Pregnant and lactating healthcare staff.
- Healthcare staff who had not received a COVID-19 vaccine.

3.5 Data Collection Tool

A structured questionnaire was designed for data collection and consisted of four main parts:

Part 1: Demographic Characteristics
Information regarding age, gender, marital status, occupation, educational level, and years of experience.
Part 2: Information Sources
Questions related to the main sources from which participants obtained information about COVID-19 vaccines.
Part 3: COVID-19 Vaccination Status
Questions regarding the number of vaccine doses received and the time intervals between doses.
Part 4: Factors Influencing Vaccine Acceptance

- Questions assessing factors affecting healthcare staff acceptance of COVID-19 vaccination.
- To evaluate the influence of demographic characteristics on COVID-19 vaccine uptake, a three-point Likert scale was used: Agree, disagree, neutral (equivocal)
- The frequency and percentage of responses were calculated for each category.

3.6 Questionnaire Validity and Reliability

The questionnaire was reviewed by experts in relevant fields to assess its content validity, clarity, and reliability Necessary modifications were made according to their recommendations.

3.7 Data Analysis

Data was entered and analyzed using appropriate statistical methods. Descriptive statistics, including frequencies and percentages, were used to summarize the data. For questions with dichotomous responses (Yes/No), frequencies and percentages were calculated.

3.8 The Chi-square

The test was used to assess the association between study variables and COVID-19 vaccine acceptance. A p-value of less than 0.05 ($p < 0.05$) was considered statistically significant.

4 Results and Discussion

As shown in Table 1, which indicated that (128–64%) of the participants from the age group between 20–25 years while the lowest number from the age group above 40 years (6–3%), (166–83%) from them receiving vaccine working in Azadi-teaching hospital (120–60%).

Table 1. Distribution of the participants according to demographic characteristics

Age	Number	%
20–25	128	**64%**
26–30	48	24%
31–35	8	4%
36–40	10	5%
>40	6	3%
Receiving vaccine		
Yes	166	**83%**
No	34	17%
Working place		
Azadi-teaching hospital	120	60.0%
Kirkuk general hospital	80	40.0%
Sex		
Male	123	61.5%
Female	97	48.5%
Educational level		
Primary (school)	132	66%
Secondary (institute)	48	24%
Tertiary (collage)	20	10%
Marital status		
Married	47	23.5%
Not married	153	76.5%
Working experience		
< 5 years	81	40.5%
5–10 years	102	51.0%
>10 years	17	8.5%
Residence		
Urban	173	86.5%
Rural	27	13.5%

Concerning the sex, (123–61.5%) from the participants were male in comparison to female who constitute (97–66%), not married (153–76.5%). For working experience years, the highest number and percent among those who have 5–10 years (102–51.0%), living in urban areas (173–86.5%), with primary level of education (132–66%).

As shown in Table 2 that only vaccinated staff were included among the 166 vaccinated staff members included in the study, most participants had received two doses of a

Table 2. Distribution of participants according to vaccine dose

Vaccine dose	Number	%
1	34	20.5%
2	116	**69.8%**
3	16	9.7%
Total	166	100. %
Type of vaccine:		
Pfizer	150	**90.4%**
Sino pharm	8	4.8%
AstraZeneca	16	4.8%
Total	166	100. %
Period in between:		
2 weeks interval	20	21.4%
21 days	100	**60.2%**
3 months	18	10.8%
3–4 months	12	7.2%
>4 months	16	9.6%
Total	166	100.0%

COVID-19 vaccine (69.8%), while 20.5% had received one dose and 9.7% had received three doses. Pfizer was the predominant vaccine administered, accounting for 90.4% of vaccinations, whereas Sinopharm and AstraZeneca each accounted for 4.8%. Regarding the interval between vaccine doses, many participants (60.2%) received their doses 21 days apart, followed by 12.0% with a 2-week interval, 10.8% with a 3-month interval, 9.6% with an interval greater than 4 months, and 7.2% with an interval of 3–4 months.

Table 3. COVID-19 distribution of participants

Source	Number	%
Ministry of Health	106	53%
Social Media	32	16%
News	10	5%
WHO	36	18%
Friends & Family	16	8%
Total	200	

From Table 3, more than half of the participants (53.0%) reported the Ministry of Health as their primary source of information. The World Health Organization (WHO) was the second most common source, reported by 18.0% of participants. Social media accounted for 16.0% of information sources, while friends and family contributed 8.0%. News media were the least reported source, accounting for only 5.0% of participants. This finding indicates that official health authorities, particularly the Ministry of Health and WHO, were the main sources of COVID-19 vaccine information among the study participants.

Table 4. The attitude of the health staff towards Covid 19 vaccine acceptance

Item		Number	percentage
Additional information about covid 19-concerns and reluctance among staff workers	Yes	164	82%
	No	6	3%
	Don't know other	30	15%
		Total:200	
factors influence in acceptance of the Covid-19 vaccine	Ministry of Health	48	24%
	Higher numbers of Covid-19 positive case	30	15%
	Negative complications	54	27%
	The increasing number of people died	68	34%
		Total:200	
The main reason for accepting the Covid-19 vaccine	Vaccine doses counted	36	18%
	Condition of Health	38	19%
	Recommendation for family & friends	82	41%
	Vaccine producing country	44	22%

As shown in Table 4 which presents the attitudes of health staff toward COVID-19 vaccine acceptance. Most participants (82.0%) reported a need for additional information regarding COVID-19 vaccines because of concerns and reluctance among healthcare workers, while only 3.0% indicated that additional information was unnecessary and 15.0% were uncertain. Regarding factors influencing vaccine acceptance, the increasing number of COVID-19-related deaths was the most influential factor (34.0%), followed by concerns about negative complications of the disease (27.0%). Recommendations from the Ministry of Health influenced 24.0% of participants, whereas the increasing number of positive COVID-19 cases influenced 15.0%. Concerning the main reasons for accepting the COVID-19 vaccine, recommendations from family and friends were the most reported reason (41.0%). This was followed by the vaccine-producing country (22.0%), personal health condition (19.0%), and the number of vaccine doses required (18.0%). These findings suggest that social influence, particularly recommendations from family and friends, played an important role in vaccine acceptance among health staff, while the severity of the pandemic, reflected by mortality rates, was a major factor encouraging vaccination.

Table 5. Effect of demographic parameters on Covid-19 vaccine uptake

Demographic parameters		Strongly agree	Disagree	Equivocal	p. value
Age group	20–25	70(35%)	10(5%)	48(24%)	0.652
	26–30	28(14%)	4(2%)	16(8%)	0.376
	31–35	2(1%)	2(1%)	4(2%)	0.252
	>35	6(3%)	4(2%)	6(3%)	0.359
Receiving vaccine	Yes	136(68%)	14(7%)	16(8%)	0.036
	No	20(10%)	10(5%)	4(2%)	

As shown in Table 5 which illustrates the association between demographic characteristics and COVID-19 vaccine uptake. Regarding age groups, the highest proportion of participants who strongly agreed with COVID-19 vaccine uptake was observed among those aged 20–25 years (35.0%), followed by participants aged 26–30 years (14.0%). However, no statistically significant association was found between age group and vaccine uptake, as all p-values were greater than 0.05. In contrast, vaccine uptake was significantly associated with vaccination status. Among participants who had received the vaccine, 68.0% strongly agreed with vaccine uptake, compared with 10.0% among those who had not received the vaccine. The association between receiving the vaccine and attitudes toward vaccine uptake was statistically significant (χ^2 test, $p = 0.036$). These findings indicate that age was not significantly related to attitudes toward COVID-19 vaccine uptake, whereas vaccination status showed a significant association with participants' acceptance of COVID.

5 Conclusion

Most health care workers were vaccinated and a high acceptance rate for covid – 19 vaccinations with Pfizer type with a period 21 days between each dose. The future recommendations: (1) Further advanced studies can be done on large populations; (2) Educational training programs about the Covid-19 vaccine and its advantages; (3) More effective communication methods to reach the target about the successful of taking Covid-19 vaccine; (4) Increase trust and awareness by health authorities and Governments; (5) Improve communication between all health sectors.

References

1. Uddin, M. et al.: SARS-CoV-2/COVID-19: viral genomics, epidemiology, vaccines, and therapeutic interventions. Viruses. **12**(5), 526 (2020). https://doi.org/10.3390/v12050526

2. Mohamud, J., Hossain, J., Ahmed, M., Mitra, D., Nath, B., Olekhnovitch, M.D., et al.: The natural history, pathobiology, and clinical manifestations of SARS-CoV-2 infections. J. Neuroimmune Pharmacol. **15**(3), 359–386 (2020). https://doi.org/10.1007/s11481-020-09944-5
3. Irshad, A.: The race to treat COVID-19: potential therapeutic agents for the prevention of COVID-19. Eur. J. Med. Chem. **221**, 113157 (2021). https://doi.org/10.1016/j.ejmech.2021.113157
4. Shi, M., Guo, R., Wang, Q., Zhou, J.R.: Screening and molecular modeling evaluation of food peptides to inhibit key targets of COVID-19 virus. Biomolecules. **11**(2), 330 (2021). https://doi.org/10.3390/biom11020330
5. Li, Y.L., Hu, Z., Zhao, Q., Alias, H., et al.: Understanding COVID-19 vaccine demand and hesitancy: a nationwide online survey in China. PLoS Negl. Trop. Dis. **14**, 1–22 (2020). https://doi.org/10.1371/journal.pntd.0008961
6. Amanat, F., Krammer, F.: SARS-CoV-2 vaccines: status report. Immunity. **52**(4), 583–589 (2020). https://doi.org/10.1016/j.immuni.2020.03.007
7. Wibawa, T.: COVID-19 vaccine research and development: ethical issues. Trop. Med. Int. Health. **26**(1), 14–19 (2021). https://doi.org/10.1111/tmi.13503
8. Polack, F.P., Thomas, S.J., Kitchin, N., Absalon, J., Gurtman, A., Lockhart, S., et al.: Safety and efficacy of the BNT162b2 mRNA COVID-19 vaccine. N. Engl. J. Med. **383**(27), 2603–2615 (2020). https://doi.org/10.1056/NEJMoa2034577
9. Cabezas, C. et al.: Associations of BNT162b2 vaccination with SARS-CoV-2 infection, hospital admission and death with COVID-19 in nursing homes and healthcare workers in Catalonia: prospective cohort study. BMJ. **374**, n1868 (2021). https://doi.org/10.1136/bmj.n1868
10. Geng, M., Peng, Y., Meng, L., Li, X., et al.: Molecular immune pathogenesis and diagnosis of COVID-19. J. Pharm. Anal. **10**(2), 102–108 (2020)
11. Yuki, K., Fujiogi, M., Koutsogiannaki, S.: COVID-19 pathophysiology: a review. Clin. Immunol. **215**, 108427 (2020). https://doi.org/10.1016/j.clim.2020.108427
12. Cao, W., Li, T.: COVID-19: towards understanding of pathogenesis. Cell Res. **30**(5), 367–369 (2020)
13. Liu, Y.C., Kuo, R.L., Shih, S.R.: COVID-19: the first documented coronavirus pandemic in history. Biomed. J. **43**(4), 328–333 (2020). https://doi.org/10.1016/j.bj.2020.04.007
14. Merza, M.A., Aswad, S.M., Sulaiman, H.M., Abdulah, D.M., Rasheed, W.S., Taib, N.I.: Clinical and epidemiological characteristics and outcomes of coronavirus disease 2019 patients in a large longitudinal study. Int. J. Health Sci. **15**, 29 (2021)
15. Merza, M.A., Al Mezoris, A.A., Mohammed, H.M., Abdulah, D.M.: COVID-19 outbreak in Iraqi Kurdistan: the first report characterizing epidemiological, clinical, laboratory, and radiological findings of the disease. Diabetes Metab. Syndr. Clin. Res. Rev. **14**, 547–554 (2020). https://doi.org/10.1016/j.dsx.2020.04.047
16. Grabowski, F., Kochańczyk, M., Lipniacki, T.: The spread of SARS-CoV-2 variant omicron with a doubling time of 2.0–3.3 days can be explained by immune evasion. Viruses. **14**(2), 294 (2022). https://doi.org/10.3390/v14020294
17. Seong, H., Hyun, H.J., Yun, J.G., Noh, J.Y., Cheong, H.J., Kim, W.J., et al.: Comparison of the second and third waves of the COVID-19 pandemic in South Korea: importance of early public health intervention. Int. J. Infect. Dis. **104**, 742–745 (2021). https://doi.org/10.1016/j.ijid.2021.02.004
18. Shekhar, R., Sheikh, A.B., Upadhyay, S., Singh, M., Kottewar, S., al Mir, H.e.: COVID-19 vaccine acceptance among health care workers in the United States. Vaccine. **9**(2), 119 (2021). https://doi.org/10.3390/vaccines9020119

19. Brandal, L.T., MacDonald, E., Veneti, L., Ravlo, T., Lange, H., Naseer, U., et al.: Outbreak caused by the SARS-CoV-2 omicron variant in Norway, November to December 2021. Eurosurveillance. **26**(50), 2101147 (2021). https://doi.org/10.2807/1560-7917.ES.2021.26.50.2101147
20. Ghazi, H.F., Jwad Taher, T.M., Alfadhul, S.H.A., Al-Mahmood, S., Hassan, S.H.A., Hamoudi, S.A., et al.: Acceptance of COVID-19 vaccine among the general population in Iraq. Iraqi Natl. J. Med. **3**(1), 93–99 (2021)
21. Ibrahim, A., Tahir, D., Shamal, R., Piro, S.S., Abdullah, R.Y., et al.: COVID-19 vaccine acceptance, hesitancy and refusal among the Iraqi Kurdish population. Int. J. Health Sci. **16**(1), 10–16 (2022)
22. Khamis, A., Mohamed, E., Nimir, M., Ahmed, M., Jawad, M., et al.: Acceptability of COVID-19 vaccination among healthcare workers in Sudan. East Mediterr. Health J. **29**(5), 5–10 (2023). https://doi.org/10.26719/emhj.23.061
23. Sulaiman, K.H.: COVID-19 vaccination acceptance among visitors of Erbil primary health care centers. Diyala J. Med. **26**(1), 127–139 (2024)
24. Jasim, M.T., Noraldeen, M.Y., Al-Kadi, N.A.: Comparison of some biochemical parameters in COVID-19 patients and vaccinated individuals in Kirkuk, Iraq. NTU J. Pure Sci. **1**(4), 1–7 (2023). https://doi.org/10.56286/ntujps.v1i4.301
25. Breesam, A.A., Nooruldeen, M.Y.: Evaluation of the immunochromatography assay's diagnostic performance for quickly detecting the presence of COVID-19 antigen in patients with positive PCR results. NTU J. Pure Sci. **1**(4), 35–40 (2022). https://doi.org/10.56286/ntujps.v1i4.309

CatBoost Regression and Feature Engineering for Forecasting Spare Parts Demand in the Heavy Equipment Industry

Diva Kurnianingtyas[1(✉)], Nathan Daud[1], Muhammad Farhan Dwi Putra Soeyanto[1], and Tutut Herawan[1,2,3]

[1] Faculty of Computer Science, Universitas Brawijaya, 65154 Malang, East Java, Indonesia
divaku@ub.ac.id

[2] Faculty of Computer Science and Information Technology, Universiti Malaya, 50603 Kuala Lumpur, Malaysia

[3] AMCS Research Center, Yogyakarta, Indonesia

Abstract. Spare parts inventory management requires demand forecasting to plan stock levels and support ongoing operational activities in the heavy equipment industry. However, spare parts demand often exhibits intermittent characteristics, where many items have zero demand in several periods and nonzero demand values fluctuate substantially. This condition makes forecasting more complex than normal demand forecasting and reduces the reliability of conventional methods when demand is sparse, irregular, and highly variable. This study evaluates CatBoost Regression with statistical feature engineering to forecast the demand for 186 spare parts in the heavy equipment industry using historical demand data from August to November 2021. The feature engineering process includes demand-based and intermittency-related features, such as mean demand, standard deviation, maximum demand, minimum demand, zero ratio, average demand interval (ADI), coefficient of variation squared (CV^2), and trend. CatBoost is compared with Moving Average, Linear Regression, Random Forest, LightGBM, and XGBoost. The results show that CatBoost achieves the best performance in terms of RMSE (3.2840) and R^2 (0.8147), indicating its ability to capture nonlinear relationships between engineered demand features and future demand. Meanwhile, Moving Average obtains slightly better MAE and WAPE values, suggesting that simple methods may still be competitive for low and stable demand items. Feature importance analysis indicates that mean demand, recent demand, maximum demand, and trend are the most influential predictors. Overall, the findings show that CatBoost with statistical feature engineering is a promising approach for forecasting spare parts demand with sparse and intermittent characteristics, particularly when the demand pattern contains nonlinear variation across items.

Keywords: Artificial intelligence · Spare part demand forecasting · Intermittent demand · CatBoost Regression · Feature engineering

N. Mohd Sharef et al. (Eds.): CAMP 2026, CCIS 3094, pp. 124–136, 2027.
https://doi.org/10.1007/978-981-92-4305-1_10

1 Introduction

Demand forecasting is a critical component of spare parts inventory management because it supports procurement planning, stock control, maintenance readiness, and continuity of industrial operations. In the heavy equipment industry, spare parts availability is closely related to equipment reliability because damaged or worn components must be replaced to avoid operational disruption (Kim et al. 2023). When spare parts are unavailable, maintenance activities may be delayed, and equipment downtime may increase. Conversely, excessive inventory can create unnecessary holding costs (Ifraz et al. 2023). Therefore, forecasting spare parts demand is important for balancing service availability and inventory efficiency.

Forecasting spare parts demand is challenging because demand is often intermittent, sparse, and highly variable. In intermittent demand, many observation periods contain zero demand, while nonzero demand may occur suddenly and with different magnitudes. This pattern differs from normal demand because the forecasting model must learn not only the demand quantity but also the irregular occurrence of demand. Previous studies have shown that intermittent and sporadic demand remain difficult to forecast because limited historical information, low demand frequency, and sudden demand changes reduce prediction accuracy (Adur Kannan et al. 2020; Boylan and Syntetos 2021; Fan et al. 2023).

Traditional statistical forecasting methods such as Moving Average, ARIMA, and Exponential Smoothing are useful for demand patterns that are relatively stable and have sufficient historical observations. However, these methods may become less adaptive when the dataset contains many zero values, abrupt increases, or irregular decreases. For example, Moving Average can provide simple and competitive estimates for stable spare parts demand, but it may respond slowly to sudden upward or downward changes (Sherliyanda et al. 2025). Similarly, ARIMA and Exponential Smoothing have been applied to spare parts inventory forecasting, but their performance is generally more suitable when demand does not exhibit strong intermittent behavior (Sutomo and Hasibuan 2021; Budiarto et al. 2024).

Machine learning approaches have been increasingly used to address the nonlinear and irregular characteristics of spare parts demand. Studies on intermittent and sporadic demand forecasting have evaluated models such as Random Forest, XGBoost, LightGBM, and LSTM and have shown that machine learning can improve forecasting performance when demand contains complex relationships among historical features (Li 2025; Wang et al. 2025; da Silva et al. 2026; Fu et al. 2026). In addition, demand classification using average demand interval (ADI), coefficient of variation squared (CV^2), and ABC analysis has been recommended to better understand spare parts demand patterns before forecasting (da Silva et al. 2026). These studies indicate that feature-based learning is important for intermittent demand because the model requires information about demand magnitude, sparsity, and variation.

CatBoost is a gradient boosting algorithm that is effective for tabular data and nonlinear regression problems. Previous studies have shown that CatBoost can produce competitive performance compared with other machine learning models because its ordered boosting mechanism can improve generalization and reduce prediction shift (Ding et al. 2020; Hancock and Khoshgoftaar 2020; Vishwakarma et al. 2024). Nevertheless, the use of CatBoost for intermittent spare parts demand forecasting remains relatively limited, especially in the context of heavy equipment spare parts, where many items have low demand frequency and sudden demand changes.

Based on these conditions, this study proposes a CatBoost Regression-based forecasting model using demand history and statistical features derived from feature engineering. The added features are designed to represent demand magnitude, demand variation, sparsity, and recent trends so that the model can learn intermittent demand patterns more effectively. The contribution of this study is threefold. First, it evaluates CatBoost Regression for sparse and intermittent spare parts demand in the heavy equipment industry. Second, it integrates demand-based and intermittency-related statistical features, including ADI, CV^2, zero ratio, and trend. Third, it compares CatBoost with Moving Average, Linear Regression, Random Forest, LightGBM, and XGBoost to identify its relative forecasting performance. Through this evaluation, the study provides practical insight into the potential of CatBoost and feature engineering for supporting spare parts demand forecasting in industrial inventory management.

2 Methods

2.1 Data Collection

This study uses spare part demand data obtained from a heavy equipment company in Indonesia. Demand data for each spare part was collected from August to November 2021. Each data set includes 186 spare part types, with varying demand for each component. The dataset exhibits an intermittent demand pattern, indicated by a majority of zero demand values in certain periods. This data pattern presents a challenge in the forecasting process. Features used in the dataset include part number, part name, and total sales. The part number feature represents the unique identity of each spare part, while the total sales feature is used as the main variable in the demand forecasting process. The data is then integrated and processed with machine learning to build a predictive model using the Python-based CatBoost Regression algorithm.

2.2 Data Preparation

The purpose of data preparation is to ensure the data has a consistent structure. If the data is structured, the forecasting process runs smoothly and achieves high accuracy. In the data preparation process, this study used the Pandas library for data manipulation and integration and NumPy for numerical calculations and statistical processing. Both libraries are available in the Python programming language.

This study conducted three data preparation stages. First, spare part demand data from August to October were integrated based on part number attributes. Second, the part number and total sales attributes were standardized to facilitate the machine learning process. Third, missing values were handled by converting null values into zero. This treatment was used because not all spare parts are demanded every month; therefore, a missing demand record in the integrated monthly data represents a zero-demand condition. After the data structure was standardized, feature engineering was performed by creating additional statistical features based on historical demand. These features were used to help the model learn demand behavior not only from raw monthly demand values but also from the characteristics of demand variation, sparsity, and change for each spare part.

Eight statistical features were created, namely mean demand, standard deviation, maximum demand, minimum demand, zero ratio, average demand interval (ADI), coefficient of variation squared (CV^2), and trend. The mean demand feature calculates the average demand across observed periods, while standard deviation represents the level of demand variation for each item. The maximum and minimum demand features identify the highest and lowest historical demand values. The zero-ratio feature measures the proportion of observed periods with zero demand and represents the sparsity level of each spare part item. ADI measures the average interval between nonzero demand occurrences, whereas CV^2 represents the squared coefficient of demand variation. Finally, the trend feature captures changes in demand between periods to indicate whether demand tends to increase or decrease. These features were designed to help the model capture both demand magnitude and intermittent demand behavior, following common intermittent demand characterization using ADI and CV^2 (Boylan and Syntetos 2021).

2.3 CatBoost

CatBoost regression is an algorithm used to predict demand based on historical spare parts demand. This study uses CatBoost because the method can handle data with nonlinear patterns and unstable demand variations found in industrial demand data. CatBoost is also capable of learning relationships between features more effectively through a gradient boosting approach (Zhang et al. 2022). This study implemented the CatBoost library in Python, supported by the Pandas and NumPy libraries for data manipulation and Scikit-learn for model evaluation.

There are two stages to implementing CatBoost in this study (see Fig. 1). First, the CatBoost model parameters were initialized. Second, the model was trained and tested using the prepared monthly demand features. The implementation process was designed to evaluate the baseline capability of CatBoost Regression when combined with statistical feature engineering for spare parts demand forecasting.

2.3.1 Parameter Initialization

Several key parameters in CatBoost need to be initialized for demand forecasting. This study uses the iterations, learning rate, depth, loss function, and random seed parameters. The iterations parameter determines the number of boosting trees built during training. The learning rate controls each decision tree's contribution to the final prediction. The depth parameter determines tree depth in each boosting process and affects the model's ability to capture nonlinear relationships between historical demand features. The loss function parameter is used as an evaluation function during model training. Root Mean Square Error (RMSE) was selected because it measures the difference between predicted and actual demand values. The random seed parameter was used to maintain reproducibility. In this study, CatBoost was implemented using default parameters from

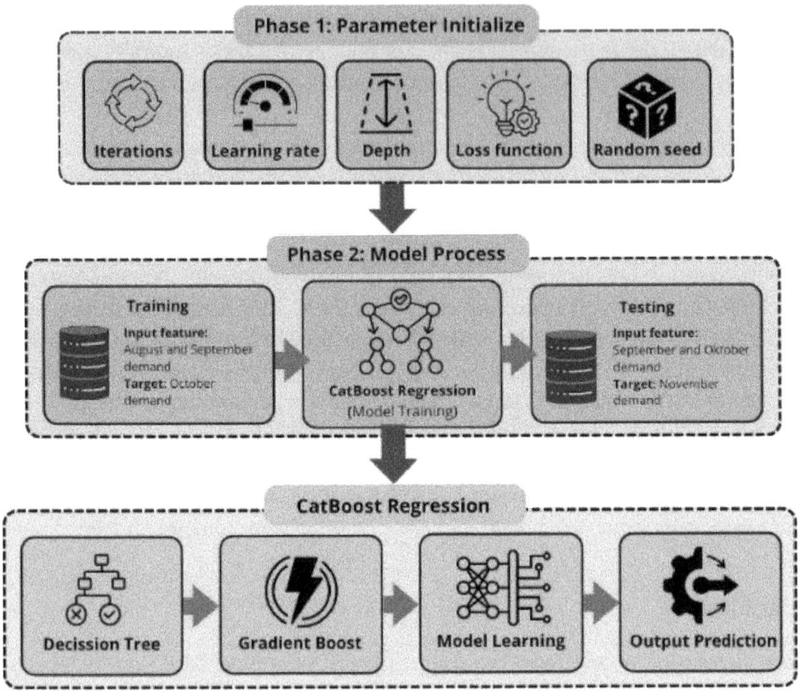

Fig. 1. CatBoost pipeline in this study

the Python CatBoost library to evaluate its baseline capability under the same forecasting setup as the comparison models.

2.3.2 Model Process

CatBoost Regression performs a training and testing process to simulate the prediction of demand in the next period based on recent historical demand. In the training stage, demand data from August (S1) and September (S2), together with statistical features formed from these periods, were used as input features, while demand data for October served as the target variable. After the training process, the model was tested using demand data from September (S1) and October (S2) with their corresponding statistical features as input, while November demand served as the forecasting target. The forecasting results were then compared with the actual November demand values to evaluate the ability of CatBoost to predict spare parts demand.

2.4 Regression Evaluation

The forecasting models were evaluated using five regression metrics: Root Mean Squared Error (RMSE), Mean Absolute Error (MAE), Coefficient of Determination (R^2), Mean Absolute Percentage Error (MAPE), and Weighted Absolute Percentage Error (WAPE). These metrics were selected to assess model performance from different perspectives, including average absolute error, sensitivity to large errors, relative percentage error, and the ability of the model to explain variation in actual demand. Better forecasting performance is indicated by lower MAE, RMSE, WAPE, and MAPE values, while a higher R^2 value indicates a stronger ability to follow actual demand patterns. All metric calculations were implemented using Scikit-learn and NumPy in Python.

3 Results and Discussion

3.1 Exploratory Data Analysis

Before training CatBoost, exploratory data analysis was conducted to understand the characteristics of the dataset. This analysis is important because spare parts demand generally exhibits unstable patterns, many zero values, and high demand variation across items. This study used the Pandas, Matplotlib, and Seaborn libraries in Python to support statistical analysis and data visualization. Based on the EDA results in Table 1, the average monthly demand ranges from 1 to 3 units per spare part, while the standard deviation is relatively high compared with the average demand. This indicates that demand distribution across items is uneven and contains substantial variation.

Table 1. Exploratory data analysis

Metric	Value
Number of parts	186
Total demand in August	332
Total demand in September	474
Total demand in October	351
Total demand in November	460
Mean demand in November	2.473118
Std demand in November	7.649188
Number of Zero Demand items in November	148
Percentage of Zero Demand items in November (%)	79.56989
Average of ADI	2.553763
Average of CV^2	0.25595
Average of Zero Ratio	0.776882

The demand for spare parts in November was very low and tended to approach zero. The distribution pattern formed a right-skewed distribution, which is a distribution that extends to the right side because some items had high demand, such as 53C0051 and 53C0052 with 59 units, as shown in Fig. 2. On the other hand, items with part numbers 06B0226, 08A5864, and 10A2658 had no demand in the November period. This condition indicates that some spare parts are only used for specific maintenance needs or are slow-moving items. This often becomes a challenge in forecasting because the model must be able to distinguish items with active demand from sporadic demand (Boylan and Syntetos 2021).

The next analysis measures the frequency of demand and the level of demand variation between periods. Based on the ADI and CV^2 calculations visualized in Fig. 3, the ADI values for some items are close to one, indicating that demand for spare parts still occurs in most observation periods, albeit in small amounts. Meanwhile, the relatively low CV^2 values for some items indicate that demand variation between periods remains quite stable. However, several items have high CV^2 values, indicating significant demand fluctuations. For example, there are items with an ADI value of around 2 and a CV^2 close to 2, indicating a very unstable and infrequent demand pattern.

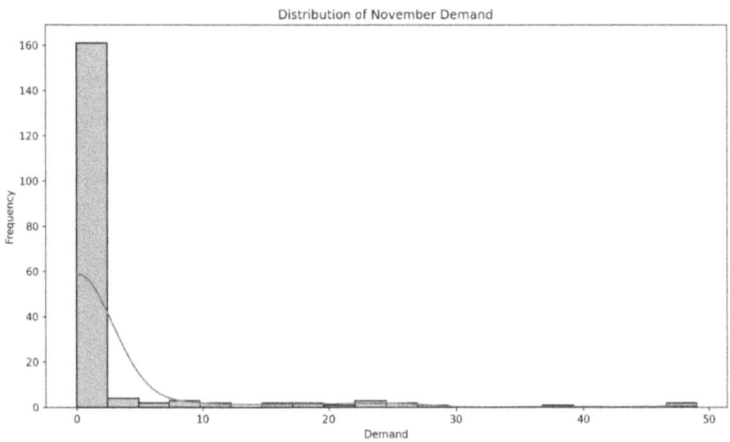

Fig. 2. Distribution of November demand

Overall, the EDA results indicate that the dataset has complex spare part demand characteristics due to an imbalanced demand distribution, many zero-demand observations, and demand variation across items. This supports the use of feature-based machine learning approaches such as CatBoost, because the model can learn nonlinear relationships between demand-based features and future demand (Ding et al. 2020; Hancock and Khoshgoftaar 2020; Vishwakarma et al. 2024).

3.2 Forecasting Results

To evaluate the ability of CatBoost Regression to learn spare parts demand patterns, the forecasting results were compared with the actual demand values. Based on Table 2, CatBoost obtained an MAE value of 1.290323, indicating an average absolute prediction error of approximately one to two units. The RMSE value of 3.284044 shows greater sensitivity to several items with large prediction errors, which is expected in intermittent demand data with sudden changes. The WAPE and MAPE values of 52.17391% and 34.13297%, respectively, indicate that forecasting sparse and irregular spare parts

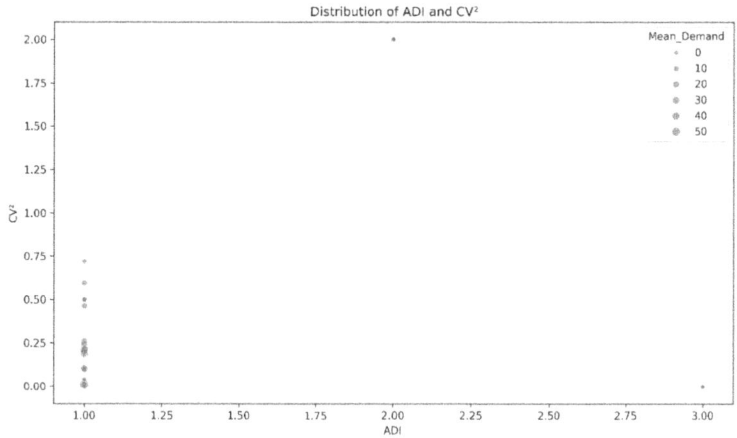

Fig. 3. Distribution of ADI and CV^2

demand remains challenging. Nevertheless, the R^2 value of 0.814677 shows that CatBoost can explain approximately 81% of the variation in actual demand based on the engineered historical features.

Table 2. CatBoost model evaluation

Metrics evaluation	Value
MAE	1.290323
RMSE	3.284044
R^2	0.814677
WAPE (%)	52.17391
MAPE (%)	34.13297

The forecasting results produced by CatBoost are close to the actual values for many spare part items, especially those with low to medium demand. This can be observed in Fig. 5, where several prediction points are located near the diagonal line representing perfect prediction. For actual demand values in the range of 1–5 units, many predictions remain in a relatively similar range. However, several items still show high prediction errors. For example, spare part item 40C5856 has an actual demand of 18 units, but the model predicts only 1 unit, resulting in an error of 17. Conversely, spare part item SP125335 has an actual demand of 3 units, but the model predicts 17 units, resulting in an error of 14. These cases indicate that abrupt demand changes remain difficult to forecast. The right-skewed error distribution in Fig. 4 confirms that most items have small errors, while a few sporadic items produce large prediction errors.

In summary, CatBoost can learn demand patterns reasonably well for spare parts with low demand and relatively stable behavior. However, several items with extreme demand

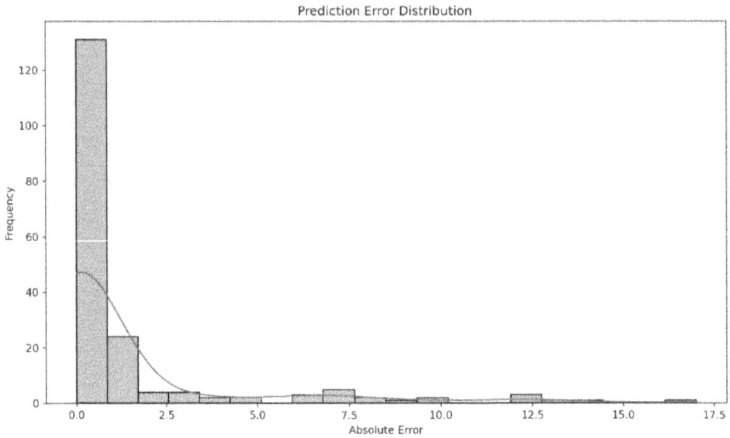

Fig. 4. Prediction error distribution

swings still produce large errors. This finding shows that industrial spare parts demand forecasting remains complex, even when boosting-based machine learning models and statistical feature engineering are applied.

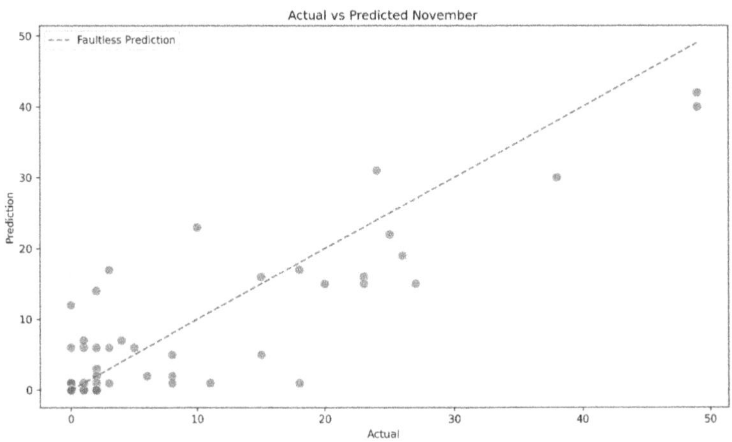

Fig. 5. Comparison of actual and prediction demand in November

3.3 Feature Importance Analysis

In addition to model evaluation, feature importance analysis is useful for identifying which features contribute most to the CatBoost forecasting process. Based on the analysis results shown in Fig. 6, the feature with the largest contribution is mean demand, with an importance value of 26.2483. This indicates that historical average demand is the strongest information source in the forecasting process. In spare parts forecasting, items with higher average demand tend to be predicted as having higher future demand than items with lower average demand. Recent demand (S1), maximum demand, and trend also contribute substantially, showing that CatBoost relies on both demand magnitude and recent demand movement. Meanwhile, standard deviation, minimum demand, CV^2, ADI, and zero ratio have lower importance values, although these features still help the model capture unstable and sparse demand characteristics. For example, item 31B0173 has a demand of 10 units in September and 0 units in October, with a CV^2 value of 2.0000. CatBoost predicted November demand at 12 units, while the actual demand was 0 units. This case shows that high-demand variation can still lead to prediction errors.

The conclusion obtained from the feature importance results is that CatBoost utilizes demand-based features, such as mean demand, S1, and maximum demand, more than sparsity features like ADI and zero ratio. This finding is supported by the fact that most spare part items in the dataset have low demand, while only a few items experience spikes in demand. Therefore, the average demand and historical demand features are the primary information for the model in determining the next month's demand estimate.

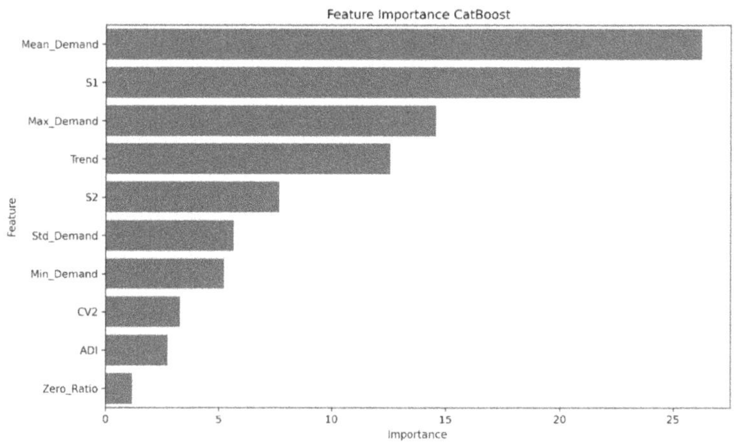

Fig. 6. The importance of the features used by CatBoost

3.4 Discussion

CatBoost achieved competitive forecasting performance on spare parts demand data with intermittent characteristics. Based on the evaluation results, CatBoost produced MAE, RMSE, and R^2 values of 1.2903, 3.2840, and 0.8147, respectively. The R^2 value indicates that the model can explain approximately 81% of the variation in actual demand using the engineered historical features. In addition, the model produced predictions that were close to the actual values for many low- and medium-demand items. This result suggests that CatBoost can learn nonlinear relationships between statistical demand features and future spare parts demand.

Compared with other methods, CatBoost produced the best performance in RMSE and R^2 compared with Linear Regression, Random Forest, LightGBM, and XGBoost.

Based on Table 3, CatBoost is more effective in reducing large prediction errors than most comparison models, which suggests that its boosting mechanism is useful for complex demand patterns (Ding et al. 2020; Hancock and Khoshgoftaar 2020). However, the Moving Average produced slightly better MAE and WAPE values than CatBoost. This result does not contradict the usefulness of CatBoost. Rather, it indicates that simple forecasting methods can remain competitive for many low-demand and relatively stable items. CatBoost is more advantageous when the evaluation emphasizes RMSE and R^2, which are more sensitive to large errors and overall demand variation.

Table 3. Comparison of models based on evaluation metrics

Methods	Metrics evaluation				
	MAE	RMSE	R^2	WAPE (%)	MAPE (%)
Moving average	1.27957	3.78878	0.75333	51.73913	22.69802
CatBoost	1.290323	3.284044	0.814677	52.17391	34.13297
Linear regression	1.35484	3.46255	0.79398	54.78261	36.39418
Random forest	1.46774	3.76957	0.75583	59.34783	34.43378
LightGBM	1.61828	4.53718	0.64626	65.43478	41.75856
XGBoost	1.62903	4.49671	0.65254	65.86957	36.24835

The intermittent demand characteristics significantly affect forecasting performance in this study. Most spare parts have low demand and do not appear consistently in each period. This condition creates an imbalanced demand distribution and makes forecasting more difficult than normal demand forecasting (Adur Kannan et al. 2020; Fan et al. 2023). Some spare parts also have sporadic demand patterns with sudden increases or decreases between periods, which explains why the model performs well for many items but still produces high errors for specific parts.

The results should be interpreted in light of several methodological considerations. First, the available demand history covers August to November 2021, so future studies with longer observation periods are needed to examine seasonality, maintenance cycles, and long-term replacement trends. Second, CatBoost was evaluated using default parameters; therefore, hyperparameter tuning can be explored in future work to determine whether the model performance can be further improved. Third, the current model comparison provides descriptive evidence based on the available testing setup. Formal statistical significance tests would be more appropriate when multiple rolling-origin test windows are available. Despite these considerations, the results demonstrate that CatBoost with statistical feature engineering has practical potential for spare parts procurement, stock control, and inventory planning in the heavy equipment industry. Future research can also incorporate external variables such as maintenance schedules, equipment utilization, equipment age, operating hours, and failure records to better anticipate sudden spare parts demand (da Silva et al. 2026).

4 Conclusion

CatBoost Regression can be used to forecast spare parts demand with sparse and intermittent characteristics. In this study, CatBoost achieved MAE, RMSE, and R^2 values of 1.2903, 3.2840, and 0.8147, respectively, and showed the strongest performance in terms of RMSE and R^2 among the evaluated models. These results indicate that CatBoost can learn relationships between engineered historical demand features and target demand, especially when the demand pattern contains nonlinear variation and high fluctuation. This suggests that engineered historical demand features can help capture complex demand behavior. Moving Average remained competitive in terms of MAE and WAPE, indicating that simple methods may still be suitable for low and stable demand items. This suggests that method choice should reflect demand stability. Analysis showed that mean demand, recent demand, maximum demand, and trend were the main features influencing the forecasting results. This suggests that these features should receive priority in forecasting design. The findings highlight that intermittent demand remains a major challenge because several spare parts still produce high prediction errors. Nevertheless, statistical feature engineering provides useful predictors for supporting demand forecasting and can contribute to spare parts procurement, inventory control, and logistics planning in the heavy equipment industry. Future studies can extend the history of demand, evaluate hyperparameter tuning, apply rolling-origin validation, and add external variables such as maintenance schedules, operational activities, and equipment conditions.

Acknowledgments. This study was supported by the Faculty of Computer Science, Universitas Brawijaya, through funding assistance for reputable international scientific publication (Decision No: 00219/DST/UN10.A0501/B/PJ/2026).

References

Adur Kannan, B., Kodi, G., Padilla, O., et al.: Forecasting spare parts sporadic demand using traditional methods and machine learning-a comparative study. SMU Data Sci. Rev. **3**, 9 (2020)

Boylan, J.E., Syntetos, A.A.: Intermittent Demand Forecasting: Context, Methods and Applications. John Wiley & Sons (2021)

Budiarto, D.D., Miftahudin, M., Riwurohi, J.E.: Application of exponential smoothing method for forecasting spare parts inventory at heavy equipment distributor company. Eduvest J. Univers. Stud. **4**, 959–976 (2024)

da Silva, L.F., da Silva Christo, E., Costa, K.A., Tavares, H.P.: Combining demand classification and forecasting models in spare parts inventory management for the energy sector. Int. J. Adv. Manuf. Technol. **144**, 1–12 (2026)

Ding, J., Chen, Z., Xiaolong, L., Lai, B.: Sales forecasting based on catboost. In: 2020 2nd International Conference on Information Technology and Computer Application (ITCA), pp. 636–639. IEEE (2020)

Fan, L., Liu, X., Mao, W., et al.: Spare parts demand forecasting method based on intermittent feature adaptation. Entropy. **25**, 764 (2023)

Fu, W., Bian, H., Chen, J., Jing, S.: Tree-based machine learning intermittent demand forecasting for spare parts in electric vehicle manufacturing. World Electr. Veh. J. **17**, 127 (2026)

Hancock, J.T., Khoshgoftaar, T.M.: CatBoost for big data: an interdisciplinary review: JT Hancock, TM Khoshgoftaar. J. Big Data. **7**, 94 (2020)

Ifraz, M., Aktepe, A., Ersöz, S., Çetinyokuş, T.: Demand forecasting of spare parts with regression and machine learning methods: application in a bus fleet. J. Eng. Res. **11**, 100057 (2023)

Kim, J.-D., Kim, T.-H., Han, S.W.: Demand forecasting of spare parts using artificial intelligence: a case study of KX tanks. Mathematics. **11**, 501 (2023)

Li, W.: Spare parts demand forecasting and inventory optimization based on machine learning algorithms. In: Machine Intelligence and Digital Applications: Proceedings of the 2nd International Conference (MIDA 2025), Ningbo, China, 18-20 April 2025, pp. 509–518. SAGE Publications, London (2025)

Sherliyanda, D., Irawan, M.D., Nasution, A.B.: Implementation of the single moving average method in forecasting sales of motorcycle spare parts. J. Dinda Data Sci. Inf. Technol. Data Anal. **5**, 99–106 (2025)

Sutomo, A., Hasibuan, S.: Comparison of ARIMA, linear trend and single exponential smoothing for short-term forecasting of power plant spare parts inventory. In: Proceedings of the 11th Annual International Conference on Industrial Engineering and Operations Management (2021)

Vishwakarma, D.K., Kumar, P., Yadav, K.K., et al.: Evaluation of CatBoost method for predicting weekly pan evaporation in subtropical and sub-humid regions. Pure Appl. Geophys. **181**, 719–747 (2024)

Wang, W., Yang, Z., Zhao, T.: International trade demand forecasting model based on transformer-XGBoost-LightGBM. Artif. Intell. Appl. (2025)

Zhang, F., Fleyeh, H., Bales, C.: A hybrid model based on bidirectional long short-term memory neural network and Catboost for short-term electricity spot price forecasting. J. Oper. Res. Soc. **73**, 301–325 (2022)

Data Governance and Knowledge Management: Higher Education Performance Analytics and Knowledge-Driven Decision Making

A Proactive Institutional Intelligence Framework: Transitioning from Reactive Reporting to an Integrated Knowledge Management Ecosystem

Shahrinaz Ismail[1(✉)], Nordiana Jamaludin[2], and Siti Haryani Shaikh Ali[3]

[1] School of Computer Sciences, Universiti Sains Malaysia, Penang, Malaysia
shahrinaz@usm.my

[2] Faculty of Business Management and Professional Studies, Management and Science University, Shah Alam, Selangor, Malaysia

[3] Malaysian Institute of Information Technology, Universiti Kuala Lumpur, Kuala Lumpur, Malaysia

Abstract. The implementation of Outcome-Based Education (OBE) in Malaysian higher education focuses on the systematic measurement of Course Learning Outcomes (CLOs) and Program Learning Outcomes (PLOs). Through a comparative analysis of seven diverse institutions, this study identifies that while automated systems are increasingly used, the current "As-Is" processes are often reactive, hampered by manual data entry, semester-end "batch" processing, and fragmented "data silos". To address these challenges, this research proposes a Proactive Institutional Intelligence Framework designed to transition institutions into an integrated Knowledge Management Ecosystem. This study proposes a "To-Be" solution that utilizes the Extract-Transform-Load (ETL) framework to ensure continuous data flow, scalability, and data veracity. Emphasis is also made on the importance of sound data governance to secure the sharing of personalized attainment data with future employers, as well as suggesting the integration of Natural Language Processing (NLP) to incorporate qualitative student feedback into the analytics process. The proposed framework aims to extend the simple attainment calculation to predictive analysis that supports institutional quality assurance and academic excellence.

Keywords: Institutional Intelligence · Knowledge Management Ecosystem · Outcome-Based Education · Learning Outcome Attainment

1 Introduction

In Malaysia, Outcome-Based Education (OBE) has become a core requirement for higher education institutions to implement, ensuring graduates' employability in the global market. Entailing this, there is a need to measure the students' competencies based on their learning outcome attainment (Mokhtar et al., 2016). At the heart of OBE lies the

N. Mohd Sharef et al. (Eds.): CAMP 2026, CCIS 3094, pp. 139–153, 2027.
https://doi.org/10.1007/978-981-92-4305-1_11

tracking of Course Learning Outcomes (CLOs) and Program Outcomes (POs), which act as the primary metrics for attainment measurement, leading to institutional accreditation. As universities move from manual spreadsheets to web-based platforms, the learning outcomes attainment analysis evolves as well, providing opportunities to perform Continuous Quality Improvement (CQI) as part of the overall academic quality assurance.

A typical scenario at a Malaysian university would start with a collection of raw grades from various forms of assessments, i.e., coursework and final exam. As planned based on the Scheme of Work (SOW), i.e., the blueprint that outlines core skills required for a student to master in a course, these assessment grades are entered into a system and saved in a database. The best practice is to have the CLOs and POs ready for analysis as they are mapped to the respective assessments with the guide from the SOW. The attainment is commonly standardized across the university with a threshold that is applied to all programs.

While academic frameworks, from data collection to reporting, are well-defined, the standardization of Extract-Transform-Load (ETL) alignment required for comprehensive analytics remains somewhat unclear. Current research prioritizes quantitative compliance and often leaves behind the qualitative information that could be integrated into the system for better analysis (Namasivayam et al., 2013). However, there is no denial that the qualitative information is gathered and exists in the university's overall system, waiting to be utilized.

This study aims to investigate the situation across multiple higher education institutions in Malaysia concerning practices of CLO and PLO attainment analysis. With the results, the gap found in the investigation is addressed by proposing an institutional intelligence framework that could be standardized across the institutions, supporting both academic excellence and institutional quality assurance.

2 Related Works

This section is structured according to the two aspects of the study: higher education performance analytics and related theories of data governance. Despite the various studies conducted on higher education performance analytics, the scope is not limited to learning outcome attainment, since different institutions place different importance on what to measure as higher education performance. Research on data governance, on the other hand, is not specific to higher education performance analytics, especially since there are many aspects of "data" to govern in an institution. Data governance in practice is summarized at the end of this section.

2.1 Higher Education Performance Analytics

Research into Malaysian higher education evaluates various assessment models and instructional strategies to measure Program Outcome (PO) and Course Learning Outcome (CLO) attainments within the Outcome-Based Education (OBE) framework. Ibrahim et al. (2025) established that both cumulative and culminating models are reliable for systematic analysis, although culminating models often lack the continuous monitoring required for early student intervention. In technical domains, Zakwan et al. (2026) identified persistent disparities between lower-order knowledge application and higher-order problem-solving in Statics courses, while Salim et al. (2024) utilized mixed-method approaches to identify barriers, such as time constraints and syllabus complexity in Business Accounting. Specialized instructional methods, like the Awakening to Languages (ATL) approach for French (Mohd Ridzuan et al., 2024) and the use of modern computer modeling in engineering (Ovinis et al., 2018), have shown success in boosting student motivation and cognitive attainment, respectively. Despite these advancements, many studies remain limited by their reliance on single cohorts or specific programs, indicating a need for more diverse, multi-cohort longitudinal research. Nevertheless, an attempt has been made in a recent study by performing a comparative analysis on five different case institutions to propose a knowledge-based multi-agent reinforcement learning (MARL) (Ismail et al., 2025) as a solution for higher education performance analytics, but not specific to measuring the attainment of learning outcomes.

In response to the need to support the Continuous Quality Improvement (CQI) cycle, institutions have increasingly turned to automated measurement systems and advanced computational technologies to manage the learning outcome attainment data. Systems such as the Learning Outcome Attainment Measurement System (LOAMS) (Mokhtar et al., 2016) and the End-of-Semester Assessment Tool (ESAT) (Gamboa et al., 2013) have been implemented to reduce administrative burden for faculty while providing quantitative data for institutional reviews. However, bridging the gap in attainment evaluation requires shifting from purely quantitative data measurements toward a more holistic approach. As Namasivayam et al. (2013) highlighted, Malaysian universities must integrate qualitative assessments, including student portfolios and self-reflection. Moving forward, the literature emphasizes the need for data mining and sentiment analysis to better correlate learning behaviors with student retention and graduation rates. The knowledge-based MARL suggested by Ismail et al. (2025) could be incorporated as a solution, as it also highlighted the use of Natural Language Processing (NLP) to support the qualitative assessment of the data.

2.2 Data Governance and Related Theories in Practices

In terms of data governance, current research into institutional analytics and cybersecurity highlights the application of diverse theoretical frameworks to drive technological adoption and organizational growth. In higher education, several studies focus on developing maturity models for learning analytics (LA), such as the HELA-CMM (Šimić et al., 2025) and MMALA (Freitas et al., 2024), which utilize design science research and the Higher Education Reference Model (HERM) to guide systematic adoption. Additionally, institutions are pursuing analytics maturity through Davenport's stages and DELTA/DELTA Plus models (Webber & Zheng, 2019) to enhance data-informed decision-making, while others examine the responsible integration of Generative AI in assessments using the SAMR model (Slotnick et al., 2025) and the People-Process-Technology (PPT) framework (Duong et al., 2024). Popular theories like the Technology Acceptance Model (TAM) and Diffusion of Innovations (DOI) are used to explore predictive analytics (Nadpurajah, 2025), while Socio-Technical Systems theory is applied to conceptualize cybersecurity awareness within healthcare settings (Duong et al., 2024).

Digital performance appraisal systems are also being analyzed under Contingency Theory to improve faculty assessment and institutional performance (Hashim & Azman, 2025).

Existing research highlights several gaps that limit the effectiveness of the technological rollouts. As an example, while persisting in system integration, the current maturity models for learning analytics often overlook the complex organizational structures of higher education institutions. This is observed at smaller institutions that often struggle with insufficient resources for broad analytics initiatives. In Malaysia specifically, there is a clear shortage of empirical research on digital performance systems and even less longitudinal data to measure long-term success.

Over the past five years, higher education has shifted toward systematic, maturity-based analytics designed to drive institutional excellence and student success. Among the evidence found is the implementation of automated data acquisition via APIs (Application Programming Interfaces) that replace manual transfers and overcome the fragmentation of legacy systems, ensuring data veracity and interoperability across student, faculty, and financial records (Slotnick et al., 2025). Several institutions are found transforming retrospective reporting to predictive and prescriptive analytics, utilizing maturity models like HELA-CMM (Šimić et al., 2025) and MMALA (Freitas et al., 2024) as guides. These advanced models are critical for identifying at-risk students for early intervention and improving retention, while also supporting operational intelligence in certain areas.

Robust data governance and ethics are essential to this framework, requiring accountability systems, strict adherence to privacy regulations like FERPA and GDPR (Webber & Zheng, 2019), and "human-in-the-loop" oversight for Generative AI to mitigate algorithmic bias (Slotnick et al., 2025). Effective governance ensures that business intelligence initiatives are strategically aligned with institutional goals rather than remaining isolated technical projects (Romero & Garzon, 2025). To support this transition, knowledge management strategies leverage the People-Process-Technology (PPT) paradigm to foster a data-informed culture that balances statistical insights with professional experience (Duong et al., 2024).

3 Methodology

A comparative analysis is conducted across seven institutes of higher learning in Malaysia. These institutions are from three different categories: two public institutions (i.e., TI, TM); two government-linked institutions (i.e., LK, PT); and three private institutions (i.e., PA, MS, TL). The seven case institutions have the strengths in the number of students and/or being the leading universities in the country. The names of the institutions are kept confidential in this study to adhere to the research ethics.

The following standardized terminology is used in this study:

- Course Learning Outcome (CLO) is used instead of Course Outcome (CO)
- Program Learning Outcome (PLO) is used instead of Program Outcome (PO)
- CLO/PLO attainment is used instead of CLO/PLO achievement
- Course Leader is used instead of Course Lecturer or Lecturer
- Program Leader is used instead of Head of Program
- Assessment system is used instead of multiple names given by the universities

- CLO/PLO attainment system is used instead of multiple unique names given by the case universities

Table 1 summarizes the comparison made across the seven case institutions, described in terms of the assessment process (AP), CLO attainment analysis process (CAAP), PLO attainment analysis process (PAAP), and document source. Anonymous names are assigned to the case institutions for confidentiality, in which case identifications (i.e., Case IDs) are used instead of the full name.

Table 1 shows that all case institutions implement proper systems for the assessment process (AP), as the starting point of data acquisition for learning outcome attainment analysis. The assessment process is followed by the CLO attainment analysis process (CAAP), which is integrated with the PLO attainment analysis process (PAAP) in all case institutions, except TM. With proper mapping in the systems, the PLO attainment could be auto-calculated once the CLO attainment analysis is completed, as performed in five case institutions, i.e., TM and PT. PT has a different way of defining the PLO, unlike the rest of the case institution, hence the system merely needed to capture the data as retrieved from the assessment system.

Table 1 Comparative Analysis across Case Institutions (n = 7)

Case ID	Assessment Process (AP)	CLO Attainment Analysis Process (CAAP)	PLO Attainment Analysis Process (PAAP)	Document Source
LK	Assessor grades and enters the marks in the assessment system. Missing marks and anomalies in grades are documented in academic report, then verified by peer assessor.	The Course Leader sets the CLO-assessment mapping in the system before the start of each semester. CLO attainment system extracts the marks from the assessment system. The system calculates CLO attainment based on a minimum threshold of 50%. Significant findings are summarized in CLO attainment system, then verified by Program Leader.	The system calculates PLO attainment based on the accumulated CLO attainment across courses. The record is accessible by the Program Leader and academic administrators.	AP: Data captured in the system, assessment report CAAP: Data captured in the system, CLO attainment report PAAP: Data calculated in the system

(*continued*)

Table 1 (*continued*)

Case ID	Assessment Process (AP)	CLO Attainment Analysis Process (CAAP)	PLO Attainment Analysis Process (PAAP)	Document Source
PA	Assessor grades and enters the marks in the assessment system. Missing marks and anomalies in grades are documented in academic report, then verified by peer assessor.	The Course Leader sets the CLO-assessment mapping in the system at the end of each semester during the CLO attainment reporting process. CLO attainment system extracts the marks from the assessment system. The system calculates CLO attainment based on a minimum threshold of 50%. Significant findings are documented in CLO attainment report, then verified by Academic Leader.	The system calculates PLO attainment based on the accumulated CLO attainment across courses. The record is accessible by the Program Leader and academic administrators.	AP: Data captured in the system, assessment report CAAP: Data captured in the system, CLO attainment report PAAP: Data calculated in the system
MS	Assessor grades and enters the marks in the assessment system. Missing marks and anomalies in grades are documented in academic report, then verified by peer assessor.	The Course Leader sets the CLO-assessment mapping in the system before the start of each semester. CLO attainment system extracts the marks from the assessment system. The system calculates CLO attainment based on a minimum threshold of 40% or 50% depending on the program. Significant findings are documented in CLO attainment report, then verified by Program Leader.	The system calculates PLO attainment based on the accumulated CLO attainment across courses. The record is accessible by the Program Leader and academic administrators.	AP: Data captured in the system, assessment report CAAP: Data calculated in spreadsheet, CLO attainment report PAAP: Data calculated in spreadsheet

(*continued*)

Table 1 (*continued*)

Case ID	Assessment Process (AP)	CLO Attainment Analysis Process (CAAP)	PLO Attainment Analysis Process (PAAP)	Document Source
TI	Assessor grades and enters the marks in the assessment system. Missing marks and anomalies in grades are documented in academic report, then verified by peer assessor.	CLO attainment system extracts the marks from the assessment system. The system calculates CLO attainment based on a minimum threshold of 50%. Significant findings are summarized in CLO attainment system, then verified by Program Leader.	The system calculates PLO attainment based on the accumulated CLO attainment across courses. The record is accessible by the Program Leader and academic administrators.	AP: Data captured in the system, assessment report CAAP: Data captured in the system, CLO attainment report PAAP: Data calculated in the system
TM	Assessor grades and enters the marks in the assessment system. Missing marks and anomalies in grades are documented in academic report, then verified by peer assessor.	The Course Leader creates the CLO-assessment mapping in the system before the start of each semester. Assessment marks are exported from the system for the CLO attainment analysis spreadsheet template to produce the CLO attainment report and graphs. *Significance: Students provide feedback to support the CLO attainment analysis for each course*	The process is manually performed in the spreadsheet template.	AP: Data captured in the system, assessment report CAAP: Data calculated in spreadsheet, CLO attainment report PAAP: Data calculated in spreadsheet *Significance: Auto ➔ Manual ➔ Manual*

(*continued*)

Table 1 (*continued*)

Case ID	Assessment Process (AP)	CLO Attainment Analysis Process (CAAP)	PLO Attainment Analysis Process (PAAP)	Document Source
TL	Assessor grades and enters the marks in the assessment system. Missing marks and anomalies in grades are documented in academic report.	CLO attainment system extracts the marks from the assessment system. Significant findings are summarized in CLO attainment system, then verified by Program Leader.	The system calculates PLO attainment based on the accumulated CLO attainment across courses. The record is accessible by the Program Leader and academic administrators.	AP: Data captured in the system, assessment report CAAP: Data captured in the system, CLO attainment report PAAP: Data calculated in the system
PT	Assessor grades and enters the marks in the assessment system. Missing marks and anomalies in grades are documented in academic report, then verified by peer assessor.	CLO attainment system extracts the marks from the assessment system. The system calculates CLO attainment based on a minimum threshold of 40%.	Since all CLOs for a course are mapped to the same PLO, the calculation of PLO attainment is easily captured and calculated across courses. The record is accessible by the Program Leader and academic administrators.	AP: Data captured in the system, assessment report CAAP: Data captured in the system, CLO attainment report PAAP: Data calculated in the system *Significance: Students can view their accumulated individual PLO attainment*

4 Results and Findings

Based on the concept of Process-People-Technology (PPT), the results in Table 1 provided insights into the significant findings summarized in Table 2, as observed in the relevant case institutions (shown as Case ID in the table).

Table 2 Significant Findings from the Case Institutions ($n = 7$)

Significant Findings	Case ID
Criteria for measuring CLO and PLO attainments	
CLO attainment minimum threshold of 50%	LK, PA, TI

(*continued*)

Table 2 (*continued*)

Significant Findings	Case ID
CLO attainment minimum threshold of 40%, except for one professional-certified program that sets a minimum threshold of 50%	MS
CLO attainment minimum threshold of 40%	PT
No indication of the minimum threshold for the CLO attainment	TM, TL
Process of measuring CLO and PLO attainments	
The system calculates PLO attainment based on the accumulated CLO attainment across courses	LK, PA, MS, TI, TL
CLOs for a course are mapped to the same PLO, causing the PLO attainment analysis to require less effort in calculation	PT
Manual process of tabulating the CLO and PLO attainment analysis	TM
Supporting Qualitative Data in CLO and PLO attainments	
CLO and PLO attainment analysis is supported by qualitative data from the students' feedback	TM
The supporting qualitative data is from the Course Leaders' reviews in the attainment reports	LK, PA, MS, TI, TL, PT

Further elaboration on the findings presented in Table 2 is summarized under the components of process, people, and technology, as follows.

Process

It is observed that one university (TM) still practices the manual process of exporting marks from the assessment system into a spreadsheet template for the sole purpose of analysing CLO and PLO attainments.

In terms of CLO and PLO attainments calculation, one university (MS) has a different threshold than the standard 50%, particularly for professional courses, e.g., Accounting. This may cause complexity if the courses are shared across programs. In addition to this, one university (PT) sets the threshold to 40% for all programs instead of the standard 50% used by other universities. Since the threshold is the same across programs, unlike MS, which has a few programs with different thresholds, the CLO and PLO attainments calculation for PT is not as complex as MS.

Different mapping methods are also found across the case universities. As an example, PT sets all CLOs of a course to map to one PLO, making the PLO analysis simpler than at other universities. On the other hand, TL recently revamped its CLO-PLO mapping to the assessments, partly due to the realization that their previous mapping was different than the required standard. Nevertheless, this study assumes that the CLO-PLO mapping is correct, regardless of the methods adopted by the case universities.

People

Students' involvement is observed at two universities: TM students as active participants in providing feedback on their opinion of their own CLO attainment every semester; and PT students as passive participants in receiving the final analysis of their individual CLO and PLO attainments accumulatively every semester. The difference between these two cases is that the former (TM) involves students during the CLO/PLO reporting, whereas the latter (PT) involves students as the final receivers of the CLO/PLO reporting results. Although the former (TM) involves them during the CLO/PLO reporting, there is no evidence that the students receive the final CLO/PLO reporting results; their opinion may be deemed inconsequential.

In terms of internal staff, different names are assigned to the person-in-charge of the program at different case universities. As an example, Program Leader is commonly used compared to Head of Program, hence this study standardizes the name to Program Leader. Despite this, PA has two levels of people-in-charge, namely Program Leader and Academic Leader, but the Academic Leader is the one who verifies and approves the CLO/PLO attainment reports. This is observed to be true when the course is served by an academic team that the Program Leader does not belong to, hence the Academic Leader is in charge as the supervisor of the Course Leader from the team.

Technology

All universities are at a mature stage of assessment system implementation, which is the starting point of data capture for CLO and PLO attainments. Nevertheless, not all universities implement a proper or complete system for the CLO/PLO attainment analysis. The most "complete" system is implemented by PT, in which the final analysis is shared with the individual students, closing the loop of reporting to the end-beneficiaries, i.e., students. Nevertheless, there is no evidence that the highest level of attainment, i.e., Program Educational Objective (PEO), is shared with other stakeholders, and the discussion on the PEO attainment is often kept within members of the strategic level, e.g., academic board and senate.

Although the expansion of the CLO/PLO attainment system makes it possible to measure the PEO attainment, no evidence could be retrieved during this study on the existence of automated PEO attainment. This shows the importance of data security and confidentiality as the analysis goes to a higher level of the university hierarchy system. However, the complexity of the calculation of PEO attainment could be a huge challenge in terms of algorithm and coding validation, due to the size of the big data, as the number of courses, assessments, CLOs, and PLOs grows over time. It is not surprising if this challenge affects the capacity of the university's data storage and the capability of the university's technical and system development teams.

This study assumes that the CLO/PLO attainment system is separately developed and maintained to avoid possible data and algorithm clashes with the assessment system. However, there is a possibility that the case universities actually implement one huge information system that covers both systems. Again, this depends on the capacity and capability of the university's infrastructure.

5 The Common as-Is Process of CLO/PLO Attainment System

The summary above provides an insight into the As-Is process as practiced at the common higher education institutions in Malaysia. Figure 1 shows the overall view of the As-Is process, indicated with the concepts of People-Process-Technology (PPT) and Extract-Transform-Load (ETL).

As shown in Fig. 1, the current CLO/PLO attainment framework in Malaysian universities is initiated at the "People" level, involving two primary groups of stakeholders. Course Leaders, typically the academic staff or lecturers, act as the primary data providers by inputting raw student marks into the Assessment System. Program Leaders and Academic Administrators operate at the management level, utilizing the CLO/PLO attainment system to monitor overall student performance. This dual-track interaction highlights a common challenge: the reliance on manual data entry and a clear division between those who generate the data and those who analyze it for institutional quality assurance.

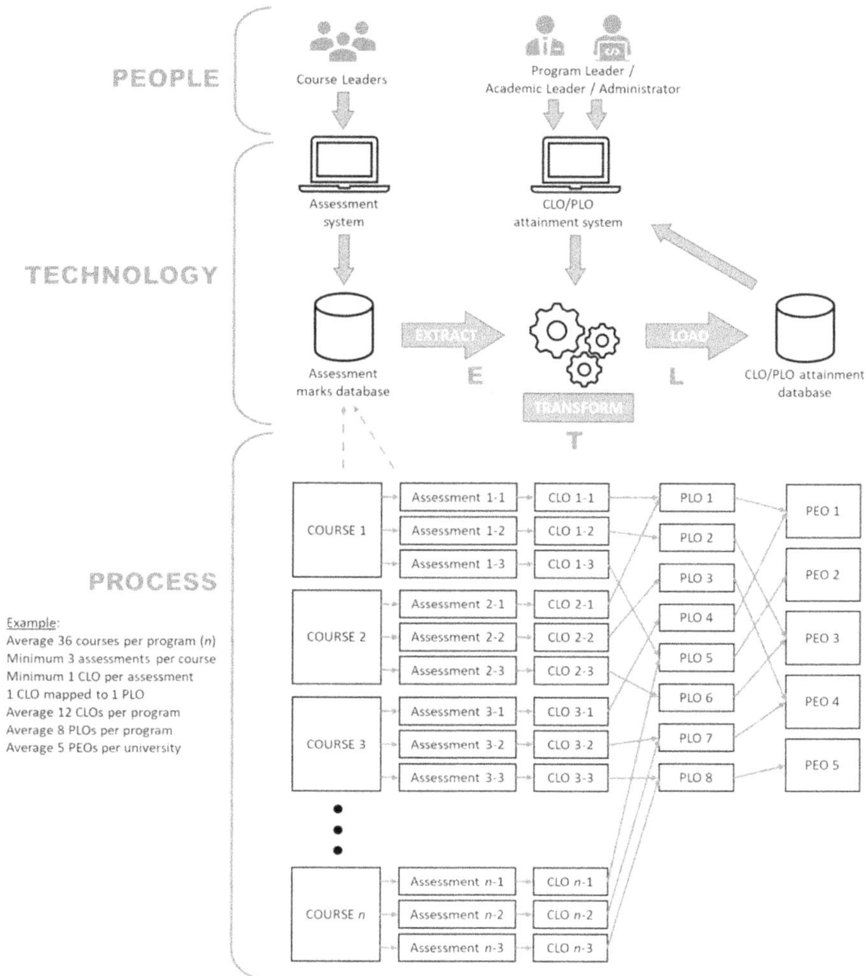

Fig. 1 As-Is process of CLO/PLO attainment system

From the technical standpoint, i.e., "Technology", the system is built on a standard data warehousing concept of Extract-Transform-Load (ETL). It starts with the extraction of raw assessment data from the grade database. This data then undergoes a transformation phase of complex mathematical calculations applied to align raw scores with predefined learning outcomes. The results of this process are loaded into a specialized CLO/PLO attainment database. This feedback loop allows for the generation of reports, but it often operates as a "batch" process at the end of a semester, which can limit the ability to provide just-in-time academic interventions.

The "Process" segment illustrates the hierarchy of Outcome-Based Education (OBE) as practiced in common Malaysian institutions. This bottom-up flow starts at the course level, where individual assessments (e.g., quizzes, projects, exams) are mapped to specific Course Learning Outcomes (CLOs). These CLOs are subsequently aggregated and mapped to broader Program Learning Outcomes (PLOs). The final stage involves aligning PLOs with Program Educational Objectives (PEOs), which reflect the graduates' long-term career achievements. This multi-tiered mapping is essential for regulatory compliance, yet it introduces a high level of complexity in data management and traceability.

An analysis of the current process reveals significant operational bottlenecks and data management challenges. As Fig. 1 suggests, an average program involves 36 courses and multiple assessments per student, leading to a massive volume of data points that must be precisely mapped across three different levels of outcomes. The separation of the assessment marks database from the attainment system creates a "silo" effect, which often necessitates manual synchronization. Consequently, the "As-Is" state is frequently reactive; by the time the ETL cycle is complete, and the PLO attainment results are visible to administrators, the academic cycle ends, leaving little room for proactive Continuous Quality Improvement (CQI).

In addition to these findings, it is common across the case institutions to gather students' feedback at the end of the semester of their studies, as part of the requirements by the Ministry of Higher Education (MoHE). The students' evaluation form is often attached to support the reporting of the learning outcome attainment analysis, but the structure of the evaluation may not be directly aligned with the need for the attainment analysis. In other words, a separate "system" is implemented, which may or may not be governed by the same department/unit that governs the CLO/PLO attainment analysis, i.e., academic quality assurance. Nevertheless, the existence of this evaluation feedback could represent the qualitative data to support the learning outcome attainment analysis, provided a suitable mechanism could be in place.

6 Discussion on the To-Be Solution: A Proactive Institutional Intelligence Framework

The question that drives this section is "Who should benefit from the final PLO attainment analysis?" Answering this, the final beneficiaries after the students graduate should be the industry or the students' future employers. If personalized PLO attainment for individual students is not yet shared in the current practice, how could it be made possible? The fundamental goes back to data governance. With sound data governance in place, further sharing with the right people within the education ecosystem, i.e., industry and future employers, can be securely done.

Figure 2 illustrates the expansion (To-Be solution) of the learning outcome attainment analysis and reporting at higher education institutions. This solution comprises the elements of People, Process, and Technology, aligned to the knowledge management processes of knowledge exchange, knowledge acquisition and storage, and knowledge retrieval. The implementation features a robust ETL framework, ensuring scalability as data volume, complexity, and velocity grow.

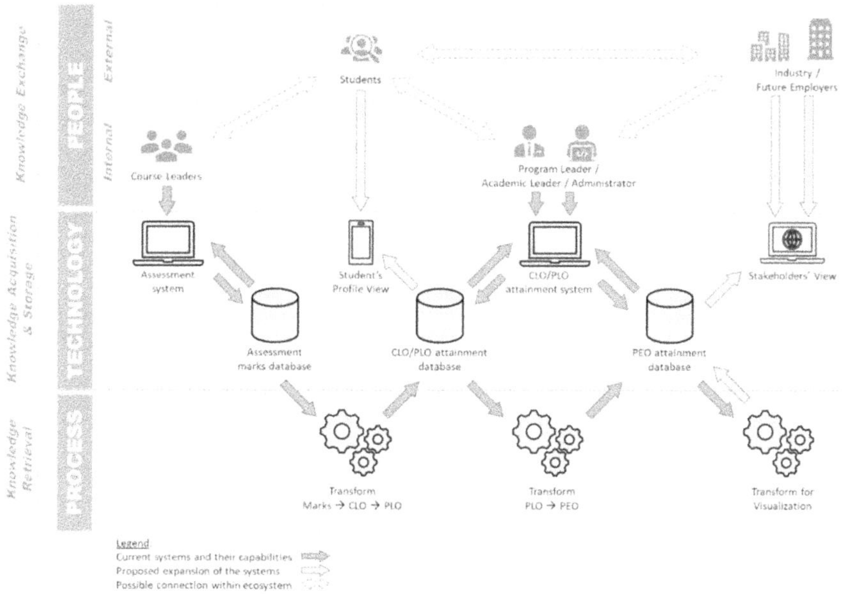

Fig. 2 To-be solution of the proactive institutional intelligence framework

Coined as "Proactive Institutional Intelligence", this solution directly addresses the shift away from end-of-semester manual checks and, in some cases, beginning-of-semester manual triggers, which are currently reactive in nature. Since being reactive enforces the internal staff to initiate the data acquisition, knowledge retrieval, and other required processes, the efficiency could be compromised. Since the information systems are already made available in most cases of higher education institutions, smart integration could be deployed to make the solution proactive. The databases are no longer isolated but part of a continuous flow, as the integration takes place.

The institutional intelligence framework is not in a silo, but needs to be implementable within a "Knowledge Management Ecosystem". The ecosystem connects the three vertical pillars, i.e., Exchange, Acquisition, and Retrieval, shown on the left side of the diagram in Fig. 2.

Not shown in Fig. 2 is the possible expansion to include NLP to analyse the qualitative feedback from the students in the assumed separate system made available by the universities. NLP could facilitate the mapping of comments from the feedback form to the learning outcome attainment results. This integration across platforms could facilitate quality insights to the stakeholders as the analysis is visualized in a thoughtful and integrated manner.

7 Conclusion

The systematic measurement of Course Learning Outcomes (CLOs) and Program Learning Outcomes (PLOs) is essential for the success of Outcome-Based Education (OBE) and the continuous quality improvement (CQI) cycle in Malaysian higher education. While the comparative analysis of seven case institutions reveals that many have successfully automated their assessment processes, significant challenges remain, including data silos, varying attainment thresholds, and a predominantly reactive "batch" processing approach at the end of semesters. Closing these gaps means leaving behind the era of manual data management. Ultimately, the objective is to adopt a streamlined ETL process that does not treat quantitative results and qualitative student perspectives as separate silos, integrating them into a single, functional ecosystem.

The proposed Proactive Institutional Intelligence Framework tackles these hurdles by moving toward a Knowledge Management Ecosystem, where the focus shifts from static reports to a continuous stream of data. At its core, this model relies on rigorous data governance. This ensures that a student's personal achievement data remains secure yet portable, allowing it to be shared with potential employers to boost graduate outcomes. Furthermore, the future of this framework lies in Agentic AI (artificial intelligence). By leveraging multi-agent reinforcement learning (MARL) and NLP, institutions can begin to automate the more nuanced qualitative analysis, providing the real-time, predictive data needed for both student success and high-level quality assurance.

References

Duong, D., Sarbazhosseini, H., Alsheikh, M.A., Ma, W.: Conceptualising people, process, technology, data, governance and continuous improvement (2PTDGC) as a framework to explore the cybersecurity awareness and process of Australian medical practices. In: Australasian Conference on Information Systems, Canberra (ACIS), Canberra, 4–6 December 2024 (2024) https://aisel.aisnet.org/acis2024/127

Freitas, E. et al.: MMALA: developing and evaluating a maturity model for adopting learning analytics. J. Learn. Anal. **11**(1), 67–86 (2024). https://doi.org/10.18608/jla.2024.8099

Gamboa, R., Namasivayam, S., Al-Atabi, M., Singh, R.: Quantitative measurement of students PO attainments for Taylor's university engineering programmes. Procedia. Soc. Behav. Sci. **103**, 753–762 (2013)

Hashim, N.H., Azman, N.: Reconfiguring faculty performance appraisal for sustainable institutional growth and academic excellence. J. Sustain. Sci. Manage. **20**(12), 2793–2817 (2025)

Ibrahim, A., Mohamad Nor, M.R., Ahmad, H., Mohamed Latiff, M.F.P.: Reliability of program outcome attainment evaluation based on cumulative model and culminating model analysis. Jurnal Kejuruteraan. **27**(1), 527–540 (2025). https://doi.org/10.17576/jkukm-2025-37(1)-39

Ismail, S., Ismail, M.H., Jamaludin, N., Hitam, N.A.: Knowledge-based multi-agent reinforcement learning for ambiguity mitigation in course moderation process. In: Uden, L., Ting, I.H. (eds.) Knowledge Management in Organisations. KMO 2025. Communications in Computer and Information Science 2562, pp. 341–352. Springer, Cham (2025). https://doi.org/10.1007/978-3-031-95898-4_24

Mohd Ridzuan, M.U., Mat Isa, C.M., Zakaria, N.N.N., Jaafar, F.: Attainment of course learning outcome using awakening to languages approach among hotel management students in Malaysia. Environ. Behav. Proceed. J. (E-BPJ). **9**(SI21), 3–10 (2024)

Mokhtar, S.A., Zulfadli, A.S.M.S.: Functioning modules for learning outcomes attainment measurement system in Universiti Kuala Lumpur. In: 2016 International Conference on Information and Communication Technology (ICICTM), Kuala Lumpur, 16–17 May 2016, pp. 77–83 (2016). https://doi.org/10.1109/ICICTM.2016.7890781

Nadpurajah, L.: Strategies to Improve Higher Education Institutional Performance through Predictive Analytics Implementation, Walden Dissertations and Doctoral Studies Collection. Walden University (2025)

Namasivayam, S., Al-Atabi, M., Chong, C.H., Choong, F., Hosseini, M., Gamboa, R.A.: A blueprint for measuring programme outcomes. J. Eng. Sci. Technol. **8**, 46–55 (2013) Special Issue on Engineering Education

Ovinis, M., Karuppanan, S., Anwar Sulaiman, S., Sri Melor, P., Zulhilmi Paiz, M., Urquia, A.: A comparative analysis of attainment of program outcomes for courses with and without the use of modern tools. MATEC Web Conf. **225**, 06022 (2018). https://doi.org/10.1051/matecconf/201822506022

Romero, K.M., Garzon, C.A.: Critical success factors for business intelligence implementation in public universities. Suma de Negocios. **16**(34), 68–79 (2025). https://doi.org/10.14349/sumneg/2025.V16.N34.A7

Salim, M., Zainal Abidin, A.H., Ramlee, H.H., Azlan, Z.N., Zainal, M.H.: Analysis of course learning outcomes attainment for business accounting in diploma of Islamic finance: a case study of college professional MARA bandar Melaka. Eval. Stud. Soc. Sci. (ESS). **5**(2), 68–79 (2024). https://doi.org/10.37134/esss.vol5.2.5.2024

Šimić, D., Begičević Ređep, N., Rako, S., Kadoić, N., Van Petegem, W., Rienties, B., et al.: HELA-CMM: capability maturity model for adoption of learning analytics in higher education. Int. J. Educ. Technol. High. Educ. **22**(1) (2025)

Slotnick, R.C., et al.: GenAI in Assessment: A Practical Guidebook for Massachusetts Public Higher Education. Massachusetts Department of Higher Education, Boston (2025)

Webber, K.L., Zheng, H.: Data-informed decision making and the pursuit of analytics maturity in higher education. IHE research projects series 2019-010 (2019). https://ihe.uga.edu/rps/2019010

Zakwan, F.A.A., Ibrahim, N., Ismail, S., et al.: Advancing accreditation and sustainable education goals through outcome attainment analysis: a case study of statics education in Malaysia. Discov Educ. **5**(1) (2026). https://doi.org/10.1007/s44217-026-01302-6

AI-Assisted Decision Support for Academic Risk Detection in Mega-Universities: A Conceptual Framework for Universitas Terbuka

Muhammad Rif'an[1(✉)], Vivi Indra Amelia Nasution[2], Iwan Susanto[1], Mira Aryani[1], Dwi Astuti Aprijani[1], Tora Akadira[2], and Suci Dwijayanti[3]

[1] Information Systems, Universitas Terbuka, Tangerang Selatan, Indonesia
m.rifan@ecampus.ut.ac.id
[2] Public Administration, Universitas Terbuka, Tangerang Selatan, Indonesia
[3] Electrical Engineering, Universitas Sriwijaya, Palembang, Indonesia

Abstract. Mega-universities play a strategic role in widening access to higher education, but their large scale, open admission policies, geographically dispersed students, and reliance on distance education make early detection of academic risk challenging. Using Universitas Terbuka (UT) as the guiding case, this paper proposes a conceptual framework for AI-assisted decision support in academic risk detection. The framework was developed through literature-informed synthesis, contextual problem mapping, risk construct definition, architecture derivation, and validation-oriented refinement. It focuses on three interrelated risks: non-registration risk, course failure risk, and delayed graduation risk. The framework emphasizes that academic risk in open and distance learning should be interpreted contextually: non-registration may reflect financial or administrative constraints, while delayed graduation may reflect intentional slow pacing among working adult learners. The proposed architecture consists of four layers: data sources, data platform, AI and analytics, and decision support applications. These layers are supported by operational triage, human-in-the-loop decision support, localized regional contextualization, cohort-drift monitoring, feedback loops, and responsible AI governance. The paper also outlines minimal data requirements and a planned evaluation strategy covering predictive performance, fairness, usability, workflow fit, intervention outcomes, and governance compliance. Rather than reporting a completed deployment, the paper contributes a context-sensitive roadmap for proactive, explainable, and responsible academic risk support in mega-university distance education.

Keywords: Artificial intelligence · Learning analytics · Early warning system · Academic risk detection · Decision support system · Open and distance learning

1 Introduction

Mega-universities and open and distance learning (ODL) institutions play an essential role in widening access to higher education, especially in countries with large, geographically dispersed, and socioeconomically diverse populations. Daniel defined

N. Mohd Sharef et al. (Eds.): CAMP 2026, CCIS 3094, pp. 154–169, 2027.
https://doi.org/10.1007/978-981-92-4305-1_12

mega-universities as institutions that deliver mass higher education through distance education at relatively low unit cost [1]. However, their scale and openness also create persistent challenges in student retention, academic progression, and completion [2–6].

Universitas Terbuka (UT), Indonesia's national open and distance learning university, reflects this model. Its open admission policy, large student population, distributed regional service structure, and reliance on mediated learning interactions make academic risk difficult to monitor using conventional approaches. In campus-based universities, risk may be detected through classroom attendance, direct advisor interaction, or visible changes in student behavior. In a mega-university ODL context, risk signals are more fragmented across registration systems, payment records, academic data, learning management systems, assessment activity, student support channels, and regional field knowledge.

Academic risk in this context is broader than low grades or weak LMS engagement. For UT and similar open universities in emerging economies, non-registration in the following semester is a major risk. Yet non-registration may be caused not only by academic disengagement, but also by immediate financial constraints, delayed tuition payment, limited access to payment channels, administrative barriers, or temporary disruption in work and household income. Similarly, delayed graduation should not automatically be treated as academic failure, because many ODL students are working adults who intentionally take fewer courses each semester. Therefore, risk detection must distinguish true academic risk from financial delay, intentional slow pacing, and temporary local disruption.

Learning analytics and early warning systems have shown that institutional records, LMS traces, and assessment data can be used to predict course failure, dropout, and disengagement [10]–[14]. However, many existing systems were developed for campus-based or moderate-scale online learning contexts and often emphasize predictive accuracy more than operational decision-making. For a mega-university, the main challenge is not only producing risk scores, but also determining which risks should trigger automated reminders, which require advisor follow-up, and how local knowledge from regional offices should contextualize central analytics.

Recent developments in artificial intelligence, explainable AI, fairness-aware modelling, and human-in-the-loop decision support offer new opportunities for academic risk detection [15–19]. AI can help identify risk patterns across large and heterogeneous datasets, while explainability can help advisors and institutional leaders understand why a student is flagged. However, AI-assisted risk detection must be governed responsibly because it involves personal and academic data, potential bias, and decisions that may affect access to student support. In the Indonesian context, this also requires alignment with Law No. 27 of 2022 on Personal Data Protection.

Despite the growing literature, several gaps remain. First, limited guidance exists on how to design AI-assisted academic risk detection for mega-university ODL contexts. Second, non-registration is often treated mainly as academic disengagement, although in emerging-economy open universities it may be strongly shaped by financial and administrative constraints. Third, few frameworks define an operational triage strategy that separates low-level risks suitable for automated institutional safety nets from critical risks requiring human intervention. Fourth, open-admission systems need mechanisms to detect cohort shift when new students differ substantially from historical training data. Fifth, centralized AI analytics must be connected with localized regional knowledge, such as telecommunications outages, local disasters, or payment-channel disruptions.

This paper addresses these gaps by proposing a conceptual framework for AI-assisted decision support for academic risk detection in mega-universities, using UT as the guiding case. The framework is developed through literature-informed synthesis, contextual problem mapping, risk construct definition, architecture derivation, and validation-oriented refinement. It focuses on three risk categories: non-registration risk, course failure risk, and delayed graduation risk. The proposed architecture consists of four layers: data sources, data platform, AI and analytics, and decision support applications, supported by human-in-the-loop practices, operational triage, cohort-drift monitoring, localized regional contextualization, feedback loops, and responsible AI governance.

The contribution of this paper is threefold. First, it reframes academic risk detection in mega-university ODL settings by integrating academic, financial, behavioral, administrative, and regional factors. Second, it proposes a layered AI-assisted decision support framework that connects data integration, predictive analytics, explainability, triage, and human-centered intervention workflows. Third, it outlines practical requirements for implementation and evaluation at UT, including payment-aware non-registration detection, regional override mechanisms, study-pace-aware delayed graduation risk, and evaluation across technical, human, organizational, and governance dimensions. Rather than reporting a completed deployment, the paper offers a context-sensitive roadmap for proactive, explainable, and responsible academic risk support at scale.

2 Related Work

2.1 Mega-Universities, ODL, and Academic Risk

Mega-universities provide large-scale access to higher education through distance education, flexible learning systems, and technology-mediated academic services [1, 2]. Their inclusive mission is especially important in countries where conventional campus-based higher education cannot fully address geographic, economic, and demographic barriers. However, large-scale ODL institutions also face persistent risks related to retention, course completion, registration continuity, and long-term progression [3–6].

In ODL settings, academic risk is multidimensional. It includes course failure, disengagement from learning activities, interruption of semester registration, and delayed completion. These risks are often harder to detect than in campus-based settings because student behavior is distributed across institutional systems rather than directly observed in classrooms. Prior studies have emphasized that persistence in distance education is influenced by self-regulation, academic readiness, employment status, service quality, and institutional support [4–7]. For open universities in emerging economies, however, academic risk must also be interpreted in relation to financial constraints, payment timing, administrative barriers, digital access, and regional conditions.

2.2 Learning Analytics and Early Warning Systems

Learning analytics and early warning systems have shown that institutional records, LMS activity, and assessment data can help identify students at risk of failure or dropout

[10–14]. LMS indicators such as page views, quiz participation, assignment submission, and forum activity can provide early signals of engagement and performance [10]. Early alert systems have also demonstrated the value of combining institutional and learning activity data to support timely intervention [11–14].

Despite these contributions, many early warning systems focus primarily on predictive accuracy, feature importance, or dashboard design. Less attention is given to the institutional process through which risk predictions are prioritized and acted upon. For mega-universities, this operational issue is critical because advisors and support staff may be responsible for very large numbers of students. Risk detection must therefore be linked to triage, intervention capacity, and clear responsibility for follow-up.

A second limitation is that many models rely heavily on academic and LMS engagement indicators. These indicators are useful for course failure detection, but they may be insufficient for non-registration risk in open universities. A student may be academically capable but fail to register because of tuition payment constraints, administrative delay, or local service disruption. Conversely, a working adult may show slow credit accumulation because they intentionally follow a part-time study pattern. These cases require a framework that distinguishes academic difficulty from financial delay, intentional slow pacing, and contextual disruption.

2.3 AI, Explainability, Fairness, and Human-in-the-Loop Support

Artificial intelligence and machine learning provide opportunities to detect complex risk patterns across large and heterogeneous educational datasets [15, 16]. Models such as regularized regression, random forests, and gradient boosting can process academic histories, registration behavior, LMS engagement, and support-service interactions. However, AI-based risk detection also raises concerns about transparency, fairness, and accountability.

Explainable AI is important because educators and advisors need to understand why a student is flagged and what action may be appropriate [16]. In an academic risk system, explanations should identify practical contributing factors, such as unpaid tuition near a deadline, repeated course failure, missed early assessment, or a sudden decline in LMS activity. Fairness monitoring is also necessary because models trained on historical data may produce unequal errors across regions, programmes, age groups, employment categories, or students with different levels of digital access [17].

Human-in-the-loop decision support is essential in education because student support requires professional judgement and contextual understanding [18, 19]. AI should guide attention rather than replace human decisions. In UT's context, regional staff should be able to contextualize or override central risk scores when local knowledge indicates that low engagement or delayed registration is caused by disaster, telecommunications outage, payment-channel disruption, or administrative barriers.

2.4 Responsible AI Governance and Research Gap

Responsible AI governance provides the institutional basis for trustworthy academic risk detection. It requires clear rules on data access, purpose limitation, data minimization, accountability, fairness monitoring, audit logging, and appropriate use of risk classifications [16–19]. For Universitas Terbuka, governance should also be aligned with Indonesia's Law No. 27 of 2022 on Personal Data Protection. Risk scores should be used to support students, not to stigmatize, penalize, or restrict them.

The literature provides a strong foundation for academic risk detection, but several gaps remain. First, limited work has translated learning analytics and AI-assisted risk detection into the operational context of mega-university ODL systems. Second, non-registration risk is often treated as academic disengagement, although it may be strongly shaped by financial and administrative constraints. Third, few frameworks define triage mechanisms that separate automated safety-net interventions from human escalation. Fourth, cohort shift remains underdeveloped in open-admission systems where new student intakes may differ from historical training data. Fifth, centralized analytics are rarely connected to localized regional decision support and override mechanisms.

This paper addresses these gaps by proposing a context-sensitive conceptual framework that integrates academic, behavioral, financial, administrative, and regional-contextual data; distinguishes non-registration, course failure, and delayed graduation risk; and embeds triage, explainability, fairness monitoring, cohort-drift detection, regional contextualization, and responsible AI governance into a unified decision support architecture.

3 Framework Development Methodology

This study develops a conceptual framework rather than reporting a completed system implementation. The framework was constructed through a literature-informed and context-sensitive design process. This approach is appropriate because the objective is to synthesize concepts from mega-university studies, open and distance learning, learning analytics, early warning systems, explainable AI, human-in-the-loop decision support, and responsible AI governance into an operational model for academic risk detection at Universitas Terbuka.

The framework development process consisted of five stages.

First, relevant literature was synthesized to identify key concepts, prior findings, and limitations in academic risk detection. The synthesis focused on ODL persistence, learning analytics, early warning systems, AI in higher education, explainability, fairness, and responsible AI governance [1–19]. This stage showed that predictive models can identify student risk, but institutional impact depends on whether predictions are interpretable, actionable, and aligned with support workflows.

Second, the academic risk problem was mapped to the context of UT. This mapping considered UT's open admission system, large and geographically dispersed student body, heterogeneous learner profiles, distributed regional offices, and fragmented institutional data. It also highlighted that risk is not always academic in origin. Non-registration may be shaped by financial constraints, delayed tuition payment, administrative barriers, or local access problems. Delayed graduation may reflect intentional slow pacing, especially among working adult learners.

Third, the framework defined three academic risk constructs: non-registration risk, course failure risk, and delayed graduation risk. Non-registration risk refers to the probability that a student will not register in the following semester and should be interpreted using academic, financial, administrative, and regional indicators. Course failure risk refers to the probability that a student will fail a registered course and can be estimated from academic history, course characteristics, assessment activity, LMS engagement, and tutorial participation. Delayed graduation risk refers to the probability that a student will not complete a programme within an expected time frame, but it should be adjusted according to declared study intent, expected study pace, employment condition, and semesterly course load.

Fourth, the risk constructs and institutional requirements were translated into a layered architecture consisting of data sources, data platform, AI and analytics, and decision support applications. The data sources layer captures academic, registration, payment, LMS, assessment, support-service, curriculum, and regional-context data. The data platform layer supports integration, identity resolution, feature engineering, access control, pseudonymization, and audit logging. The AI and analytics layer estimates risk, generates explanations, monitors fairness, and detects cohort drift. The decision support layer presents actionable outputs for programme leaders, advisors, tutors, regional offices, support units, and institutional leadership.

Fifth, the framework was refined against four criteria: conceptual coherence, contextual fit, operational actionability, and responsible AI requirements. This refinement led to the inclusion of four additional elements beyond a standard predictive analytics architecture: an operational triage mechanism, cohort-drift monitoring, localized regional decision support with override capability, and student study-intent adjustment for delayed graduation risk.

The resulting framework should therefore be understood as a design-oriented conceptual framework. It does not claim to validate a specific predictive model or report measured intervention outcomes. Instead, it provides a structured roadmap for designing, piloting, and evaluating AI-assisted academic risk detection in a mega-university context.

4 Problem Definition and Design Goals

In a mega-university ODL context, academic risk detection is not only a technical prediction problem, but also an institutional decision support problem. The system must identify students who need timely support while avoiding overgeneralization, false alarms, and inappropriate automated decisions. For Universitas Terbuka, academic risk signals are distributed across academic records, registration and payment systems, LMS activity, assessment data, student support channels, and regional field knowledge.

This paper focuses on three interrelated forms of academic risk.

The first is non-registration risk, defined as the probability that a student will not register in the following semester. In open universities serving adult learners in emerging economies, non-registration should not be treated only as academic disengagement. It may also be caused by inability to pay tuition on time, delayed payment, administrative barriers, payment-channel problems, or temporary work and household income disruption. Therefore, non-registration risk should be modelled using academic, LMS, registration, payment, and regional-contextual indicators.

The second is course failure risk, defined as the probability that a student will fail a registered course in the current semester. This risk is more directly related to academic history, course characteristics, assessment participation, LMS engagement, tutorial activity, and submission timeliness. Course failure risk matters because repeated failure can reduce motivation, delay credit accumulation, increase financial burden, and contribute to later non-registration.

The third is delayed graduation risk, defined as the probability that a student will not complete a programme within an expected time frame. However, in ODL institutions this risk must be interpreted carefully. Many students are working adults who intentionally take fewer courses each semester. The framework therefore proposes that delayed graduation risk should be adjusted using self-declared study intent, expected academic pace, employment condition, historical course load, and programme requirements. This prevents the system from flagging intentional slow pacing as academic failure.

The central problem can therefore be formulated as follows: how can a mega-university design an AI-assisted decision support system that estimates academic risk early enough for meaningful intervention, distinguishes academic risk from financial delay and intentional slow pacing, presents risk information in an understandable and actionable form, allows regional contextualization, and operates under responsible AI governance?

To address this problem, the framework is guided by six design goals.

First, it requires scalable and integrated data infrastructure that connects student records, registration, payment, LMS, assessment, support-service, curriculum, and regional data using common identifiers and standardized definitions. Second, it requires context-sensitive risk modelling so that different risks are interpreted according to their causes, time horizons, and appropriate interventions.

Third, the framework requires operational triage. Low-level risks should trigger automated safety nets such as SMS reminders, payment notifications, registration alerts, and learning activity nudges. Medium-level risks may require targeted monitoring, while high and critical risks should be escalated to advisors, tutors, support units, or regional offices. This distinction is necessary because mega-universities cannot manually review all risk signals.

Fourth, the framework requires explainable and human-in-the-loop decision support. AI-generated risk scores should guide attention rather than determine final decisions. Stakeholders should be able to see the key contributing factors behind each risk classification, such as unpaid tuition near a deadline, repeated course failure, missed assessment, or sharp decline in LMS activity. Advisors and regional staff should also be able to validate, contextualize, correct, or override risk classifications when justified by field knowledge.

Fifth, the framework requires cohort-drift monitoring. In open-admission systems, new student cohorts may differ from historical training data in age, employment profile, programme choice, regional composition, payment behavior, digital access, or academic preparation. The system should therefore monitor feature distributions, flag model uncertainty, recalibrate thresholds, or require more conservative human-supervised use when substantial divergence is detected.

Sixth, the framework requires responsible AI governance and data protection. Academic risk detection involves personal, academic, administrative, and behavioral data. The system must therefore define clear rules for purpose limitation, data minimization, role-based access, audit logging, fairness monitoring, accountability, and appropriate use of model outputs. In the Indonesian context, this governance should be aligned with Law No. 27 of 2022 on Personal Data Protection. Risk information should be used to support students, not to stigmatize, penalize, or exclude them.

Together, these design goals ensure that academic risk detection is not reduced to a risk-scoring exercise. Instead, it becomes an institutional decision support process that connects data, AI, human judgement, regional context, governance, and student-centered intervention.

5 Proposed Conceptual Framework

5.1 Overall Architecture

The proposed framework is an AI-assisted decision support architecture for academic risk detection in mega-university ODL contexts. It consists of four core layers: data sources, data platform, AI and analytics, and decision support applications. These layers are supported by cross-cutting components: human-in-the-loop practices, operational triage, localized regional contextualization, cohort-drift monitoring, feedback loops, and responsible AI governance (Fig. 1).

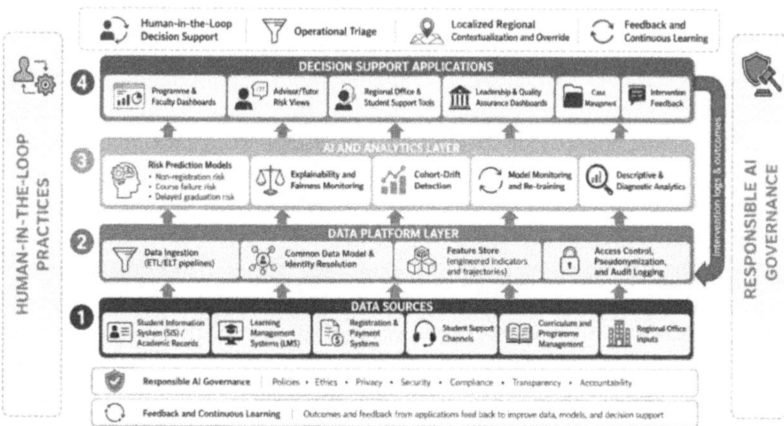

Fig. 1 Conceptual framework for AI-assisted academic risk detection and decision support in Universitas Terbuka.

The framework follows the operational flow of academic risk detection at scale. Student-related data are collected from institutional systems and regional inputs, integrated into an analytics platform, transformed into risk-relevant features, processed by AI and analytics models, and delivered through role-specific decision support applications. Human users then interpret the outputs, select interventions, record outcomes, and provide feedback for continuous improvement. The architecture is not intended to automate academic decisions, but to prioritize attention and support timely, explainable, and context-sensitive student support.

5.2 Data Sources Layer

The data sources layer includes academic information systems, registration systems, payment systems, LMS platforms, assessment systems, tutorial records, student support channels, curriculum databases, and regional office inputs. Academic records support the detection of course failure and delayed graduation. Registration and payment data are essential for non-registration risk because students may fail to register not only due to disengagement, but also because of delayed payment, payment-channel barriers, or temporary financial constraints. LMS and assessment data provide engagement and performance signals, while support-service and regional data add contextual information that may not appear in central systems.

The data platform layer integrates these sources into a structured analytics environment. Its main functions include data ingestion, identity resolution, common data modelling, feature engineering, access control, pseudonymization, and audit logging. A canonical student identifier is needed to link academic, registration, payment, LMS, assessment, support-service, and regional-context data across semesters. Feature engineering transforms raw records into indicators such as payment timing, registration continuity, repeated course attempts, LMS activity, submission timeliness, credit accumulation, and declared study intent.

5.3 AI and Analytics Layer

The AI and analytics layer estimates non-registration risk, course failure risk, and delayed graduation risk. The framework does not prescribe a single algorithm. Instead, it supports a model strategy that starts with interpretable baselines, such as logistic regression or regularized linear models, and may later include tree-based ensemble methods if they provide meaningful improvement. This approach balances predictive performance with interpretability, governance, and institutional trust.

The layer also provides explanations, monitors fairness, and detects cohort drift. Explanations should identify practical contributing factors, such as unpaid tuition near a deadline, repeated failure in a required course, missed early assessment, low LMS engagement, or credit accumulation below the student's declared study plan. Fairness monitoring should examine whether error rates differ across programmes, regions, age groups, employment categories, socioeconomic proxies, or digital access conditions. Cohort-drift monitoring is needed because open-admission cohorts may differ from historical training data in age, employment profile, programme choice, regional composition, payment behavior, or digital access. When substantial drift is detected, the system should flag uncertainty, recalibrate thresholds, retrain models, or increase human review.

5.4 Decision Support Applications

The decision support layer converts analytics outputs into actionable information for different institutional roles. Programme and faculty dashboards should show aggregate risk patterns across programmes, courses, cohorts, and regions. Advisor and tutor views should provide prioritized student lists, key contributing factors, intervention history, and recommended follow-up. Regional office and student support tools should combine central risk scores with local knowledge, such as disaster impacts, telecommunications outages, payment-channel disruption, or administrative barriers. Leadership and quality assurance dashboards should provide aggregate indicators for strategic planning and service improvement.

A key feature of this layer is localized decision support. Regional staff should be able to annotate, contextualize, adjust, or override central AI-generated risk classifications when justified by local evidence. For example, if students in a specific district show low LMS activity due to a telecommunications outage, regional staff should be able to mark the case as a contextual exception. These annotations should be logged and fed back into the data platform to improve future interpretation.

5.5 Operational Triage Mechanism

The operational triage mechanism defines how different risk levels trigger different responses. This is essential because mega-universities cannot manually review all risk signals. Low-level risks can be addressed through automated safety nets, while high and critical risks require human review and case management (Table 1).

Table 1 Operational triage strategy for AI-assisted academic risk detection

Risk level	Example signals	Primary response	Responsible actor
Low	Unpaid invoice early in registration period; brief LMS inactivity; upcoming assignment deadline	Automated SMS, email, or system notification	System
Medium	Declining engagement; missed early assessment; repeated late payment pattern	Targeted reminder and monitoring	Advisor or support unit
High	Payment near deadline unpaid; repeated course failure; long LMS inactivity	Human follow-up and intervention planning	Advisor, tutor, or regional office
Critical	Combined financial, academic, engagement, and contextual risk; affected by disaster or local disruption		Regional office, programme unit, support unit

This triage logic allows the institution to use automated interventions for routine risks while reserving human attention for cases that require judgement, empathy, and contextual interpretation.

5.6 Responsible AI Governance and Feedback

Responsible AI governance surrounds all layers of the framework. It defines the purpose of risk detection, permitted data use, access rights, accountability, model review, audit logging, fairness monitoring, and safeguards against punitive use. In the Indonesian context, the framework should align with Law No. 27 of 2022 on Personal Data Protection. Individual risk scores should be accessible only to authorized staff and used to support students, not to stigmatize, penalize, or exclude them.

The framework also includes feedback and continuous learning. Interventions, advisor notes, regional annotations, student responses, and outcomes should be recorded and periodically reviewed. These feedback data help evaluate whether risk detection leads to meaningful support and whether models, thresholds, explanations, triage rules, and workflows need improvement.

6 Data Requirements and Planned Evaluation for Universitas Terbuka

6.1 Minimal Data Requirements for a Pilot

A realistic pilot at Universitas Terbuka should begin with a minimal but operationally meaningful dataset. The objective is not to integrate all institutional systems at once, but to establish a reliable foundation for detecting non-registration risk, course failure risk, and delayed graduation risk. The dataset should therefore include student background and programme data, longitudinal academic records, registration and payment data, LMS engagement data, support-service and regional-contextual data, and student study intent.

Student background and programme data include a canonical student identifier, cohort, programme, region or regional office, broad location, age group, prior education, and employment category if available. These variables support cohort analysis and fairness monitoring. Academic records include semester registrations, registered courses, grades, credit accumulation, cumulative GPA, repeated course attempts, academic status changes, inactive periods, and completion status. These data are essential for modelling course failure and delayed graduation risk (Table 2).

Registration and payment data are critical for non-registration risk. Relevant elements include registration history, course-load selection, billing status, payment status, payment timing, late-payment history, and proximity to registration or payment deadlines. These variables help distinguish academic disengagement from financial or administrative delay. LMS engagement data include weekly login frequency, content access, quiz participation, assignment submission, submission timeliness, forum activity, and early-semester activity trends.

Support-service and regional-contextual data may include helpdesk topics, call-center records, chatbot interactions, counselling referrals, complaints, and annotations

Table 2 Minimal data requirements for the pilot implementation

Data group	Example elements	Main risk supported
Student background and programme data	Student identifier, cohort, programme, region, age group, prior education, employment category	Fairness monitoring, cohort analysis
Academic records	Grades, GPA, credits, course attempts, academic status, inactive periods	Course failure, delayed graduation
Registration and payment data	Registration history, billing status, payment status, payment timing, late-payment pattern	Non-registration
LMS engagement data	Weekly login, content access, quiz activity, assignment submission, forum activity	Course failure, disengagement
Support-service and regional context	Helpdesk topics, complaints, counselling referrals, regional disruption notes	Non-registration, course failure, contextual risk
Study intent and academic pace	Full-time or part-time intent, expected course load, preferred study pace	Delayed graduation interpretation

from regional offices about disasters, telecommunications outages, payment-channel disruption, or local administrative problems. Finally, study intent data should capture whether students intend to study full-time, part-time, or flexibly, together with their expected course load. This information is important for interpreting delayed graduation risk without incorrectly flagging intentional slow pacing.

6.2 Phased Data Integration Strategy

The pilot should follow a phased integration strategy. The first phase may focus on one or two high-enrolment programmes and a limited historical window, such as three to four semesters. This phase should prioritize academic records, registration data, payment data, and aggregated LMS indicators to establish common identifiers, data definitions, baseline features, and initial models.

The second phase may add support-service records and regional contextual inputs, because student risk may first appear through administrative, financial, technical, or local service problems. The third phase may expand to more programmes, regions, and student groups while strengthening governance, fairness monitoring, cohort-drift detection, and role-specific dashboards. This gradual approach reduces implementation risk and allows UT to test data quality, model validity, staff acceptance, and workflow fit before scaling institution-wide.

6.3 Planned Evaluation Strategy

Evaluation should examine the framework as an institutional decision support process, not only as a predictive model. Technical evaluation should define the prediction target, observation window, prediction window, and unit of analysis for each risk category. Non-registration risk should be evaluated as the likelihood of failing to register in the following semester. Course failure risk may use the student-course as the unit of analysis. Delayed graduation risk should be evaluated against expected progression adjusted by declared study intent.

Model performance should be assessed using ROC AUC, precision, recall, F1-score, calibration, confusion matrices, and recall at operationally meaningful thresholds. Calibration is important because risk probabilities should be interpretable by decision-makers. Fairness evaluation should examine whether false positives and false negatives differ across programmes, regions, age groups, employment categories, socioeconomic proxies, and digital access conditions where such data are available and lawful to use.

Because UT admits large open-admission cohorts, evaluation should also include cohort-drift and robustness analysis. Key feature distributions in new cohorts should be compared with historical training data, including programme, region, age group, employment profile, payment behavior, early LMS activity, and course-load selection. If substantial drift is detected, the system should flag uncertainty, recalibrate thresholds, retrain models, or apply predictions more conservatively with greater human review. Robustness testing should also examine missing data, delayed updates, and inconsistent records.

Human-centered evaluation should assess whether advisors, tutors, programme leaders, student support units, regional offices, and institutional leaders find the system understandable, useful, and compatible with their work. Prototype dashboards and risk reports can be tested through usability sessions, scenario-based walkthroughs, surveys, and interviews. The evaluation should examine clarity of explanations, trust, perceived usefulness, cognitive workload, and fit with existing workflows.

Workflow and organizational evaluation should measure whether the institution can act on risk information. Important indicators include the number of students flagged, proportion receiving automated reminders, proportion escalated to human review, response time, advisor workload, regional office workload, intervention completion rate, and documentation of follow-up actions. The localized decision support component should be evaluated by tracking regional annotations, override decisions, reasons for override, and whether these contextual inputs improve risk interpretation.

Impact evaluation should examine whether interventions contribute to student success. Possible indicators include registration completion, reduced late registration, improved assignment submission, improved course pass rates, increased student response to support messages, reduced repeated course failure, and improved credit accumulation. At the pilot stage, UT may use pre-post comparisons, matched comparison groups, or phased rollout across programmes or regions. The evaluation should also monitor unintended effects, such as excessive alerts, staff overload, student anxiety, or inappropriate use of risk labels.

Governance evaluation should assess compliance with institutional policy and Indonesian personal data protection requirements. It should review whether data processing has a clear purpose, whether only necessary data are used, whether access is role-based, whether risk outputs are logged, and whether staff understand appropriate use of risk information. Risk scores should be used to support students, not to punish or restrict them.

6.4 Implementation Challenges

Several implementation challenges should be anticipated. First, data fragmentation and inconsistent identifiers may limit the quality of longitudinal student profiles. Second, payment, LMS, and support-service data may be incomplete, delayed, or stored in different formats. Third, staff capacity may constrain how many risk alerts can be followed up manually. Fourth, regional offices may vary in their ability to provide contextual annotations or conduct outreach. Fifth, users may distrust AI outputs if explanations are unclear or if the system appears to increase surveillance rather than support.

Governance readiness is also a challenge. UT will need clear policies on access rights, data retention, audit trails, model review, fairness monitoring, and escalation procedures. Technical teams and academic units must work together so that the system remains aligned with institutional workflows and student support values. For this reason, implementation should begin with a limited pilot, iterative evaluation, and continuous stakeholder engagement before wider deployment.

7 Conclusion

This paper proposed a conceptual framework for AI-assisted decision support for academic risk detection in mega-universities, using Universitas Terbuka as the guiding case. The framework addresses three interrelated academic risks: non-registration risk, course failure risk, and delayed graduation risk. It emphasizes that these risks should be interpreted contextually, since non-registration may be shaped by financial and administrative constraints, while delayed graduation may reflect intentional slow pacing among working adult learners.

The framework consists of four core layers: data sources, data platform, AI and analytics, and decision support applications. These layers are supported by operational triage, human-in-the-loop decision support, localized regional contextualization, cohort-drift monitoring, feedback loops, and responsible AI governance. This design connects central analytics with practical student support by distinguishing low-level risks suitable for automated reminders from high and critical risks requiring human review and case management.

The main contribution of this paper is a context-sensitive roadmap for moving from fragmented retrospective reporting toward proactive, explainable, and responsible academic risk support at scale. In the Indonesian context, the framework should be aligned with Law No. 27 of 2022 on Personal Data Protection and should ensure that risk scores are used to support students rather than to penalize or stigmatize them.

As a conceptual study, this paper does not report empirical implementation or measured intervention outcomes. Future work should pilot the framework in selected UT programmes or regions, evaluate predictive performance and fairness, test staff-facing dashboards, and examine whether AI-assisted interventions improve registration continuity, course completion, and student progression.

Acknowledgments. The authors acknowledge Universitas Terbuka for providing the institutional context that motivated the development of this conceptual framework. The authors also acknowledge the broader body of research on open and distance learning, learning analytics, early warning systems, explainable AI, and responsible AI governance that informed the design of the proposed framework.

Competing Interests The authors have no competing interests to declare that are relevant to the content of this paper.

Ethics Approval This paper presents a conceptual framework and does not report primary research involving human participants or animals. Therefore, formal ethics approval was not required. Future empirical implementation or pilot testing involving student-level data should follow institutional ethics procedures, applicable data governance policies, and Indonesian personal data protection requirements.

References

1. Daniel, J.S.: Mega-Universities and Knowledge Media: Technology Strategies for Higher Education. Kogan Page, London (1996)
2. Berberoğlu, N., Berberoğlu, B.: Grouping the mega university countries according to their similarities. Procedia. Soc. Behav. Sci. **174**, 2153–2159 (2015). https://doi.org/10.1016/j.sbspro.2015.01.870
3. Bozkurt, A.: The historical development and adaptation of open universities in Turkish context: the case of Anadolu university as a giga university. Int Rev Res Open Distrib Learn. **20**(4), 36–59 (2019). https://doi.org/10.19173/irrodl.v20i4.4086
4. Simpson, O.: Student retention in distance education: are we failing our students? Open Learn. **28**(2), 105–119 (2013). https://doi.org/10.1080/02680513.2013.847363
5. Hart, C.: Factors associated with student persistence in an online program of study: a review of the literature. J. Interact. Online Learn. **11**(1), 19–42 (2012)
6. Radovan, M.: Should I stay, or should I go? Revisiting student retention models in distance education. Turk Online J Distance Educ. **20**(3), 29–42 (2019). https://doi.org/10.17718/tojde.598191
7. Belawati, T., Zuhairi, A.: The practice of a quality assurance system in open and distance learning: a case study at universitas Terbuka Indonesia (the Indonesia Open University). Int Rev Res Open Distrib Learn. **8**(1), 1–15 (2007). https://doi.org/10.19173/irrodl.v8i1.340
8. Zuhairi, A., Maria, R.D., Mir, K.: Implementing quality assurance system for open and distance learning in three Asian open universities: Philippines, Indonesia and Pakistan. Asian Assoc Open Univ J. **15**(3), 297–320 (2020). https://doi.org/10.1108/AAOUJ-07-2020-0047
9. Budiman, R.: A Longitudinal Study of Student Achievement and Dropout in a Distance English Writing Course. University of Dundee, Dundee (2015)
10. Macfadyen, L.P., Dawson, S.: Mining LMS data to develop an early warning system for educators: a proof of concept. Comput. Educ. **54**(2), 588–599 (2010). https://doi.org/10.1016/j.compedu.2009.09.008

11. Jayaprakash, S.M., Moody, E.W., Lauría, E.J.M., Regan, J.R., Baron, J.D.: Early alert of academically at-risk students: an open source analytics initiative. J Learn Anal. **1**(1), 6–47 (2014). https://doi.org/10.18608/jla.2014.11.3
12. Akçapınar, G., Altun, A., Aşkar, P.: Using learning analytics to develop early-warning system for at-risk students. Int. J. Educ. Technol. High. Educ. **16**, 40 (2019). https://doi.org/10.1186/s41239-019-0172-z
13. Bañeres, D., Rodríguez, M.E., Guerrero-Roldán, A.E., Karadeniz, A.: An early warning system to detect at-risk students in online higher education. Appl. Sci. **10**(13), 4427 (2020). https://doi.org/10.3390/app10134427
14. Bañeres, D., Rodríguez-González, M.E., Guerrero-Roldán, A.E., Cortadas, P.: An early warning system to identify and intervene online dropout learners. Int. J. Educ. Technol. High. Educ. **20**, 3 (2023). https://doi.org/10.1186/s41239-022-00371-5
15. Zawacki-Richter, O., Marín, V.I., Bond, M., Gouverneur, F.: Systematic review of research on artificial intelligence applications in higher education—where are the educators? Int. J. Educ. Technol. High. Educ. **16**, 39 (2019). https://doi.org/10.1186/s41239-019-0171-0
16. Khosravi, H., et al.: Explainable artificial intelligence in education. Comput Educ Artif Intell. **3**, 100074 (2022). https://doi.org/10.1016/j.caeai.2022.100074
17. Kesgin, K., Kiraz, S., Kosunalp, S., Stoycheva, B.: Beyond performance: explaining and ensuring fairness in student academic performance prediction with machine learning. Appl. Sci. **15**(15), 8409 (2025). https://doi.org/10.3390/app15158409
18. Memarian, B., Doleck, T.: Human-in-the-loop in artificial intelligence in education: a review and entity-relationship (ER) analysis. Comput Hum Behav Artif Humans. **2**(1), 100053 (2024). https://doi.org/10.1016/j.chbah.2024.100053
19. Papagiannidis, E., Mikalef, P., Conboy, K.: Responsible artificial intelligence governance: a review and research framework. J. Strateg. Inf. Syst. **34**(2), 101885 (2025). https://doi.org/10.1016/j.jsis.2024.101885
20. Carter, J.W., Scott, J.T., Barrett, J.D.: The emotional cost of AI chatbots in education: who benefits and who struggles? Comput Hum Behav Artif Humans. **5**, 100181 (2025). https://doi.org/10.1016/j.chbah.2025.100181
21. Mazzullo, E., Bulut, O., Wongvorachan, T., Tan, B.: Learning analytics in the era of large language models. Analytics. **2**(4), 877–898 (2023). https://doi.org/10.3390/analytics2040046
22. Ahmed, W., et al.: Machine learning-based academic performance prediction: a regression approach. PLoS One. **20**(7), e0307645 (2025). https://doi.org/10.1371/journal.pone.0307645
23. Nasution, V.I.A., Budiman, R., Istianda, M.: Digital governance for a sustainable agenda: lessons learnt from universitas Terbuka. In: Proceedings of the 7th Open Society Conference 2025 (OSC 2025), pp. 54–68 (2025). https://doi.org/10.2991/978-2-38476-505-8_6
24. Nasution, V.I.A., Yuliani, I., Akadira, T., Saleha, E.: Transforming governance: Indonesia's digital leap from bureaucracy to big data in the era of society 5.0. In: Proceedings of the 6th Open Society Conference (OSC 2024), pp. 190–198 (2024). https://doi.org/10.2991/978-2-38476-340-5_17
25. Nasution, V.I.A.: Dynamic capabilities for transforming governance in open and distance learning University in Indonesia: lesson from SALUT DGCN. Jurnal Pendidikan Humaniora. **13**(1), 26 (2025). https://doi.org/10.17977/um011v13i12025p26-43

Autonomous Knowledge Agents for Institutional Performance Analytics in Higher Education: A Social Presence-SECI Framework with AI Tutoring

Siti Haryani Shaikh Ali[1(✉)] and Shahrinaz Ismail[2]

[1] Malaysian Institute of Information Technology, Universiti Kuala Lumpur, Kuala Lumpur, Malaysia
sharyani@unikl.edu.my
[2] School of Computer Sciences, Universiti Sains Malaysia, Penang, Malaysia

Abstract. Online learning disrupts the SECI knowledge creation spiral by eliminating the social conditions on which students' tacit knowledge sharing depends. This paper argues that social presence is the condition that lets the spiral keep working in online settings, and that AI personal tutors, built as autonomous knowledge agents, can generate and sustain the kind of presence each stage of knowledge conversion needs, so that students reach genuine learning outcomes rather than surface content. Drawing the SECI model, social presence theory, and AI tutoring research together, the paper proposes the Social Presence–SECI with Autonomous Intelligence (SPS-AI) Framework. Its central claim is that each SECI stage calls for a different type of social presence: affective-relational at Socialisation, cognitive co-presence at Externalisation, cohesive-informational at Combination, and motivational-coaching at Internalisation. An AI tutor can shift between these at a scale human instruction cannot match, and while doing so it generates the interaction evidence institutions need to monitor learning outcomes. The analytics key performance indicators mapped to the SECI phases are also proposed.

Keywords: SECI Model · Social Presence · AI Personal Tutor · Autonomous Knowledge Agents · Student Learning Outcomes · Online Higher Education · KPIs

1 Introduction

Online and hybrid learning had given students more access to course content than at any earlier point in the history of education. While students gained easy access to more content, the quality and depth of how they actually processed and understood that content had not kept pace. Researches consistently show that moving instruction onto digital platforms did not, on its own, reproduce the conditions for the iterative, socially embedded knowledge construction that effective higher education depends on (Guzik et al., 2025; Zatuchin, 2024). The problem is not access to content. It is the social and cognitive

N. Mohd Sharef et al. (Eds.): CAMP 2026, CCIS 3094, pp. 170–183, 2027.
https://doi.org/10.1007/978-981-92-4305-1_13

conditions under which students process, interrogate, and turn that content into durable competency. The SECI model of knowledge creation (Nonaka and Takeuchi, 1995) had set out this distinction with some precision, naming four modes of knowledge conversion, namely Socialisation, Externalisation, Combination, and Internalisation, each of which needs social interaction that online environments are poorly equipped to provide. Of the four, Socialisation suffers most, since it depends on the spontaneous, co-present sharing of tacit experiential knowledge. Researches had shown that its disruption cascades through the rest of the spiral: students who have not shared tacit understanding cannot externalise it well, cannot synthesise it, and cannot internalise it as durable competency (Guzik et al., 2025; Barreto and Abarca, 2025). Institutionally, this shows up as the familiar gap between content-completion metrics and genuine learning outcome attainment, a gap that had become one of the more pressing concerns for quality assurance officers and academic managers.

Against these gaps, AI personal tutoring systems, or autonomous knowledge agents, capable of adaptive and personalising interaction with individual students across the full cycle of knowledge creation, had offered a pathway of real theoretical and practical significance. Where such systems were well designed, Lin et al. (2023) and Baumgart and Mamlouk (2021) had shown that much of the adaptive responsiveness, the personalised scaffolding, and the motivational support of effective one-to-one human tutoring could be replicated. Yet the integration of these capabilities into a theoretically coherent whole had not followed. Crompton et al. (2023) had found AI in higher education to be used overwhelmingly for assessment, for prediction, and for the management of student learning, and only rarely for the deeper work of supporting knowledge creation. Hence, the knowledge-creation framework within which these scattered functions might had remained mostly absent from what institutions actually deployed. The notion of agent-mediated personal knowledge management, whereby software agents resided within the learning environment and mediated the knowledge-creation processes of individual learners on their behalf and in response to them, had been advanced as one coherent answer to this difficulty (Ismail and Ali, 2018), and it is this notion that the present paper had taken up, arguing that its integration with the SECI model and with social presence theory supplied the existing AI tutoring systems had required.

The present paper concerned with the development of the Social Presence–SECI with Autonomous Intelligence (SPS-AI) Framework, a conceptual architecture that integrated the SECI model, social presence theory, and AI tutoring research into a unified framework for the design and evaluation of AI personal tutors as autonomous knowledge agents serving student learning outcome attainment in online higher education. The framework also proposes a set of analytics Key Performance Indicators (KPIs) mapped to each SECI phase, through which institutions would be furnished with operationalised metrics for evaluating the effectiveness of SPS-AI implementations at both the individual student and the cohort level. These additions were supported by recent empirical researches in learning analytics, intelligent tutoring systems, and AI system architecture in education (Crompton et al., 2023; Lin et al., 2023; Ouyang and Chang, 2023; Mahamad et al., 2025).

2 Background and Rationale

The problem this paper addresses is that existing AI tutoring systems do not account for the knowledge-creation and social-presence dimensions of online learning, and this gap has consequences for whether they support genuine learning rather than surface content acquisition. Researches had shown that AI tutors in higher education mostly follow a content-delivery and assessment model. They treat learning as information transfer and neglect the dynamic, socially embedded processes through which tacit competency is actually built (Barreto and Abarca, 2025; Lin et al., 2023). The result is systems that score well on narrow metrics while doing little for the deeper cognitive and social work that produces transferable competency. A second problem is fragmentation. Knowledge management, educational technology, and intelligent systems had each built productive literatures, but largely apart from one another, rarely meeting on the specific question of student knowledge creation in AI-mediated environments. A third is practical: there were no operationalised analytics metrics by which an institution could check whether an SPS-AI implementation was producing the outcomes it promised. Hence, there is a need for a framework that brings the three traditions together on AI tutor design, and that hands institutions the analytics KPIs to build, monitor, and evaluate such systems.

Three objectives shaped the study, and they belong together as one programme of inquiry. The first is to synthesise the literature on the SECI model, social presence theory, and AI tutoring and autonomous-agent systems, so as to find the points where these traditions converge and on which an integrated framework can rest. From that synthesis the paper builds the SPS-AI Framework itself, a conceptual architecture for AI personal tutors working as autonomous knowledge agents, with the social presence requirements and agent responsibilities spelled out for each SECI phase. It then proposes a set of analytics KPIs, for each phase and for institutional monitoring, each with an operational definition, and a data source. Taken together, this is conceptual and theoretical work of the kind educational technology has long recognised as legitimate (Jaakkola, 2020). The findings are expected to help instructional designers, AI system developers, academic managers, and quality assurance practitioners understand the conditions, architectural and social, under which AI personal tutors can support student learning outcome attainment online.

3 Related Work

3.1 The SECI Model in Student Learning

The SECI model describes knowledge creation as a continuous spiral of four conversion modes (Nonaka and Takeuchi, 1995). Socialisation transfers tacit knowledge between people through shared experience and co-presence. Externalisation turns that tacit knowledge into explicit, articulable form through dialogue, metaphor, and reflection. Combination synthesises discrete explicit pieces into larger structures through recombination and documentation. Internalisation absorbs explicit knowledge back into tacit practice through application. Complete the cycle and it sets up a fresh round of Socialisation at a deeper level, which is what makes it a spiral and not a closed loop. Within higher education, research had tested how well the model maps onto student learning, and had found both real explanatory power and specific trouble online. Guzik et al. (2025), studying e-learning across corporate and higher education settings, had found online courses supporting Externalisation and Combination well enough, but the Socialisation phase, which needs tacit knowledge passed through shared experience, stayed the hardest to support digitally. That is inline with Zatuchin's (2024) point that each phase needs deliberate scaffolding if tacit knowledge conversion is not to collapse into rote information transfer. Barreto and Abarca (2025) had shown generative AI tools aligned with SECI pedagogy accelerating Externalisation and Combination while undermining the quality of Internalisation, and Asbari and Asbari (2025) had found Socialisation the phase most likely to collapse where structured social interaction was not designed in.

A further body of research had examined the correspondence between the four phases of the SECI spiral and the progressive development of student competency, identifying a theoretically significant alignment between Socialisation and the activation of prior experiential knowledge, Externalisation and the assessment of conceptual articulation, Combination and the demonstration of synthesis and critical thinking, and Internalisation and the transfer of learning to novel practical contexts, an alignment that had important implications for the design of AI tutoring systems intended to achieve measurable learning outcome attainment rather than content coverage. Among the problems identified was the tendency for online assessment systems to capture only the explicit knowledge outputs of the Combination phase, namely scores on tests and written assignments, while consistently neglecting the conversational and interpersonal processes through which tacit competency was developed in the Socialisation and Internalisation phases, and researches had argued that this selective capture constituted a form of learning evidence loss that systematically understated the depth of student knowledge creation (Zatuchin, 2024; Liccardo and Cerchione, 2025). Hence, there is a need for frameworks supporting the full SECI spiral within AI-mediated learning environments, rather than merely automating the assessment of explicit knowledge products, and it is hoped that the SPS-AI Framework proposed in the present paper would address this need.

3.2 Social Presence Theory and Student Knowledge Creation

Social presence is the degree to which a medium, or a communicating agent, is perceived as carrying the social, emotional, and interpersonal qualities of the parties involved. Researches had established that this perception is not a simple function of the medium's technical features. It is a socially constructed achievement, something participants actively produce as they work with the communicative affordances around them (Kreijns et al., 2021; Alsayer et al., 2024). Together with cognitive presence and teaching presence, it shapes the quality of online and blended learning, and it is significantly associated with learner satisfaction, with collaborative knowledge construction, and with a student's willingness to take the intellectual risks that higher-order knowledge creation asks for. Perceived learning and satisfaction in online courses were significantly and positively predicted by social presence (Richardson & Swan, 2003; Martin et al.,2022), and social presence was a significant positive predictor of learning engagement and self-regulation in online higher education (Miao & Ma, 2022). These findings, which are in line with Wu (2023) and Gao et al. (2024), found that social presence mediated the relationship between online learning self-efficacy and student engagement.

More recent researches had begun to examine social presence specifically in the context of human-AI agent interaction, asking whether and under what conditions an AI agent within a learning environment could be perceived as socially present in ways that supported genuine knowledge exchange and tacit knowledge externalisation. Wang et al. (2024) had found that AI agents designed to display social acknowledgement, contextual responsiveness, and collaborative framing significantly enhanced both the cognitive and social presence of students in asynchronous settings, and Xu et al. (2025) had demonstrated that pedagogical agents embedded in online learning achieved meaningful levels of social presence that significantly predicted learner engagement and knowledge co-construction. Norz et al. (2022, 2024) had found, through social network analysis of LMS log data, that social presence in online courses could be reliably quantified using indicators including network density, tie strength, and active forum participation, and that these indicators showed statistically significant patterns across the course duration, providing a methodological foundation for the real-time measurement of social presence within an analytics dashboard. Ebadi et al. (2022) had further found that learner motivation was significantly predicted by social presence and the human-likeness of the AI agent, and Wang et al. (2022) had demonstrated that the perceived cognitive and social presences of an AI coach were significantly associated with language learning outcomes. For all its growth, the said body of evidence had still not reached the question of whether an AI agent can be made to generate different kinds of social presence, each matched to what a particular SECI phase demands of it. Whether one uniform mode of presence could carry a student through the whole spiral is yet to be determined, and it is that doubt the present framework had been built to answer.

3.3 AI Tutoring Systems as Autonomous Knowledge Agents

Intelligent tutoring systems are not new, and several decades of study had left a fairly settled evidence base: in many cases their learning gains approached those of one-to-one human tutoring, and they exceeded conventional classroom instruction by a clear margin. Lin et al. (2023), reviewing the field systematically, had found that architectures drawing on machine learning (ML), natural language processing (NLP), and adaptive feedback achieved effect sizes well above conventional instruction, and that the systems doing best had three things in common, namely the real-time modelling of learner knowledge states, instructional sequencing fitted to the individual, and feedback that arrived close to the performance it addressed rather than long after it. The breadth of use is wide but shallow. Crompton et al. (2023), for instance, had found AI in higher education clustered into assessment and evaluation, prediction, assistance, tutoring, and the management of learning, each productive on its own terms yet seldom drawn together within one theoretically coherent architecture, which is the gap that the SPS-AI Framework had been built to close.

Where AI tutoring had been joined to learning analytics, the pairing had proved unusually productive for institutional monitoring, yielding richer evidence of student learning than either part could produce alone. Ouyang and Chang (2023) had found that coupling an AI performance-prediction model to learning-analytics feedback lifted student engagement, improved collaborative performance, and strengthened satisfaction, which sits well with the present framework's claim that agent-generated interaction data can serve individual learning and institutional monitoring at the same time. Other studies reached similar conclusion that an effective AI tutor needs not only good pedagogical algorithms but an analytics pipeline that can turn interaction data into something an institution can act on (Mahamad et al., 2025; Demartini et al., 2024; Sajja et al., 2023; Kochmar et al.,2021).

4 The SPS-AI Framework

4.1 Conceptual Architecture

The Social Presence–SECI with Autonomous Intelligence (SPS-AI) Framework was proposed as an integrated conceptual architecture for the design and evaluation of AI personal tutors operating as autonomous knowledge agents in online higher education, its purpose being to support student knowledge creation across all four phases of the SECI spiral through the deliberate generation and modulation of stage-specific social presence. Three claims, each drawn from the literature reviewed above, hold the framework together: that the SECI spiral is the right model for student knowledge creation in higher education, hence the whole spiral must be supported if learning outcomes are to be genuinely attained; that social presence is not generic but a stage-specific variable the different phases call for in distinct forms and intensities; and that an AI personal tutor, built as an autonomous knowledge agent able to modulate the presence it projects, is both technically and theoretically capable of supplying the range of presence types the full spiral requires in settings where face-to-face interaction is simply not available.

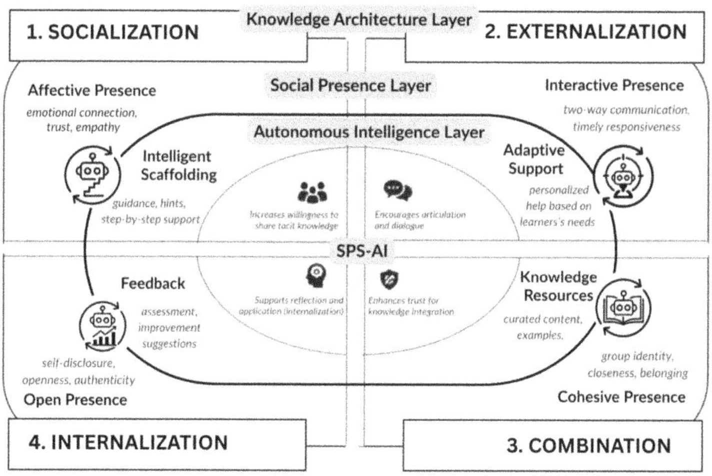

Fig. 1 The Social Presence–SECI with Autonomous Intelligence (SPS-AI) Framework

The SPS-AI Framework (Fig. 1) was organised around three functional layers that together constituted the operational architecture of the autonomous AI tutor. The first was the Knowledge Architecture Layer, which comprised the learner knowledge model, the domain knowledge representation, and the SECI phase recognition engine, a component responsible for diagnosing the student's current position within the knowledge creation spiral based on interaction data, response patterns, question types, and reflective discourse indicators. The second was the Social Presence Interface Layer, which housed the stage-specific social presence modulation engine, responsible for calibrating the tutor's communicative style, affective tone, and epistemic stance to the social presence requirements of the SECI phase in which the student was currently operating. The third was the Autonomous Intelligence Layer, which governed the AI tutor's pedagogical sequencing decisions, feedback generation, scaffolding strategies, and adaptive transitions between SECI phases, operating with a meaningful degree of autonomy from moment-to-moment instructor direction while remaining within institutionally defined parameters for student support.

4.2 Stage-Specific Social Presence Mediation

The central and most theoretically novel contribution of the SPS-AI Framework was its specification of four qualitatively distinct types of social presence, each calibrated to the knowledge-creation demands of a specific SECI phase, and its argument that an AI tutor operating as an autonomous knowledge agent must be capable of generating and transitioning between these presence types as the student progressed through the knowledge creation spiral. The first type, Affective-Relational Presence, was required at the Socialisation phase, where the student needed to share tacit knowledge, experiential insight, and pre-conceptual understanding in a form of dialogic exchange approximating the spontaneous knowledge-sharing of face-to-face co-presence, a process that researchers had found was highly sensitive to perceived warmth, trust, and interpersonal safety in the communicative environment (Wang et al., 2024; Xu et al., 2025). At this phase, the AI tutor's behaviour was characterised by conversational, non-evaluative dialogue, active listening indicators, personal acknowledgement of the student's experience, and invitation-framing rather than directive questioning, all of which had been documented as contributors to perceived social presence in AI-mediated educational environments. The second type, Cognitive Co-presence, was required at the Externalisation phase, where the student needed to articulate implicit understanding in explicit, communicable form, a process particularly vulnerable to self-censorship when the communicative environment was perceived as evaluative or unsympathetic, and the AI tutor's behaviour was characterised by Socratic questioning, collaborative hypothesis-building, non-judgmental scaffolding of imprecise articulations, and the co-construction of conceptual structures that progressively externalised the student's tacit understanding.

The third type, Informational Presence, was required at the Combination phase, where the student needed to synthesise discrete explicit knowledge artefacts into more complex and applicable knowledge structures, with the AI tutor functioning as a knowledgeable synthesis partner capable of connecting the student's emerging understanding to relevant disciplinary knowledge, alternative frameworks, and real-world application contexts, and researches had found that the most effective pedagogical agents at this phase demonstrated domain expertise alongside communicative responsiveness (Lin et al., 2023). The fourth type, Motivational-Coaching Presence, was required at the Internalisation phase, where the student was engaged in applying explicit knowledge

Table 1 Stage-Specific Social Presence Requirements and AI Tutor Roles in the SPS-AI Framework

SECI Stage	SP Type Required	AI Tutor Role	Learning Outcome
Socialization	**Affective Presence** Relational presence, emotional connection, trust, empathy	**Intelligent Scaffolding** Warm dialogic exchange; mimics peer empathy; non-evaluative listening	Activates tacit sharing; builds epistemic trust
Externalization	**Interactive Presence** Cognitive co-presence, two-way communication, and timely responsiveness	**Adaptive Support** Socratic prompting; scaffolds articulation of implicit knowledge	Converts tacit competency to an assessable explicit form
Combination	**Cohesive Presence** Relational Presence, group identity, closeness, belonging	**Knowledge Resources** Synthesis partner: connects student ideas to domain knowledge	Builds structured, transferable knowledge artefacts
Internalization	**Open Presence** Motivational/Coaching Presence, self-disclosure, openness, authenticity	**Feedback** Adaptive scenarios; personalised feedback; re-injects social presence	Embeds explicit knowledge as a durable student competency

to practical contexts in a process of experiential absorption inherently vulnerable to motivational withdrawal in the absence of social encouragement and performative feedback, and this fourth presence type repre-sented the critical design challenge identified by Barreto and Abarca (2025), who had found that AI-mediated learning enhanced Externalisation and Com-bination but consistently undermined Internalisation quality, a finding that the SPS-AI Framework attributed to a failure of social presence recalibration, where AI tutors designed for informational responsiveness had failed to transition to the motivational, adaptive, and personally affirming communicative style that Internalisation required. Table 1 summarises the stage-specific social presence requirements and AI tutor roles across the full SECI spiral.

5 Analytics Key Performance Indicators for SPS-AI Implementation

5.1 Rationale for a Multi-Phase KPI Architecture

To put the framework to institutional use, this paper proposes a structured set of Key Performance Indicators, so that an institution can judge an SPS-AI implementation with the rigour it applies to any other quality metric. Researches had found institutions capturing explicit indicators, namely assessment scores, completion rates, and time-on-task, which are easy to track but theoretically miss the depth of the processes that build the competencies a learning outcome is meant to reflect (Sharples, 2023; Lin et al., 2023). An AI tutor supporting the full SECI spiral can do better. It can yield measures of tacit-sharing engagement from Socialisation data, articulation quality from Externalisation dialogue, synthesis depth from Combination performance, and competency application from Internalisation assessment, which together give a multi-dimensional picture of one student's knowledge creation and support intervention that is both targeted and timely. This is based by the work of Ouyang and Chang (2023) who had found AI performance prediction joined to analytics feedback raising engagement and collaborative performance, and Sajja et al. (2023), who had shown GPT-4 analytics tools quantifying engagement, mapping progression, and evaluating instruction in real time.

The SPS-AI KPI architecture was accordingly organised across phase-specific KPIs, with three to four metrics assigned to each SECI phase, each of them operationally defined, and anchored to a specific data source. Among the important implications of this structure was its alignment with the distinction between process-based monitoring of knowledge creation trajectories and outcome-based monitoring of learning attainment, a distinction that Sharples (2023) had argued was essential for the responsible deployment of AI in education and that Crompton et al. (2023) had identified as underrepresented in existing AI-in-education implementations. Table 2 presents the full SPS-AI KPI framework, comprising twelve metrics across the four SECI phases and the institutional dashboard level.

Table 2 Stage-Specific SPS-AI analytics KPI

SECI Stage	Metric	Operational Definition	Data Source
Socialization	**Tacit Sharing Index (TSI)** Gao et al. (2024)	Ratio of personal-experience utterances to total interactions in a session; captures willingness to disclose pre-conceptual knowledge	NLP sentiment + discourse classifier on chat logs
	Social Presence Quotient – Affective (SPQ-A) Alsayer et al. (2024)	Weighted average of emotional acknowledgement cues (empathy markers, self-disclosure signals) in AI–student dialogue	Dialogue act classifier; affective computing model
	Epistemic Trust Score (ETS) Wang et al. (2024)	Frequency of student-initiated sharing after AI invitation prompts, relative to directive questions	Interaction log; response-type classifier

(*continued*)

Table 2 (*continued*)

SECI Stage	Metric	Operational Definition	Data Source
Externalization	**Conceptual Articulation Quality (CAQ)** Kochmar et al. (2021)	Automated rubric score for clarity, specificity, and conceptual completeness of student written responses	NLP-based automated essay scoring engine
	Scaffolding Responsiveness Rate (SRR) Gao et al. (2024)	Proportion of AI scaffold prompts that elicit a substantively extended student	LMS logs; response-length classifier
	Social Presence Quotient – Cognitive (SPQ-C) Alsayer et al. (2024) Martin et al. (2022)	Perceived cognitive co-presence from student survey items adapted from the CoI instrument	End-of-session 5-item Likert survey
Combination	**Knowledge Synthesis Depth (KSD)** Baumgart & Mamlouk (2021)	Graph-based integration score; valid links between student-generated concepts and the domain knowledge graph	Concept map analyser; knowledge graph alignment engine
	Analytic Trust Index (ATI) Xu et al. (2025)	Student-reported confidence in AI-generated knowledge synthesis outputs, post-Combination	3-item validated trust instrument (post-task)
	Knowledge Velocity (KV) Ouyang & Chang (2023)	Rate of movement through Combination sub-tasks per unit time, relative to cohort baseline	LMS timestamp analysis
Internalization	**Competency Transfer Score (CTS)** Lin et al. (2023)	Performance on a novel application scenario not seen during Combination; assesses transfer of internalised knowledge	Adaptive assessment engine; rubric-scored scenario tasks
	Motivational Persistence Indicator (MPI) Barreto & Abarca (2025)	Voluntary re-engagements with the AI tutor after initial task completion, within an internalisation window	LMS session logs

(*continued*)

Table 2 (*continued*)

SECI Stage	Metric	Operational Definition	Data Source
	Social Presence Quotient – Motivational (SPQ-M) Alsayer et al. (2024) Ebadi et al. (2022)	Student-rated encouragement, coaching tone, and authentic engagement from the AI during Internalisation	5-item end-of-unit survey

5.2 Rationale for a Multi-Phase KPI Architecture

The twelve KPIs in Table 2 were operationalised through a combination of automated computational measurement, adapted validated survey instruments, and rule-based threshold monitoring. The Social Presence Quotient metrics (SPQ-A, SPQ-C, SPQ-M) were adapted from the Community of Inquiry survey instrument, with item selection informed by the validation study of Alsayer et al. (2024), who had found that social presence was most strongly predicted by affective expression. The Tacit Sharing Index and Scaffolding Responsiveness Rate were adapted from Gao et al. (2024), who had identified social interaction quality as a significant predictor of online learning efficiency. The Conceptual Articulation Quality metric drew on the automated essay scoring literature, with Kochmar et al. (2021) having demonstrated that NLP-based personalised feedback in ITS shows improvement in student performance outcomes through rubric-scored automated assessment. The Knowledge Synthesis Depth metric was adapted from Baumgart and Mamlouk (2021), while The Competency Transfer Score was derived from adaptive assessment principles established in the ITS literature (Lin et al., 2023). The Analytic Trust Index was adapted from AI trust measurement instruments, given the finding of Xu et al. (2025) that pedagogical agents achieved meaningful levels of social presence significantly predicting learner engagement when trust in the agent's outputs was maintained.

6 Discussion

The framework makes three theoretical contributions across the literatures on knowledge management, educational technology, and AI tutoring. The first is to integrate social presence theory with the SECI model in an educational setting, which had not been done before. It answers Guzik et al.'s (2025) observation that online environments keep failing to support Socialisation, by saying why, namely the absence of affective-relational presence, and by showing how deliberate tutor design might fix it. The second is to treat social presence as stage-specific and type-differentiated rather than as the single composite that instruments like the Community of Inquiry survey measure. That matters for both design and measurement: an aggregate social presence score can mislead if it hides variation across phases. Yang et al. (2022) lend support here, having found the five dimensions of social presence relating differently to cognitive learning, affective learning, and motivation. The third is to reframe AI personal tutors as autonomous knowledge agents whose main job is to generate and modulate social presence in service of spiral completion, rather than to deliver content or test declarative knowledge. That makes the tutor an active participant in the student's knowledge creation, which is inline with Sharples' (2023) call for a social generative AI paradigm in education.

The analytics KPI architecture addressed a critical gap between theoretical framework and institutional practice, a gap consistent with concerns raised in the broader learning analytics literature. Researches had found that the uptake of analytics frameworks in higher education institutions was strongly contingent on the clarity and institutional relevance of the performance indicators they proposed, with systems offering precise, measurable thresholds and clear data source specifications achieving significantly higher institutional adoption rates than those offering only theoretical constructs (Demartini et al., 2024; Crompton et al., 2023). The twelve KPIs proposed in Table 2, organised across the four SECI phases were designed to meet this criterion by providing operational definitions that could be directly encoded in an analytics system, target thresholds derived from empirical benchmarks in the literature, and data source specifications aligned with standards used in contemporary LMS deployments. Among the important implications of this KPI architecture for quality assurance practice was its potential to shift institutional monitoring from retrospective, grade-based reporting to prospective, process-based tracking of knowledge creation trajectories, allowing institutions to identify students at risk of SECI spiral failure at the Socialisation or Externalisation phases rather than only at the point of final assessment, and to deploy AI tutor interventions calibrated to the specific phase at which the student's knowledge creation had stalled.

7 Conclusion

The present paper had proposed the SPS-AI Framework as a theoretically grounded architecture for AI personal tutors operating as autonomous knowledge agents in online higher education, drawing on the integrated insights of the SECI model, social presence theory, and AI tutoring research. Its novelty lay in three moves: recasting social presence as a stage-specific, type-differentiated design variable; naming four distinct presence types, each tied to the knowledge-creation demands of one SECI phase; and reframing AI personal tutors as autonomous knowledge agents whose effectiveness turns as much on the presence they can generate as on their raw computational intelligence. To these the paper had added a twelve-metric KPI architecture that gives institutions operationalised, threshold-calibrated measures of spiral progression at student and cohort level. The findings of this study are expected to help instructional designers, AI system developers, academic managers, and quality assurance practitioners to better understand the architectural and social conditions under which AI personal tutors could serve as effective partners in student knowledge creation. Further study should be made to validate the framework's claims, to refine the KPI thresholds through institutional piloting, and to support the many online higher education settings now confronting the problem of student learning outcome attainment.

References

Alsayer, A.A., et al.: Measuring social presence in online learning: a validation study. Educ. Inf. Technol. **30**, 5655–5676 (2024). https://doi.org/10.1007/s10639-024-12345-x

Asbari, M., Asbari, D.A.F.: The dynamics of student knowledge creation through SECI: a qualitative exploration in the Indonesian context. Indones. J. Manag. Econ. Res. **2**(1), 29–39 (2025)

Barreto, U., Abarca, Y.: Integration of the SECI model and ChatGPT in higher education. Heliyon. **11**, e42814 (2025). https://doi.org/10.1016/j.heliyon.2025.e42814

Baumgart, A., Mamlouk, A.M.: A knowledge-model for AI-driven tutoring systems. In: Tropmann-Frick, M., et al. (eds.) Information Modelling and Knowledge Bases XXXIII, pp. 1–12. IOS Press (2021). https://doi.org/10.3233/FAIA210474

Crompton, H., et al.: Artificial intelligence in higher education: the state of the field. Int. J. Educ. Technol. High. Educ. **20**, 22 (2023)

Demartini, C., et al.: Artificial intelligence bringing improvements to adaptive learning in education: a case study. Sustainability. **16**(3) (2024)

Ebadi, S., et al.: Examining the roles of social presence and human-likeness on Iranian EFL learners' motivation using AI technology: a case of CSIEC chatbot. Interact. Learn. Environ. **31**(2), 655–673 (2022)

Gao, F., Wang, C., Xie, H., Hong, J.: Social interaction and online learning efficiency for middle school students: the mediating role of social presence and learning engagement. Behav. Sci. **14**, 896 (2024). https://doi.org/10.3390/bs14100896

Guzik, A., Lis, A.M., Chodnicki, M.: Effective knowledge creation and transfer in e-learning using SECI model. In: Proceedings of the 26th European Conference on Knowledge Management (2025)

Ismail, S., Ali, S.H.S.: Social presence in digital community. In: 2018 International Symposium on Agent, Multi-Agent Systems and Robotics (ISAMSR), pp. 1–5. IEEE (2018)

Jaakkola, E.: Designing conceptual articles: four approaches. AMS Review. **10**(1–2), 18–26 (2020)

Kochmar, E., et al.: Automated data-driven generation of personalized pedagogical interventions in intelligent tutoring systems. Int. J. Artif. Intell. Educ. **32** (2021)

Kreijns, K., et al.: Social presence: conceptualization and measurement. Educ. Psychol. Rev. **34** (2021)

Liccardo, G., Cerchione, R.: GenAI-assisted knowledge generation: a case study. In: Proceedings of the 26th European Conference on Knowledge Management (2025)

Lin, C.-C., Huang, A.Y.Q., Lu, O.H.T.: Artificial intelligence in intelligent tutoring systems toward sustainable education: a systematic review. Smart Learn. Environ. **10**, 41 (2023). https://doi.org/10.1186/s40561-023-00260-y

Mahamad, S., et al.: Technical review: architecting an AI-driven decision support system for enhanced online learning and assessment. Future Internet. **17** (2025)

Martin, F., et al.: A meta-analysis on the Community of Inquiry presences and learning outcomes in online and blended learning environments. Online Learn. **26**(1) (2022)

Miao, J., Ma, L.: Students' online interaction, self-regulation, and learning engagement in higher education: the importance of social presence to online learning. Front. Psychol. **13**, 815220 (2022). https://doi.org/10.3389/fpsyg.2022.815220

Nonaka, I., Takeuchi, H.: The Knowledge-Creating Company. Oxford University Press (1995)

Norz, L.-M., et al.: Measuring social presence in online-based learning: an exploratory path analysis using log data and social network analysis. Internet High. Educ. **56**, 100894 (2022)

Norz, L.M., et al.: Quantifying social presence in online-based learning. Online Learn. **28**(4), 414–432 (2024)

Ouyang, F., Chang, S.: Integration of artificial intelligence performance prediction and learning analytics to improve student learning in online engineering course. Int. J. Educ. Technol. High. Educ. **20**(1), 4 (2023)

Richardson, J.C., Swan, K.: Examining social presence in online courses in relation to students' perceived learning and satisfaction. JALN. **7**(1), 68–88 (2003)

Sajja R et al: Integrating AI and learning analytics for data-driven pedagogical decisions and personalized interventions in education. arXiv preprint. (2023)

Sharples, M.: Towards social generative AI for education: theory, practices and ethics. Learn.: Res. Pract. **9**(2), 159–167 (2023). https://doi.org/10.1080/23735082.2023.2261131

Wang T et al: Generative co-learners: Enhancing cognitive and social presence of students in asynchronous learning with generative AI. https://arxiv.org/abs/2410.04365. (2024) https://doi.org/10.1145/3706598.3714259

Wang, X., et al.: Learners' perceived AI presences in AI-supported language learning: a study of AI as a humanized agent from Community of Inquiry. Comput. Assist. Lang. Learn. **37**(4), 814–840 (2022)

Wu, R.: The relationship between online learning self-efficacy, informal digital learning of English, and student engagement. Front. Psychol. **14**, 1266009 (2023). https://doi.org/10.3389/fpsyg.2023.1266009

Xu, K.M., et al.: Social presence: a key factor in embedding a pedagogical agent into online learning in primary education. Br. J. Educ. Technol. **57**(1), 227–242 (2025). https://doi.org/10.1111/bjet.70006

Yang, H., et al.: Learning in the online classroom: exploring the unique influence of social presence dimensions. Commun. Stud. **73**(3), 245–262 (2022)

Zatuchin, D.: Enhancing knowledge transformation in digital education: an analysis of the SECI model's application in course design and execution. Discov. Educ. **3**, 140 (2024). https://doi.org/10.1007/s44217-024-00229-0

Knowledge Representation and Analytical Framework for OBE Assessment in Higher TVET Institutions

Siti Haryani Shaikh Ali[1(✉)], Kamaruzzaman Ismail[2], and Shahrinaz Ismail[3]

[1] Malaysian Institute of Information Technology, Universiti Kuala Lumpur, Kuala Lumpur, Malaysia
sharyani@unikl.edu.my

[2] Malaysian France Institute, Universiti Kuala Lumpur, Bandar Baru Bangi, Malaysia

[3] School of Computer Sciences, Universiti Sains Malaysia, Penang, Malaysia

Abstract. Outcome-Based Education (OBE) in Higher Technical and Vocational Education and Training (HTVET) institutions continued to face persistent challenges in aligning assessment practices, competency attainment evidence, and institutional data systems with prescribed learning outcome hierarchies. This paper presents an integrative synthesis of knowledge representation issues and proposes a four-layer framework, namely the Knowledge Representation and Analytical Framework (KRF-HTVET) for OBE assessment in higher TVET institutions. Drawing upon ontological engineering, semantic modelling, predictive analytics, and visual analytics, the proposed framework addresses the structural disconnect between formal outcome hierarchies and the data infrastructure required to evidence graduate competency. Research found that learning analytics adoption in TVET remains critically underdeveloped, and that ontological curriculum alignment approaches are yet to be systematically applied in TVET-specific institutional contexts. The proposed KRF-HTVET framework is hoped to provide administrators, curriculum designers, and instructors with structured knowledge representation and decision-support analytics that move beyond traditional reporting toward prescriptive, real-time institutional intelligence.

Keywords: Outcome-Based Education · Knowledge Representation · TVET · Competency-based Assessment · Higher Education

1 Introduction

Outcome-Based Education (OBE) represents a fundamental reorientation of educational philosophy in which the entire architecture of a programme, from curriculum design to assessment protocols and institutional reporting, is systematically aligned with the demonstrable competencies of the graduate. The paradigm shift, from input-based to results-oriented educational models, necessitates a reconceptualisation of what institutions are designed to produce, an argument since widely adopted in higher TVET systems globally. Researches had further established that the efficacy of any OBE system is contingent upon the synchronised coherence of learning outcomes, instructional activities,

N. Mohd Sharef et al. (Eds.): CAMP 2026, CCIS 3094, pp. 184–197, 2027.
https://doi.org/10.1007/978-981-92-4305-1_14

and assessment measurements, such that every evaluative task provides direct, empirical evidence of a defined outcome (Biggs & Tang, 2007). Hashim, Ali, and Ismail (2016) had demonstrated, through the UniKL higher TVET model, that Malaysian institutions had begun operationalising this philosophy through value-based, industry-aligned curricula, and yet practical implementation had continued to expose structural, pedagogical, and technological gaps that existing assessment and analytics systems had failed to close (Sin & Hussin, 2024; Yusop, Rasul, & Yasin, 2024; Mistamiruddin & Mohamad Nasri, 2024).

The central problem confronting HTVET institutions is not a conceptual one, as the theoretical basis of OBE is well-established, but rather a representational and computational one: institutions lack the formal knowledge structures and reasoning tools needed to translate outcome hierarchies into automated, evidence-based processes. Research has found that assessment data in TVET contexts are often fragmented and inconsistently recorded, that learning analytics adoption in vocational and technical education remains critically underdeveloped, and that ontological approaches to curriculum alignment have yet to be systematically applied in TVET-specific contexts (Gedrimiene et al., 2019; Namahoot et al., 2025). Hence, there is a need for an integrated framework that formalises outcome knowledge, structures competency-based assessment, and enables predictive, visualisation-driven decision support calibrated to the institutional realities of higher TVET.

2 Research Methodology

2.1 Methodological Approach

This paper is a conceptual, theory-building contribution, and its methodology is accordingly one of synthesis rather than of primary data collection or statistical aggregation. The framework proposed in Sect. 5 was developed through an integrative synthesis of the literature, in the sense that Whittemore and Knafl (2005) gave to the integrative review, namely a method distinctive in permitting empirical, theoretical, and methodological sources to be combined within a single account. Torraco (2016) had characterised such a review as a form of scholarship in which existing literature is used to generate new knowledge rather than merely to summarise it, and Paul et al. (2023) had argued that framework-based reviews of this kind are particularly suited to theory building. The present study sits within that tradition. Its purpose was neither to estimate a pooled effect nor to chart a field exhaustively, but to draw a heterogeneous body of work into a coherent architecture for OBE assessment in higher TVET.

2.2 Source Selection and Synthesis

Literature was sought for its bearing on the conceptual components of the proposed framework, through searches of major academic databases and the tracing of citations from key works. Sources were retained on the criterion of substantive relevance to one of those components, with preference given to peer-reviewed journal articles and indexed conference papers from roughly the past decade, and with a small number of foundational works admitted for their established theoretical significance, among them Spady (1994), Anderson and Krathwohl (2001), and Biggs and Tang (2007). The retained literature is the set listed in the References.

The synthesis was purposive rather than exhaustive, and it was organised around the conceptual building blocks of the framework: the documented challenges of OBE assessment and analytics in TVET, treated in Sect. 4.1; the existing approaches to knowledge representation and to analytics, treated in Sect. 4.2; and the theoretical and applied work informing the proposed architecture, treated in Sect. 5. Within each, studies bearing on a common construct were read together and folded into interpretive statements rather than recited one by one, following the evidence-grouping logic that Chong et al. (2021) describe; and where the evidence diverged, the divergence is reported in the relevant section rather than smoothed over. It is through this thematic organisation that the literature becomes the layered architecture set out in Sect. 5.

3 Theoretical Foundations of OBE in Higher TVET

OBE traces its theoretical lineage to Spady (1994), who had proposed that educational systems be redesigned around the principle that all students can learn and succeed, provided that curriculum, instruction, and assessment are aligned with clearly defined exit outcomes. This effected a shift from the input-based model, in which institutional success was measured by resources and content coverage, toward a results-oriented architecture in which success is defined by the demonstrated competencies of the graduate. Anderson and Krathwohl (2001) had subsequently refined the cognitive taxonomy that undergirds OBE outcome writing, revising Bloom's (1956) original classification from static noun-based categories into verb-based cognitive processes and thereby enabling a more precise alignment of outcome statements with observable behaviours across the cognitive, psychomotor, and affective domains, domains of particular weight in TVET, where practical skill application sits at the centre of the educational mission. Ali (2018), reviewing curriculum-design principles, had argued that constructive alignment is the underpinning concept through which intended learning outcomes drive the design of all teaching, learning, and assessment activity, and that the synergy of formative and summative assessment is essential to the integrity of any OBE programme. Li, Zhao, He, and Deng (2024) had found, empirically, that assessment strategies and project-based learning each mediate the relationship between intended learning outcomes and graduate competence, confirming that the operational mechanisms of OBE behave as constructive-alignment theory predicts. Table 1 presents a summary of the key theoretical frameworks underpinning OBE as applied to higher TVET institutions.

The principle of constructive alignment, as established by Biggs and Tang (2007), provides the mechanism through which OBE is implemented at course and programme levels, requiring that every instructional activity and assessment task be designed to facilitate and to evidence the achievement of a defined outcome. For higher TVET this entails a tiered hierarchy of outcomes, in which Programme Outcomes, Programme Specific Outcomes, and Course Outcomes must hold internal consistency and aggregate upward, such that demonstrable achievement at course level constitutes valid evidence of attainment at programme level. Rao (2020) had argued that the integrity of this hierarchy is a precondition for the validity of any assessment or analytics system erected

Table 1 Theoretical Frameworks Underpinning OBE in Higher TVET Institutions

Framework / Theory	Author / Year	Key Contribution	Relevance to Higher TVET
OBE Paradigm	Spady (1994)	Shifts educational design from input-based delivery to measurable, results-oriented graduate competencies	Foundational justification for competency-focused TVET programme design and accreditation
Revised Bloom's Taxonomy	Anderson & Krathwohl (2001)	Recategorises educational goals using verb-based cognitive processes across cognitive, psychomotor, and affective domains	Enables precise OBE outcome writing for practical and technical TVET skill domains
Constructive Alignment	Biggs & Tang (2007)	Requires coherent synchronisation of learning outcomes, instructional activities, and assessment tasks	Provides the operational mechanism for CO-PO coherence and valid assessment design in TVET
OBE Outcome Hierarchy (PO-PSO-CO)	Rao (2020)	Defines a tiered hierarchy of programme, programme-specific, and course outcomes as the structural backbone of OBE	Prerequisite for valid analytics and attainment tracking across TVET programme levels
Higher TVET Educational Model	Hashim, Ali & Ismail (2016)	Integrates value-based, work-based learning with industry-aligned competency targets and global curriculum orientation	Demonstrates the multi-dimensional outcome representation required for TVET (technical, generic, affective skills)
TVET Assessment	Yusop et al. (2022)	Finds competency-based assessment as a dominant approach; highlights the underdevelopment of practical skills measurement	Confirms the gap between declared TVET outcomes and available assessment evidence structures

upon it. Hashim, Ali, and Ismail (2016) had further demonstrated that a higher TVET curriculum must represent at once its vocational technical competencies, the generic skills associated with Industry 4.0, entrepreneurial dispositions, and workplace-based assessment evidence, a representational complexity that exceeds what cognitive taxonomies alone are able to provide. Yusop et al. (2022), in a PRISMA-guided review of

TVET assessment, had found that competency-based assessment was the dominant approach while practical-skills measurement remained underdeveloped, confirming the gap between what TVET programmes aspire to develop and the evidence structures available to certify it. Schröder (2019) had argued that outcome-oriented qualification frameworks and praxis-oriented vocational didactics must be supported by research-driven systemic development, a reminder that the representational and analytical infrastructure of TVET is as much a policy question as a pedagogical one. This is inline with the broader observation that formal OBE structures in TVET tend to be more fully elaborated at the level of documentation and policy than at the level of operational data collection and analysis.

4 Related Work

4.1 Challenges in OBE Assessment and Analytics in Higher TVET Institutions

The implementation of OBE assessment and analytics in HTVET institutions was confronted by a pattern of interrelated challenges spanning pedagogical, structural, technological, and governance dimensions. Mistamiruddin and Mohamad Nasri (2024), in a systematic literature review following PRISMA guidelines, had identified continuous assessment and feedback and time and resource management as the most frequently cited challenges, each appearing in 80% of reviewed studies, followed by mindset and cultural change among educators (70%) and curriculum redesign (60%). Sun, Koo, and Lim (2020) had similarly found that implementing an OBE system requires proper planning, sustained investment, and systematic utilisation of attainment data for continuous quality improvement, challenges that are amplified in TVET contexts by dual accountability to both academic accreditation bodies and industry competency standards, while Phuong et al. (2024) had further identified nine specific challenges including pedagogical hurdles, lack of training and resources, institutional constraints, and technological limitations.

Among the problems identified in TVET-specific assessment contexts, the misalignment between assessment practices and learning outcomes had emerged as a particularly persistent concern. Yusop, Rasul, and Yasin (2024) had found that teachers' evaluation methods were frequently incompatible with assessment objectives and that occupational skills assessment was often disconnected from lesson-level learning outcomes, a finding that is in line with Sin and Hussin (2024), who had documented that variations in curriculum design, assessment techniques, and stakeholder engagement represented significant barriers to effective OBE implementation. Kamaruzaman et al. (2024) had further found a significant disconnect between curriculum content and industry expectations, with insufficient emphasis on soft skills development. This is a concern reinforced by Halik Bassah et al. (2023), whose qualitative study of industry expert perspectives had found that industries specifically require social and communication skills, technological literacy, and leadership, which are competencies that current TVET curricula have been found to systematically under-prioritise in assessment design. Khan et al. (2022) demonstrated that competency-based training and assessment (CBTA) produces significant improvements in practical teaching strategies and technical knowledge when the behavioural and psychological dimensions of instructors are treated as mediating variables, hence reinforcing that CBTA implementation in TVET is as much a faculty development challenge as a curriculum design.

Learning analytics adoption in TVET contexts was severely underrepresented relative to mainstream higher education. The vast majority of learning analytics research was focused on the course level in conventional higher education, with very few studies addressing VET at the system level (Gedrimiene et al., 2019), and that resistance culture had the strongest links with learning analytics adoption challenges (Alzahrani et al., 2023). Barongo et al. (2024) had further found that existing electronic assessment frameworks lacked sufficient accessibility, suffered from non-adaptive designs, and provided insufficient educator training. Ndukwe and Daniel (2020) had demonstrated that teacher data literacy is a foundational enabler that must be developed alongside any technical analytics infrastructure, and that an integrated tripartite model connecting teaching analytics, learning analytics, and learning design is required to ensure that predictive outputs translate into improved pedagogical practice. Hence, there is a need for a contextually situated framework that addresses the said challenges within higher TVET institutions in a coordinated and empirically grounded manner.

4.2 Existing Approaches of OBE Knowledge Representation and Analytics Frameworks

The application of ontological engineering to educational knowledge representation had emerged as a structurally coherent response to the problem of semantic fragmentation in OBE systems, where the relationships between course outcomes, programme outcomes, accreditation requirements, and industry competency standards are typically managed through disconnected documentation rather than formally encoded knowledge structures.

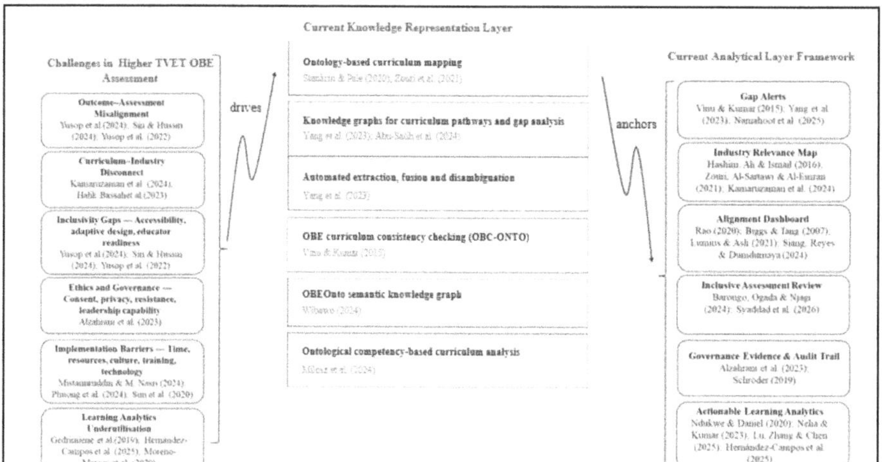

Fig. 1. Connecting OBE current challenges, current approaches, and current frameworks

As illustrated in Fig. 1, the existing body of work is organised into two interrelated layers, namely a Current Knowledge Representation Layer and a Current Analytical Layer Framework, both driven by the documented challenges in higher TVET OBE assessment. Stančin and Pale (2020) had found that ontologies are increasingly applied for curriculum modelling, describing learning domains, and structuring learning data, while Zouri, Al-Sartawi, and Al-Emran (2021) had demonstrated that ontology-based curriculum mapping supports effective knowledge representation and knowledge discovery for accreditation purposes, with ontology reuse through extension shown to be technically feasible for modular qualification frameworks of the kind required in TVET. Building upon the said body of work, the ONTO-ALIGN framework utilising a hierarchical 'Knowledge-Course-Graduation Requirements' structure was proposed (Namahoot et al., 2025), while Yang et al. (2023) had demonstrated that encoding curriculum knowledge as directed graphs of entities and relationships enables the systematic tracking of learning pathways through to graduation requirements achievement, providing a formal basis for automated gap analysis not achievable through conventional document-based curriculum management.

The automated extraction of curriculum knowledge from unstructured institutional documents had been identified as a critical prerequisite for scalable deployment of ontological approaches. Yang et al. (2023) had described the Education Entity Relationship Joint Extractor (EERJE), while Vinu and Kumar (2015), through the OBC-ONTO ontology, had demonstrated that formal outcome-based curriculum representations can support automated reasoning for consistency checking, including detection of cases where no assessment task provides evidence for a declared outcome support relationship. Wibowo (2024) had further introduced OBEOnto, a semantic knowledge graph framework designed specifically for OBE curriculum implementation, demonstrating that a general ontological model can be extended with institution-specific nodes to support data storage, retrieval, and analysis for programme accreditation purposes, and Abu-Salih et al. (2024) had further identified insufficient data for knowledge graph construction and restricted scalability across subject areas as key limitations directly relevant to multi-campus TVET environments. Taken together, the Knowledge Representation Layer depicted in Fig. 1 has established the representational infrastructure needed to transform the OBE outcome hierarchy from a documentation artifact into a reasoning-capable knowledge structure, yet it is yet to be determined whether such frameworks can be systematically adapted to higher TVET, where programme outcomes must additionally accommodate workplace assessment data, industry competency taxonomies, and sector-specific regulatory requirements.

The Current Analytical Layer Framework depicted in Fig. 1 had operationalised OBE outcome hierarchies through predictive modelling, competency-based assessment design, and visual analytics. Supported by the works of Neha and Kumar (2023), who had proposed a deep neural network-based framework that found a strong correlation between third-year and final-year performance, and Yağcı (2022) had demonstrated that educational data mining approaches achieve classification accuracy of 70–75% for final exam performance prediction even under limited data conditions, while Waheed et al. (2020) had achieved classification accuracy of 84–93% for at-risk student identification using clickstream data from virtual learning environments, confirming that legacy data and assessment-related data have significant impact on model accuracy. Among the findings most directly relevant to TVET, Lu, Zhang, and Chen (2025) achieved prediction accuracy of 95–98% and found that broader internet behaviour patterns contribute more significantly to performance prediction than platform-specific data, a finding of direct relevance to institutions where student engagement spans both campus and workplace environments. Visual analytics had been identified as essential for translating analytical outputs into institutional decision support, with Lumius and Asli (2021) proposing OBEInsights, extending the Malaysian iCGPA instrument through an interactive exploration paradigm; while Hernández-Campos et al. (2025) had found that the lack of a consistent theoretical framework and the restriction of most analyses to final grades pointed to significant gaps in analytical depth, and Ndukwe and Daniel (2020) had further established that teacher data literacy is a foundational competency that must be developed in parallel with any technical analytics infrastructure. Hence, there is yet an empirically validated framework that integrates the said Knowledge Representation and Analytical layers within the specific institutional and regulatory context of higher TVET, and the said gap is hoped to be addressed through the proposed study.

5 The Proposed Knowledge Representation and Analytical Framework (KRF-HTVET)

The proposed Knowledge Representation and Analytical Framework for OBE Assessment in Higher TVET Institutions (KRF-HTVET), as shown in Fig. 2, is structured across four integrated layers, each of which responds to a specific category of challenge identified in the preceding review of the literature. The architecture of the said framework is designed to address the fundamental limitation of existing OBE assessment systems in TVET, which is the absence of a formally encoded, machine-readable representation of the outcome hierarchy that can serve as the semantic foundation for automated reasoning, analytics, and decision support.

The Knowledge Representation Layer constitutes the foundational infrastructure of the KRF-HTVET, encoding the full OBE outcome hierarchy such as Programme Outcomes, Programme Specific Outcomes, and Course Outcomes, as a formal ontological structure using Web Ontology Language (OWL) and Resource Description Framework (RDF) technologies, following the architectural principles of Namahoot et al. (2025), the curriculum mapping ontology approach of Zouri et al. (2021), and the knowledge graph construction methodology reviewed by Abu-Salih et al. (2024). Entity types, including Graduation Requirement, Course, Knowledge Unit, Course Objective, and Major, are defined, together with relationship types, such as ‘supports’, ‘maps to’, ‘achieves’, ‘belongs to’, and ‘represents’, which collectively encode the logical dependencies between outcome levels and provide the formal basis for automated consistency checking and gap analysis. The EERJE multi-task extraction framework of Yang et al. (2023) is positioned within this layer as the mechanism for populating the ontological schema from unstructured institutional documents. In the HTVET context, this layer must additionally accommodate the representation of industry competency taxonomies, accreditation requirements from the Malaysian Qualifications Agency and sector-specific regulatory authorities, and workplace assessment evidence from on-the-job training.

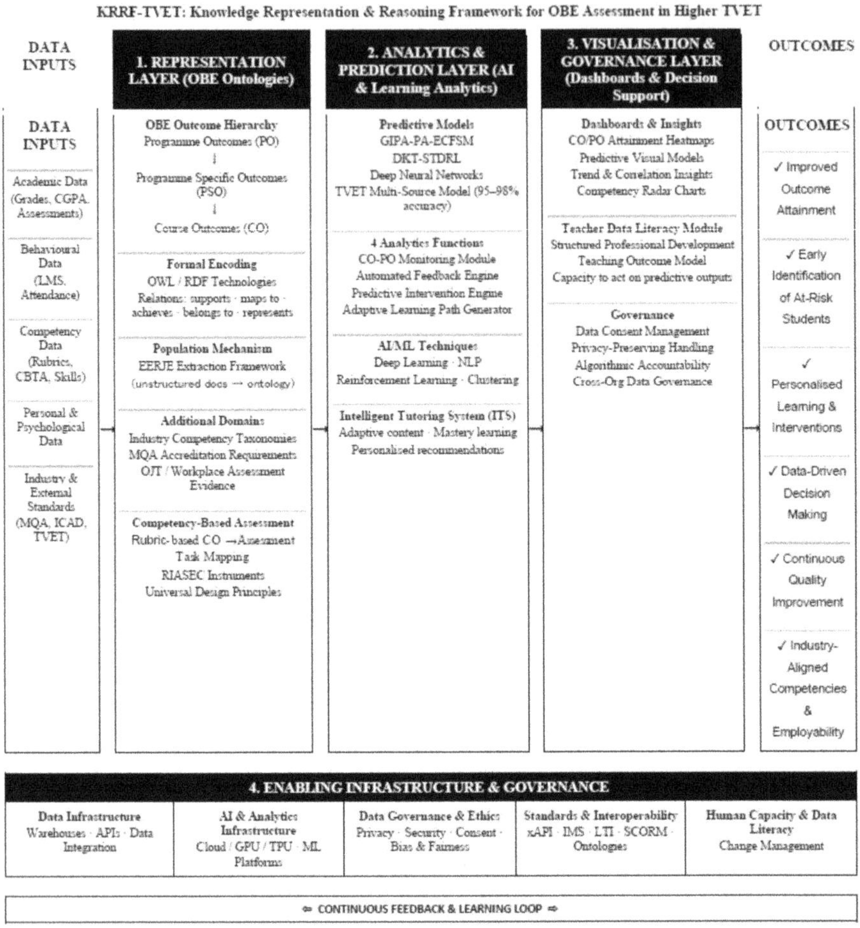

Fig. 2. The Knowledge Representation and Analytical Framework for OBE Assessment and Analytics in Higher TVET Institutions (KRF-HTVET)

The Competency-Based Assessment Layer operationalises the outcome hierarchy through structured assessment processes that generate valid, comparable evidence of competency attainment at both course and programme levels. Drawing upon the frameworks of Vargas et al. (2024) and Siang, Reyes, and Dumdumaya (2024), this layer employs a rubric-based approach in which each assessment task is formally mapped to specific course outcomes through the representational schema of the Knowledge Representation Layer, and further incorporates performance-based assessment instruments aligned with the RIASEC framework and the universal design principles recommended by Barongo et al. (2024), ensuring that competency documentation processes are both formally structured and accessible to diverse learner populations.

The Analytics and Prediction Layer transforms the semantically structured assessment data into actionable insights for instructors, programme coordinators, and institutional administrators, through predictive modelling, knowledge tracing, and outcome attainment monitoring. The enhanced correlated feature set model of Neha and Kumar (2023), the DKT-STDRL model of Lyu et al. (2023), and the deep neural network approach of Waheed et al. (2020) collectively provide the predictive architecture for early identification of at-risk students and forecasting of programme outcome attainment trajectories. The TVET-specific multi-source model of Lu et al. (2025) provides a particularly strong foundation for the operational predictive module of this layer. The four learning analytics functions identified by Caspari-Sadeghi (2023), which consisted of monitoring and analysis, automated feedback, prediction, and adaptation, are each instantiated as specific functional modules: a CO-PO Monitoring Module, an Automated Feedback Engine, a Predictive Intervention Engine, and an Adaptive Learning Path Generator informed by the ITS architecture of Syaddad et al. (2026).

The Visualisation and Governance Layer provides the institutional interface through which analytics outputs are translated into decision support and through which governance structures, ethical safeguards, and professional development pathways for sustainable analytics adoption are enacted. The visual analytics design principles of Lumius and Asli (2021) provide the foundation for the visualisation architecture. A Teacher Data Literacy Module, informed by the Teaching Outcome Model of Ndukwe and Daniel (2020) and grounded in the evidence that data-driven early identification of at-risk students, as demonstrated by Yağcı (2022) and Moreno-Marcos et al. (2020), requires instructor capacity to act upon predictive outputs, provides instructors with structured professional development support. In response to the findings of Alzahrani et al. (2023) regarding the primacy of ethical issues and resistance culture as barriers to learning analytics adoption, the Governance Layer incorporates data consent management, privacy-preserving data handling, transparent algorithmic accountability mechanisms, and cross-organisational data governance processes designed to support scalable, sustainable adoption across multi-campus higher TVET institutions.

6 Discussion: Research Gaps and Implications

The proposed KRF-HTVET framework, grounded in the integrative review of challenges and existing approaches, responds to a body of evidence that collectively established a critical structural gap in the OBE assessment and analytics landscape for higher TVET institutions. Research had found that the challenges most frequently cited by TVET practitioners, namely misalignment between assessments and outcomes, fragmented data, limited faculty capacity, and inadequate analytics systems, are precisely those that a formally structured knowledge representation framework, operating in conjunction with competency-based assessment, predictive analytics, and visualisation systems, is technically positioned to address (Mistamiruddin & Mohamad Nasri, 2024; Sin & Hussin, 2024; Gedrimiene et al., 2019; Alzahrani et al., 2023). The said observation points toward a significant implication for the field: progress in OBE implementation in higher TVET is less likely to be achieved through further elaboration of OBE theory than through the development and validation of practical representational tools that can translate existing theoretical frameworks into operational systems deployable within real institutional contexts.

Among the important implications of the proposed framework for HTVET administrators and policymakers is the recognition that the transition from descriptive, retrospective OBE reporting to prescriptive, predictive institutional intelligence requires not only investment in technical infrastructure but equally investment in the professional development of faculty and the governance structures that determine how analytics data is collected, interpreted, and acted upon. Hence, there is a need for HTVET institutions to develop integrated professional development programmes, data governance frameworks, and leadership engagement strategies in parallel with the technical implementation of the KRF-HTVET, recognising that the adoption of evidence-based predictive analytics, as Yağcı (2022) and Waheed et al. (2020) had demonstrated, depends critically on the institutional readiness to act upon analytical outputs, not merely to generate them. The findings of this paper are expected to help administrators, curriculum designers, instructors, and quality assurance officers in higher TVET institutions to understand the representational and analytical foundations of a principled OBE assessment system, and to identify the specific technical and governance investments required to transition from manual, document-based OBE management to an intelligent, data-driven approach.

7 Conclusion

This paper presented an integrative synthesis of knowledge representation issues confronting OBE assessment and analytics in higher TVET institutions and proposed the KRF-HTVET, comprising a four-layer integrated framework as a principled response. The said framework is grounded in a body of empirical evidence that had documented, across diverse national contexts, the persistent misalignment between OBE's theoretical aspirations and the operational reality of TVET assessment and data systems, manifesting concretely as curriculum-industry misalignment, assessment documentation overload, on-the-job training evidence gaps, and instructor readiness deficits, all of which find direct structural responses within the four layers of the proposed framework.

There is yet an empirically validated framework that integrates ontological knowledge representation with competency-based assessment, predictive analytics, and visualisation in a system explicitly designed for the institutional realities of higher TVET, and the said gap represents the primary contribution that the KRF-HTVET is proposed to address. Further study should be made to validate the framework through pilot implementation in higher TVET institutions, to assess the accuracy of automated extraction components across TVET-specific technical vocabularies, and to evaluate the institutional and pedagogical outcomes of analytics-informed OBE management relative to conventional approaches. The findings of such validation studies are expected to refine the architecture of the KRF-HTVET and to provide the evidence base needed for its adoption as a standard infrastructure for OBE quality assurance in higher TVET at the national and regional level. It is hoped that the proposed framework will contribute to the field of higher TVET quality assurance by providing a conceptually coherent and empirically grounded architecture that can guide both the technical development of OBE analytics systems and the institutional policy decisions that determine their deployment and use.

References

Abu-Salih, B., Al-Tawil, M., Aljarah, I., Faris, H.: A systematic literature review of knowledge graph construction and application in education. Heliyon. **10**(4), e26393 (2024). https://doi.org/10.1016/j.heliyon.2024.e26393

Ali, L.: The design of curriculum, assessment and evaluation in higher education with constructive alignment. Int. J. Soc. Sci. Educ. Stud. **4**(3), 72–78 (2018)

Alzahrani, A.S. et al.: Untangling connections between challenges in the adoption of learning analytics in higher education. Educ. Inf. Technol. **28**, 4563–4595 (2023). https://doi.org/10.1007/s10639-022-11323-x

Anderson, L.W., Krathwohl, D.R.: A Taxonomy for Learning, Teaching, and Assessing: a Revision of Bloom's Taxonomy of Educational Objectives. Longman (2001)

Barongo, A., Ogada, K., Njagi, D.: Electronic assessment framework for competency-based education amongst intellectual disability students in TVET institutions in Nairobi County. Int. J. Res. Innov. Appl. Sci. **9**(11), 138–154 (2024). https://doi.org/10.51584/IJRIAS.2024.911011

Biggs, J., Tang, C.: Teaching for Quality Learning at University, 3rd edn. Open University Press (2007)

Bloom, B.S.: Taxonomy of Educational Objectives: the Classification of Educational Goals. Handbook I: Cognitive domain. David McKay (1956)

Caspari-Sadeghi, S.: Learning assessment in the age of big data: learning analytics in higher education. Cogent Educ. **10**(1), 2162697 (2023). https://doi.org/10.1080/2331186X.2022.2162697

Chong, S.W., Reinders, H., Cheung, A.: A methodological review of systematic literature reviews in higher education: heterogeneity and homogeneity. Educ. Res. Rev. **35**, 100421 (2021). https://doi.org/10.1016/j.edurev.2021.100421

Gedrimiene, E., Silvola, A., Pursiainen, J., Rusanen, J., Muukkonen, H.: Learning Analytics in Education: Literature Review and Case Examples from Vocational Education. University of Oulu (2019)

Halik Bassah, N.S., Mohd Noor, N., Hassan, R., Zakaria, Z.: Employability skills needed for TVET graduates in Malaysia: perspective of industry expert. Online J. TVET Pract. **8**(2), 16–27 (2023)

Hashim, A.S., Ali, R., Ismail, K.: Higher TVET educational model as basis for global curriculum: the UniKL experience. J. Mod. Educ. Rev. **6**(7), 461–469 (2016). https://doi.org/10.15341/jmer(2155-7993)/07.06.2016/004

Hernández-Campos, M., González-Torres, A., García-Peñalvo, F.J.: Learning outcomes evaluation through learning analytics systems in higher education: a systematic literature review. SAGE Open. **15**(3), 1–17 (2025). https://doi.org/10.1177/21582440251347374

Kamaruzaman, M.F., Abd Majid, F., Yau, S.A., Haslee Sharil, W.N.E.: Challenges in TVET education in higher learning institutions. Int. J. Acad. Res. Bus. Soc. Sci. **14**(9) (2024). https://doi.org/10.6007/IJARBSS/v14-i9/22304

Khan, S., Ahmed, S., Ather, M., Khaskhely, M.: The competency-based training and assessment, and improvement of technical competencies and changes in pedagogical behavior. E M: Ekon. Manag. **25**(3), 177–196 (2022). https://doi.org/10.15240/tul/001/2022-3-011

Li, M., Zhao, X., He, J., Deng, Y.: The relationship between learning outcomes and graduate competences: the chain-mediating roles of project-based learning and assessment strategies. Sustainability. **16**(9), 3678 (2024). https://doi.org/10.3390/su16093678

Lu, X., Zhang, Y., Chen, J.: Multi-source behavioral data and deep learning model for student performance prediction in vocational education. J. Comput. Methods Sci. Eng. **25**(1), 112–130 (2025)

Lumius, L.D., Asli, M.F.: OBEInsights: visual analytics design for predictive OBE knowledge generation. Int. J. Adv. Comput. Sci. Appl. **12**(12), 632–674 (2021)

Lyu L, Wang Z, Yun H, Yang Z, Li Y: DKT-STDRL: spatial and temporal representation learning enhanced deep knowledge tracing for learning performance prediction. arXiv preprint (2023)

Mistamiruddin, E., Mohamad Nasri, N.: Challenges in integrating outcome-based education (OBE) in higher education institutions: a systematic literature review. Int. J. Acad. Res. Prog. Educ. Dev. **13**(4) (2024). https://doi.org/10.6007/IJARPED/v13-i4/23378

Moreno-Marcos, P.M., Alario-Hoyos, C., Muñoz-Merino, P.J., Delgado Kloos, C.: Analysis of the factors influencing learners' performance prediction with learning analytics. IEEE Access. **8**, 164331–164346 (2020). https://doi.org/10.1109/ACCESS.2020.3021476

Namahoot, C.S., Namahoot, K.S., Nuntawong, C., Sukma, N.: ONTO-ALIGN: an ontology-based framework for evaluating curriculum alignment with educational standards in computer science. Interdiscip. J. Inf. Knowl. Manag. **20** (2025). https://doi.org/10.28945/5600

Ndukwe, I.G., Daniel, B.K.: Teaching analytics, value and tools for teacher data literacy: a systematic and tripartite approach. Int. J. Educ. Technol. High. Educ. **17** (2020). https://doi.org/10.1186/s41239-020-00201-6

Neha, K., Kumar, R.: Enhancing graduate academic performance prediction and classification: an analysis using the enhanced correlated feature set model. Ingénierie des Systèmes d'Information. **28**(6), 1605–1612 (2023). https://doi.org/10.18280/isi.280617

Paul, J., Lim, W.M., O'Cass, A., Hao, A.W., Bresciani, S.: Frameworks for developing impactful systematic literature reviews and theory building: what, why and how? J. Decis. Syst. **32**(4) (2023). https://doi.org/10.1080/12460125.2022.2145948

Phuong, H.Y., Anh Thi, N., Trut Thuy, P., Thanh Thao, L., Thi Anh Thu, H., Huong Tra, N.: Challenges of implementing outcome-based assessment in Vietnamese higher education: a qualitative inquiry. J. Eff. Teach. Methods. **2**(1), 27–35 (2024). https://doi.org/10.59652/jetm.v2i1.123

Rao, N.J.: Outcome-based education: an outline. High. Educ. Future. **7**(1), 1–11 (2020). https://doi.org/10.1177/2347631119886418

Schröder, T.: A regional approach for the development of TVET systems in the light of the 4th industrial revolution. Int. J. Train. Res. **17**, 83–95 (2019)

Siang, S.E.R., Reyes, A.R.L., Dumdumaya, C.E.: Reinforcing learning management system with OBE-based learning analytics and RIASEC integration. TEM Journal. **13**(4), 2634–2646 (2024). https://doi.org/10.18421/TEM134-69

Sin, N.D.M., Hussin, M.Z.: Outcome base education implementation in technical and vocational education and training: a challenge and issue. **Int. J. Mod. Educ. 6**(22), 389–408 (2024). https://doi.org/10.35631/IJMOE.622028

Spady, W.G.: Outcome-Based Education: Critical Issues and Answers. American Association of School Administrators (1994)

Stančin, K., Pale, P.: Ontologies in education — state of the art. Educ. Inf. Technol. **25**(6), 5301–5320 (2020). https://doi.org/10.1007/s10639-020-10226-z

Sun, P.H., Koo, A.C., Lim, S.C.: The importance and challenges of outcome-based education: a case study in a private higher education institution. Int. J. Bus. Manag. **15**(5), 64–73 (2020)

Syaddad, H.N., Salim, A.A., Ishwara, L., Hasibuan, Z.A., Kurniawan, B., Supatmi, S.: A systematic literature review on intelligent tutoring systems for outcome-based education in higher education. Technol. J. **3**(1), 46–62 (2026). https://doi.org/10.62872/pd6g0a26

Torraco, R.J.: Writing integrative literature reviews: using the past and present to explore the future. Hum. Resour. Dev. Rev. **15**(4), 404–428 (2016). https://doi.org/10.1177/1534484316671606

Vargas, H., Heradio, R., Farias, G., Lei, Z., de la Torre, L.: A pragmatic framework for assessing learning outcomes in competency-based courses. IEEE Trans. Educ. **67**(2), 224–233 (2024). https://doi.org/10.1109/TE.2024.3352034

Vinu, E.V., Kumar, P.S.: Improving large-scale assessment tests by ontology based approach. In: Proceedings of the Twenty-Eighth International Florida Artificial Intelligence Research Society Conference, pp. 457–462 (2015)

Waheed, H., Hassan, S.U., Aljohani, N.R., Hardman, J., Alelyani, S., Nawaz, R.: Predicting academic performance of students from VLE big data using deep learning models. Comput. Hum. Behav. **104**, 106189 (2020). https://doi.org/10.1016/j.chb.2019.106189

Whittemore, R., Knafl, K.: The integrative review: updated methodology. J. Adv. Nurs. **52**(5), 546–553 (2005). https://doi.org/10.1111/j.1365-2648.2005.03621.x

Wibowo, A.: A semantic knowledge graph approach to support outcome-based education system. In: Proceedings of the 2024 International Seminar on Intelligent Technology and its Applications (ISITIA), pp. 245–252 (2024). https://doi.org/10.1109/ISITIA63062.2024.10667427

Yang, Y., Chen, S., Zhu, Y., Zhu, H., Chen, Z.: Knowledge graph empowerment from knowledge learning to graduation requirements achievement. PLoS One. **18**(10), e0292903 (2023). https://doi.org/10.1371/journal.pone.0292903

Yağcı, M.: Educational data mining: prediction of students' academic performance using machine learning algorithms. Smart Learn. Environ. **9**, 11 (2022). https://doi.org/10.1186/s40561-022-00192-z

Yusop, S.R.M., Rasul, M.S., Yasin, R.M.: Challenges, strengths, and relevance of integrating classroom-based assessment in technical and vocational education training. Int. J. Eval. Res. Educ. **13**(4), 2440–2449 (2024). https://doi.org/10.11591/ijere.v13i4.27900

Yusop, S.R.M., Rasul, M.S., Mohamad Yasin, R., Hashim, H.U., Jalaludin, N.A.: An assessment approaches and learning outcomes in technical and vocational education: a systematic review using PRISMA. Sustainability. **14**(9), 5225 (2022). https://doi.org/10.3390/su14095225

Zouri, M., Al-Sartawi, A., Al-Emran, M.: An ontology-based approach for curriculum mapping in higher education. In: Proceedings of the 2021 IEEE 11th Annual Computing and Communication Workshop and Conference (CCWC), pp. 0826–0831 (2021). https://doi.org/10.1109/CCWC51732.2021.9376077

Predicting MOOC Learner Types from a Single Session: A Lightweight Behavioral Classification Approach

Jamallah Mohammed Hussein Zawia(✉) and Maizatul Akmar Binti Ismail

Faculty of Computer Science and Information Technology, Universiti Malaya, 50603 Kuala Lumpur, Malaysia
jamala.zawia@um.edu.my

Abstract. Massive Open Online Courses (MOOCs) attract heterogeneous learners whose engagement patterns vary widely, yet most behavioral archetype frameworks rely on multi-session aggregation. This dependency creates a persistent cold-start problem: anonymous learners cannot be profiled until enough longitudinal data has accumulated. This paper addresses this gap with two contributions. First, the study synthesizes 18 empirical MOOC engagement studies covering 381,705 learners across 13 platforms, deriving a unified 12-archetype taxonomy, a 25-feature catalog grouped in four categories, and the Session Definition Consistency Index (SDCI = 38.9%) that quantifies the field's lack of session-reporting standardization. Second, the study specifies a three-stage pipeline for single-session learner profiling: a 30-min-gap sessionization rule justified by the literature evidence, a zero-shot DistilBERT NLI classifier over six behaviorally-grounded candidate labels, and a first-order MAML next-item recommender proposed for follow-up validation. The first two stages are evaluated on 472 sessions from a single corpus. The paper reports session statistics, the empirical label distribution, and a per-label diagnostic of the observed skew (Social/Interactive 56.1%; two of the six candidate labels with no top-1 assignments). The meta-learning stage is execution-tested only, and the reason comparative validation is deferred is set out explicitly. The closing sections list threats to validity and a follow-up agenda whose top priorities are standardized session reporting and a full benchmark study.

Keywords: MOOC · Learner archetypes · Session-based classification · Behavioral profiling · Meta-learning · Cold-start recommendation · Learning analytics

1 Introduction

Massive Open Online Courses (MOOCs) have, since their emergence in the early 2010s, evolved from large-cohort lecture replays into instrumented learning environments that capture fine-grained behavioral traces of every interaction. A substantial body of research

---

Supplementary Information The online version contains supplementary material available at https://doi.org/10.1007/978-981-92-4305-1_15.

N. Mohd Sharef et al. (Eds.): CAMP 2026, CCIS 3094, pp. 198–212, 2027.
https://doi.org/10.1007/978-981-92-4305-1_15

has used these traces to identify recurrent behavioral archetypes of MOOC learners, labels such as Auditors, Completers, Samplers and Disengagers, to support personalization, early at-risk identification, and completion-rate improvement [1–3]. The promise of these taxonomies has, however, been blunted by three structural problems visible only when the literature is examined as a whole.

First, the field lacks a shared vocabulary: closely related archetypes are reported under different names, and the same name (for example, Auditor) is operationalized differently across papers. Second, the unit of analysis varies wildly: some studies treat a whole course as the session unit, others a calendar week, others a continuous interval bounded by an inactivity threshold ranging from 3.9 to 60 min. Third, the dominant data-collection paradigm is retrospective and multi-session: archetypes are assigned after weeks of accumulated activity, so at-risk learners are identified only after they have already disengaged.

Together these problems block what should be the natural use case of an archetype framework, providing immediate, behaviourally-informed support to a newly arrived, anonymous learner. The contribution of this paper is structured around four research objectives (O1-O4) detailed below. The literature study is the empirical foundation that motivates and constrains the methodology. The pipeline is the main methodological contribution: its first two stages are specified, implemented, and evaluated on a 472-session pilot, and the third stage is specified and execution-tested only, with comparative validation deferred to Sect. 8.

1.1 Problem Statement

The challenge addressed in this paper can be stated precisely. Given an anonymous MOOC learner whose interaction history within a course consists of only the current, in-progress session, the goal is to (i) classify that learner into a behavioral archetype with enough confidence to act on it, (ii) do so using only features that are observable within the first few minutes of the session, and (iii) couple the classification to a personalized recommendation that does not require longitudinal user history. Solving this is blocked, in current MOOC research, by three concrete obstacles.

P1. Taxonomic fragmentation. Across the empirical literature surveyed (Sect. 3), 18 empirical studies report 70 distinct archetype instances under inconsistent labels.
P2. Session-definition inconsistency. Only 38.9% of empirical studies in the corpus provide an explicit definition of what they mean by a learning session, and reported inactivity thresholds vary by a factor of 15.4.
P3. Multi-session aggregation dependency. The dominant data-collection paradigm is retrospective; the single-session, cold-start setting receives less than 5% of the field's reporting effort, despite being the setting most commercial MOOC platforms need to serve every day.

1.2 Research Objectives

The paper pursues four objectives that map directly to the three problems above.

O1. Develop a unified archetype taxonomy. The 70 reported archetype instances are consolidated into a smaller set of core types defined by their behavioral signatures, not their original labels. (Addresses P1.)
O2. Identify discriminative behavioral features and minimal feature sets. The behavioral features used across the corpus are cataloged and assessed for reporting frequency and apparent discriminative power. (Addresses P1, P3.)
O3. Quantify session-definition consistency across the corpus. The Session Definition Consistency Index (SDCI) is introduced as a methodological-quality metric, computed on the included studies, and used to motivate a defensible default threshold. (Addresses P2.)
O4. Evaluate a single-session classification and meta-learned recommendation pipeline. A three-stage pipeline (sessionization, zero-shot session classification, first-order MAML next-item recommendation) is built and tested on a held-out interaction corpus; the study reports what works, what does not, and what scale is needed before deployment. (Addresses P3.)

1.3 Research Questions

The four objectives translate into four research questions answered in Sects. 3 and 5–7.

RQ1. What behavioral archetypes have been identified in MOOC research, and how can they be consolidated into a unified taxonomy with documented mapping rules?
RQ2. Which behavioral features are most commonly reported across the literature, and what is the minimum feature set required to discriminate among the core archetypes within a single session?
RQ3. To what extent do MOOC studies consistently define and operationalize the "session" as a unit of analysis, and what is the empirical distribution of session thresholds in the field?
RQ4. Can a single session's behavioral trace, classified at the session level, support meaningful personalization in the cold-start setting through a zero-shot classifier and a meta-learned recommender?

The remainder of the paper is organized as follows. Section 2 situates the work against prior taxonomies and session-modeling studies. Section 3 summarizes the literature study that provides the empirical foundation. Section 4 specifies the proposed methodology in detail. Section 5 describes the experimental setup. Section 6 reports results. Section 7 discusses methodological and practical implications and explicit threats to validity. Section 8 concludes.

2 Related Work

Early MOOC archetype research established the engagement-versus-completion plane that still organizes most taxonomies. Kizilcec et al. [1] proposed a four-type schema (Completing, Auditing, Disengaging, Sampling) from three Coursera courses, and Anderson et al. [2] independently arrived at a similar taxonomy on edX. Subsequent work has tended to reproduce these schemata on new datasets or expand them with platform-specific refinements [3–5].

A second strand of work has focused on clustering methodology. K-means remains the most common technique [6–8], but agglomerative hierarchical clustering with Ward's method [9, 10], latent class analysis [11], hidden Markov models for engagement-state transitions [12], and entropy-based feature engineering [13] all appear in the literature.

A third strand investigates session-level behavior. De Barba et al. [14] argued that the session is the natural unit at which self-regulation manifests; Maldonado-Mahauad et al. [15] applied process mining on Coursera with a 45-min threshold; Davis et al. [16] used a 60-min threshold on edX; Liao et al. [17] examined video-watching sequences at scale on a Chinese MOOC platform. Session-level metrics are reported far less frequently than course-level metrics, and the choice of inactivity threshold is rarely justified.

Several systematic reviews have surveyed adjacent literatures: Zhu et al. [18] (166 articles on MOOC learning analytics), Huang et al. [19] (PRISMA review of engagement-and-dropout factors, 21 studies), Shen et al. [20] (post-pandemic engagement, 52 articles), and Alanazi et al. [21] (70 deep-learning studies on student performance prediction). To the authors' knowledge no prior review has targeted the archetype-and-session-definition intersection, nor proposed a metric for session-definition consistency

3 Background: Synthesis of Prior MOOC Engagement Studies

This section condenses the literature study that motivates and constrains the methodology in Sect. 4. The full review protocol, extracted data, and per-study coding are released as supplementary CSV files (S1_study_metadata.csv, S2_study_population_breakdown.csv, S3_archetype_characteristics.csv, S4_comparison_matrix.csv).

3.1 Corpus Characteristics

The 18 empirical studies span 2014–2025 and thirteen platforms (Coursera, edX, FutureLearn, openSAP, China University MOOCs, OULA, XueTangX, iMooX, OpenEdu, OWL, Blackboard, Moodle, and a Custom EPFL platform), reporting 381,705 learners (Fig. 1). Coursera accounts for 39.9%, openSAP 19.0%, edX 12.3%, FutureLearn 11.5%, OULA 8.5%. Sample sizes range from 127 to 132,690 (median 7,527). Method choice (Fig. 2) splits into three families: k-means and variants [6–8, 22] (5 studies); hierarchical/Ward clustering (3 studies); and probabilistic or sequence-aware methods including LCA [11], HMM [12], lag-sequential analysis [25], N-gram modeling [23, 24], fsQCA [26], and entropy-based methods [13] (10 studies combined).

3.2 Findings Driving the Methodology

Three literature study findings directly shape the methodology in Sect. 4.

Finding 1 The field is non-comparable across studies on session-level metrics. The Session Definition Consistency Index, the proportion of empirical studies that explicitly define a session, evaluates to SDCI = 7 / 18 = 38.9% (Fig. 3). Among the five studies

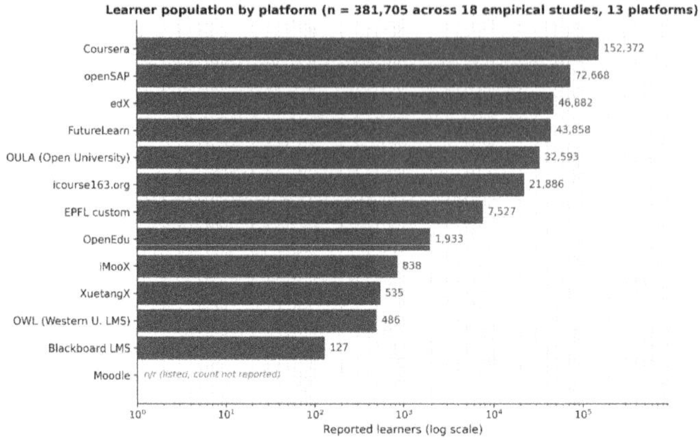

Fig. 1. Reported learner totals by platform across the 18 empirical studies (log scale)

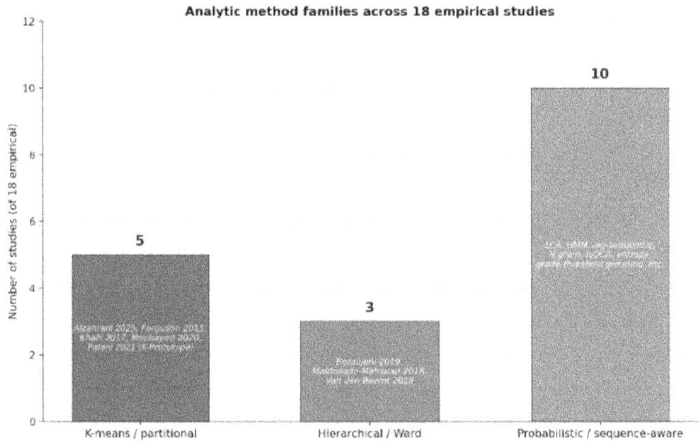

Fig. 2. Distribution of clustering and classification method families across the 18 empirical studies

that committed to an inactivity threshold, values range from 3.9 min (Maslennikova et al. 2023, via inter-activity log-curve fitting) to 60 min (Davis et al. 2017; Li 2022; Wen & Rose 2014), a 15.4x variation; median and mode are both 60 min. Section 4.2 inherits the 30-min default justified by these findings [16, 23, 27, 29].

Finding 2 Session-level features are systematically under-reported. Of 25 catalogued behavioral features across four categories (Activity Composition, Engagement Depth, Temporal Patterns, Behavioral Patterns), the most consistently reported are forum-post counts (41%) and certificate attainment (39%); session count, length, and frequency appear in only 4.3% of archetype descriptions despite being the field's stated focus (Fig. 4). This asymmetric evidence base means session-level features must be measured at pipeline runtime rather than imported from prior labels.

Fig. 3. Session-definition reporting practices across the 18 empirical studies.

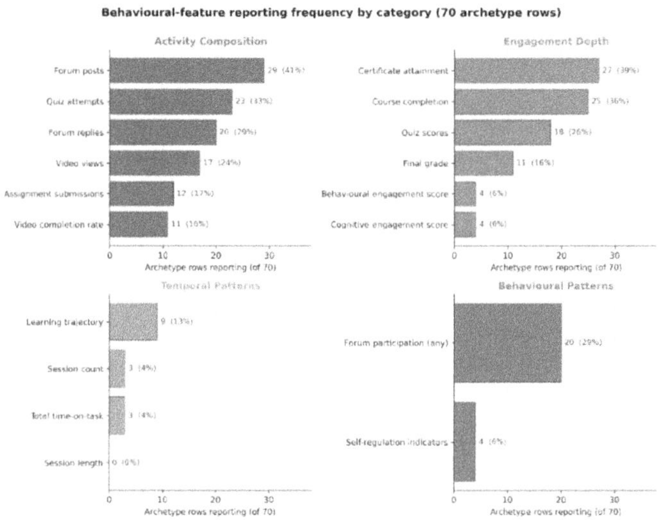

Fig. 4. Reporting frequency of behavioral features across the 70 archetype descriptions, grouped by category.

Finding 3 A 12-archetype unified taxonomy emerges from inductive consolidation. Mapping the 70 originally-reported archetype instances by semantic similarity and behavioral metric yields a 12-archetype core taxonomy (Fig. 5). Disengager is the most frequent (9 instances), followed by Medium-Achiever and Passive-Auditor (4 each), Sampler (3), Active-Completer/High-Engagement/Low-Engagement (2 each), and Assessment-Oriented, Video-Focused, Social-Engaged, Strategic-Learner, Irregular-Participant (1 each). Thirty-one of the 70 instances mapped unambiguously. The taxonomy is a proposed unification rather than a validated classifier and grounds the candidate-label set of the Sect. 4 classifier.

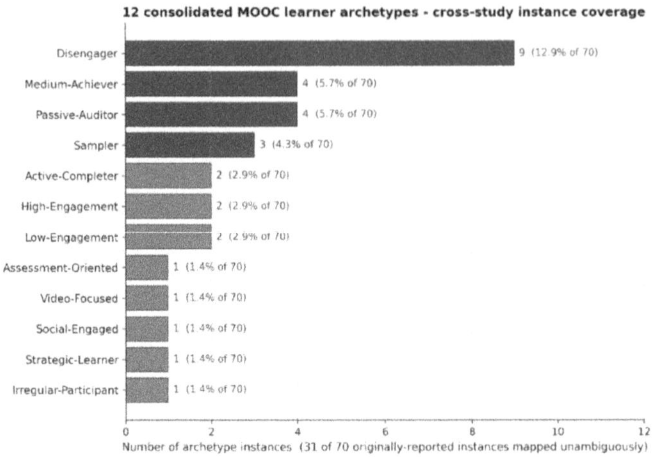

Fig. 5. Frequency of the 12 consolidated core archetypes across the included studies.

4 Methodology

This section specifies the three-stage pipeline that operationalizes Objective O4. The pipeline takes as input a sequence of raw interaction events from a single learner session and produces as output (i) a behavioral-archetype label for the session and (ii) a ranked list of next-item recommendations that adapt to the predicted label. Each design choice is grounded in the literature study findings of Sect. 3.

4.1 Framework Overview

The pipeline has three stages. Stage 1 segments raw clickstream events into discrete sessions using an inactivity-gap criterion. Stage 2 transforms each session into a natural-language description over engagement signals and classifies it with a zero-shot NLI model against behaviorally-grounded candidate labels. Stage 3 treats the predicted label as a task identifier and adapts a base next-item recommender using first-order Model-Agnostic Meta-Learning (MAML). Each component uses an off-the-shelf model architecture identified by the literature study as state-of-practice; the contribution lies in the composition and alignment with the literature study's evidence base.

4.2 Sessionization Strategy

Stage 1 segments raw events by an inactivity-gap criterion. The threshold is set to 30 min, the median of the literature study-reported thresholds (Sect. 3.2, Finding 1). A 30-min threshold is selected over the higher per-study median of 60 min because it lies between the empirical-fit threshold of Maslennikova et al. (3.9 min) and the 60-min literature defaults of Davis et al., Li, and Wen & Rose, aligns with engagement-decay observations [16], and is the value used in the educational-data-mining sensitivity literature [28]. The unit of segmentation is activity-based, mirroring the literature study's most common operationalization.

For each detected session, the pipeline computes a session-summary record consisting of (i) start and end timestamps, (ii) total event count, (iii) total duration in minutes, and (iv) an ordered list of accessed items with their associated engagement signals (rating, watch percentage, dwell time where available).

4.3 Zero-Shot Session Classifier

Stage 2 classifies each session into one of six behaviourally-grounded candidate labels using a zero-shot NLI model. The design has three components.

Candidate-Label Design The label set is Learning-Focused, Exploration, Casual Browsing, Social/Interactive, Problem-Solving Oriented, and Administrative/Logistical. Labels must be expressible in natural language (so zero-shot NLI can score them) and mappable to one or more of the 12 literature study archetypes (Finding 3). For example, Casual Browsing maps to Sampler and Passive-Auditor; Problem-Solving Oriented to Assessment-Oriented and Strategic-Learner; Social/Interactive to Social-Engaged.

Session-to-Text Encoder Each session is transformed into a short natural-language description that aggregates its engagement signals: the mean watch percentage across video events, the mean rating across rated items, the total event count, the session duration, and (where present) the modal action type. The resulting text typically fits in 80–150 tokens and is bounded at 256 tokens to control inference cost.

NLI Model Selection The pipeline uses typeform/distilbert-base-uncased-mnli as the zero-shot classifier. DistilBERT-MNLI was chosen over heavier alternatives (BART-large-mnli, DeBERTa-MNLI) because the pilot is bounded by GPU-poor inference, because zero-shot NLI in this size class is already a strong off-the-shelf baseline, and because the comparison across NLI backbones is identified in Sect. 8 as an explicit future-work item.

For each session, the classifier produces a top-1 label with a confidence score and a full distribution over the six candidates; the full distribution is retained for downstream uncertainty-aware processing.

4.4 Stage 3 (Proposed): Meta-Learning Recommender — Deferred Validation

Stage 3 specifies an adaptive next-item recommender that uses the Stage 2 label as a task identifier. Validation is deferred; the reasons are set out in the scope paragraph below. The component is built on **first-order Model-Agnostic Meta-Learning (MAML)** [30]. The intuition is that each archetype-aligned label defines a *task* with its own distribution of next-item preferences, so a few-shot adaptation step can stand in for the longitudinal user history that cold-start learners do not have. The base predictor is a compact GRU (embedding 64, hidden 128, about 384 K parameters) over a 2,000-item vocabulary. The first-order MAML loop groups sessions into tasks by their predicted label, with a minimum of eight sessions per task. Each outer step performs one inner adaptation step at learning rate 0.01 and updates meta-parameters at 3×10^{-4}, with four tasks per batch.

Scope of the Present Study The pilot dataset (Sect. 5.1) yields only three label-eligible tasks under the eight-session threshold, which is too few to support a benchmark claim. Section 6.3 therefore reports execution-trace evidence only. A full comparison against non-meta-learned baselines (popularity, recency, GRU4Rec, SASRec) on a multi-course corpus large enough to give at least six label-eligible tasks is the primary deferred experiment in Sect. 8.

4.5 Implementation and Configuration

The pipeline is implemented in Python with PyTorch for the GRU and Hugging Face Transformers for the NLI model. Sessionization and feature extraction use pandas and NumPy. The end-to-end pipeline is reproducible from raw clickstream data; the full notebook is supplied as supplementary material. Random seeds are fixed at 42 for the MAML loop. Inference for the NLI classifier uses fp16 on GPU where available and falls back to CPU otherwise. The total trainable parameter count of the GRU is approximately 384 K, which fits comfortably within the memory envelope of a Colab T4 instance.

5 Experiments

This section describes the empirical pilot used to evaluate the pipeline. The pilot is framed as a **feasibility study**: it demonstrates that each pipeline component executes end-to-end on real data and produces interpretable outputs, but it does not include comparative baselines at sufficient scale to support a benchmark claim. Limitations of the experimental scope are reported in Sect. 7.3.

5.1 Dataset

The dataset is an English-language MOOC interaction log derived from a publicly indexed source. Each event records (user_id, item_id, created_at, rating, watch_percentage). After cleaning, the dataset contains a single course span across multiple weeks, and produces **472 sessions** after applying the 30-min sessionization rule of Sect. 4.2. The dataset's engagement-signal richness is intentionally limited (no explicit forum events) to test the pipeline's behavior under sparse-signal conditions of the kind that newer or smaller MOOC platforms typically have.

5.2 Experimental Setup

The pipeline runs end-to-end in a single Colab session with a GPU runtime. Stage 1 is deterministic given the 30-min threshold. Stage 2 uses the pre-trained typeform/distilbert-base-uncased-mnli checkpoint with no task-specific fine-tuning. Stage 3 is run as an architectural feasibility check. The MAML loop is initialized from random weights and trained over the labels that pass the eight-session threshold. No comparative benchmark is run and no hyperparameter search is performed. At pilot scale the goal is to confirm that the loop completes and that adaptation reaches a usable state. Recommendation quality is not estimated here.

5.3 Evaluation Approach

Evaluation is observational. The pilot has no comparative baseline (it is single-dataset) and no ground-truth labels (the corpus has no hand-coded archetype assignments). For Stage 1, session counts are reported alongside the distribution of session duration and event count. Stage 2 results give the full top-1 label distribution and a label-by-label diagnostic of the observed skew. Stage 3 is reported only through execution traces: whether the loop completes, how many labels become task-eligible, and whether adaptation reaches a stable state. Comparative recall@k is not reported; with only a handful of eligible tasks, any delta would reflect episode sampling rather than method-level transfer. Section 8 specifies the baselines, the ground-truth labeling protocol, and the multi-corpus replication that would be needed for full validation.

6 Results

6.1 Session Statistics

Applying the 30-min sessionization rule to the dataset produces **472 sessions** from interactions of active users (defined as users with three or more sessions). Sessions vary widely in duration and density: short sessions of one or two events sit alongside long, multi-item sessions. The fact that even on a single dataset session counts and durations span such a wide range underlines Finding 1 of the literature study, that a fixed threshold across studies, while necessary for cross-study comparability, will mask substantial within-platform heterogeneity.

6.2 Zero-Shot Classification Results

The DistilBERT-MNLI classifier produces the label distribution in Figure 6: Social/Interactive 56.1% (265 sessions), Exploration 32.8% (155), Problem-Solving Oriented 10.0% (47), Casual Browsing 1.1% (5). Learning-Focused and Administrative/Logistical received no top-1 assignments under this dataset and label set. Per-label inspection shows Social/Interactive covers short and long sessions with a long upper tail, Exploration concentrates at short durations and low event counts, and Problem-Solving exhibits higher per-session activity. Detailed numerical statistics are in the supplementary notebook output.

The skew and the two empty labels are not random outputs of a noisy classifier. They follow systematically from how the candidate-label phrasing interacts with the engagement signals that the session-to-text encoder makes visible. *Social/Interactive* (56.1%) dominates because its hypothesis phrase is broad: in DistilBERT-MNLI it scores well on any session description that mentions multiple items, ratings, or repeated engagement. With no dedicated forum, comment, or peer-message tokens in the input, this label is where sessions go when there is no more specific evidence to put them anywhere else. *Exploration* (32.8%) is selected on short-to-medium sessions with a small number of distinct items, the typical pattern for early-course access. *Problem-Solving Oriented* (10.0%) is selected where event count and mean rating are higher, but in the

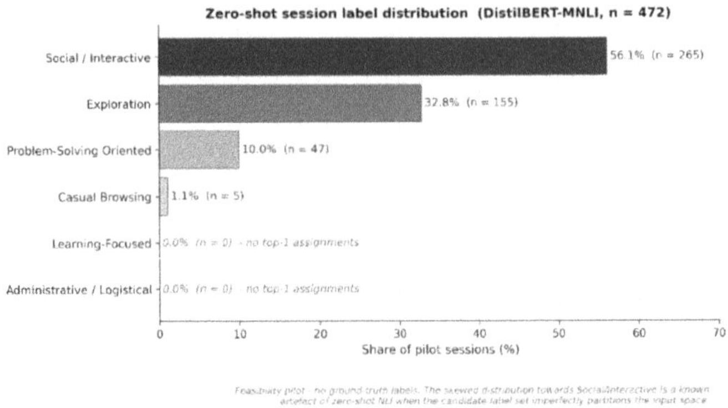

Fig. 6. Zero-shot session label distribution from the pilot (DistilBERT-MNLI; 472 sessions)

absence of explicit quiz or assessment events it is effectively measuring activity intensity rather than problem-solving. *Casual Browsing* (1.1%) is suppressed upstream by the active-user filter (users with at least three sessions). The two empty labels diagnose a label-set/signal-set mismatch: *Learning-Focused* (0%) requires study, mastery, or comprehension cues that the rating/watch-percentage/event-count encoder does not surface, so its sessions are routed to Social/Interactive; *Administrative/Logistical* (0%) requires navigation, enrollment, or account-management events, which the dataset's event schema does not contain.

Two design implications follow. The candidate-label set should be treated as a hyperparameter rather than a fixed taxonomy; a deployment-grade use of the pipeline would need a prompt-sensitivity study that jointly tunes alternative label phrasings and the NLI scoring temperature against hand-coded validation data. The two empty labels point to a gap in the input signal rather than to a real absence of those archetypes in the population. Adding forum, quiz, navigation, and certificate-attainment tokens to the session-to-text encoder, where they are available, is the simplest way to redistribute mass away from the Social/Interactive sink. Both remediations are scoped in Sect. 8.

6.3 Stage 3 (Proposed): Execution Behavior, Not Validation

The MAML loop completes end-to-end. Three of the six candidate labels (Social/Interactive, Exploration, Problem-Solving Oriented) pass the eight-session threshold; the meta-loss plateaus and inner-loop adaptation produces a ranked top-k list for a held-out anonymous session. This shows that the architecture runs and that adaptation is reachable. It does not show that the recommendations are better than those produced by simpler baselines. With only three label-eligible tasks, any reported recall@k delta would reflect episode sampling more than the method itself, so no comparative metric is reported here. Section 8 specifies what a full validation would need: popularity, recency, GRU4Rec, and SASRec baselines, HR@10 and MRR as metrics, and a corpus large enough to give at least six label-eligible tasks.

6.4 Cross-Reference with Literature Archetypes

The four labels with non-zero assignments map back to the unified taxonomy as follows: Social/Interactive → Social-Engaged; Exploration → Sampler and Passive-Auditor; Problem-Solving Oriented → Assessment-Oriented and Strategic-Learner; Casual Browsing → Low-Engagement and Disengager. As noted in Sect. 6.2, the two empty labels point to a label-set/signal-set mismatch rather than to the absence of those archetypes: the dataset does include sessions whose natural mapping would be Active-Completer or Medium-Achiever.

7 Discussion

7.1 Methodological Implications

The most actionable finding is the SDCI of 38.9%: fewer than two in five studies let a downstream researcher reproduce their session segmentation, and a 15.4x variance in inactivity thresholds quietly drives variation in nearly every session-level statistic. Without consensus on the unit of analysis, no formal meta-analysis is possible. This study proposes three remediations. First, journals and conferences in learning analytics should require explicit session definitions in methods reporting. Second, studies should report at least two sensitivity-analysis thresholds when session-level metrics are central. Third, a 30 min threshold should be the default where existing work permits: it is the field median, aligns with engagement-decay observations [16], and is the value adopted in this pipeline.

Related to the SDCI is the asymmetry of feature reporting (Sect. 3.3, Finding 2). Features that are easy to measure at course level (certificate attainment, final grade) are over-represented; features that require session-level reasoning (action pace, time-to-first-quiz, inter-action interval variance) are under-represented. The 25-feature framework contributed as part of the background deliberately includes a balanced set across the four categories so that future studies can be benchmarked uniformly.

7.2 Practical Implications for MOOC Platforms

For platforms that need to profile cold-start learners on first arrival, the zero-shot classification stage is the most immediately usable part of the pipeline. It is cheap, requires no hand-coded labels, and does not depend on a longitudinal user history; the absence of such history is the standard blocker for archetype-based personalization of first-time visitors. The proposed meta-learning stage would adapt recommendations to the predicted label, but it should not be deployed before the validation work in Sect. 8 is completed.

7.3 Limitations

Five limitations apply. (L1) Pilot scope: the empirical evidence comes from a single-corpus pilot of 472 sessions, without a comparative baseline and without hand-coded ground-truth labels. What the paper supports is feasibility evidence rather than validation. (L2) Classifier label-space: the zero-shot distribution depends on how the candidate-label phrasing interacts with the signals that the session-to-text encoder exposes (Sect. 6.2). The label set should be treated as a hyperparameter; a prompt-sensitivity study and a richer signal set would be required before any deployment claim. (L3) Meta-learning stage: Stage 3 is specified and execution-tested only. Comparative recommendation quality cannot be assessed at pilot scale because only three of the six labels reach the eight-session task-eligibility threshold; the validation protocol is in Sect. 8. (L4) Literature study methodology: single-reviewer screening with AI assistance, selection bias toward open-access studies (four eligible studies were excluded for inaccessibility), and unweighted synthesis (where a 132,690-learner study contributes equally to archetype frequency as a 127-learner study) all bound the generalizability of the SLR-derived findings. (L5) Archetype signatures: the signatures proposed for the 12 unified archetypes are design priors rather than validated classifiers, and deployment would require dataset-specific calibration.

8 Conclusion and Future Work

This paper presents a methodological contribution for single-session MOOC learner classification, grounded in a synthesis of 18 empirical MOOC engagement studies covering 381,705 learners across 13 platforms. The synthesis yields three findings that constrain methodology: the SDCI of 38.9%, the 4.3% session-level data scarcity, and the 12-archetype unified taxonomy. These findings are operationalised in a three-stage pipeline (sessionization, zero-shot classification, and a meta-learned recommender stage proposed for follow-up validation). The first two stages are evaluated on a single-corpus, 472-session pilot, with a label-by-label diagnostic of the observed zero-shot skew (Sect. 6.2). The third stage is execution-tested only; comparative benchmarking is deferred.

Five follow-up studies are scoped. (1) Cross-platform validation on at least three additional MOOC corpora (Coursera Open, edX Insights, Canvas Network) to check whether the empirical label distribution carries over. (2) Baseline benchmark at scale: the MAML recommender against GRU4Rec, SASRec, popularity, and recency baselines on held-out user splits, with HR@10 and MRR. (3) Label refinement with ground truth: redesign the candidate-label set with one-to-one correspondence to the 12 unified archetypes and validate the zero-shot output against hand-coded labels from two independent raters (target Cohen's $\kappa \geq 0.70$). (4) Intervention study: a pre-registered A/B trial in which profiled learners receive label-aware nudges in their second session, targeting a 3–5 point lift in week-1 retention. (5) Fairness and deployment audit: label-assignment disparities across demographics, device, and time zone, with a sub-200 ms 95th-percentile latency target. Together these studies are intended to take single-session profiling beyond the present feasibility evidence toward a deployment-ready cold-start solution. A secondary aim is to nudge the field toward a shared session-reporting standard, with a target SDCI of at least 80%.

Acknowledgements. The authors thank the open-data initiatives behind the platforms whose data made the included studies possible. The first author acknowledges supervisory support from the Faculty of Computer Science and Information Technology, Universiti Malaya. This research is supported by the Fundamental Research Grant Scheme (FRGS), grant number FP059-2024, funded by the Ministry of Higher Education, Malaysia.

Data Availability. All extracted study metadata, archetype characteristics, the comparison table, and the population breakdown are released as supplementary CSV files. The pilot pipeline source code is available on request from the corresponding author.

References

1. Kizilcec, R.F., Piech, C., Schneider, E.: Deconstructing disengagement: Analyzing learner subpopulations in massive open online courses. In: Proc. 3rd Int. Conf. Learning Analytics and Knowledge (LAK '13), pp. 170–179 (2013)
2. Anderson, A., Huttenlocher, D., Kleinberg, J., Leskovec, J.: Engaging with massive online courses. In: Proc. 23rd Int. Conf. World Wide Web (WWW '14), pp. 687–698 (2014)
3. Reich, J., Ruipérez-Valiente, J.A.: The MOOC pivot. Science. **363**(6423), 130–131 (2019)
4. Breslow, L., Pritchard, D.E., DeBoer, J., Stump, G.S., Ho, A.D., Seaton, D.: Studying learning in the worldwide classroom: research into edX's first MOOC. Res. Pract. Assess. **8**, 13–25 (2013)
5. Onah, D.F.O., Sinclair, J., Boyatt, R.: Dropout rates of massive open online courses: Behavioral patterns. In: Proc. EDULEARN14, pp. 5825–5834 (2014)
6. Khalil, M., Ebner, M.: Clustering patterns of engagement in massive open online courses (MOOCs): the use of learning analytics to reveal student categories. J. Comput. High. Educ. **29**(1), 114–132 (2017)
7. Moubayed, A., Injadat, M., Nassif, A.B., Lutfiyya, H., Shami, A.: E-learning: challenges and research opportunities using machine learning and data analytics. Distance Educ. **41**(2), 195–224 (2020)
8. Alzahrani, N., Almutairi, S., Alkahtani, L.: Identifying weekly student engagement patterns in e-learning via k-means clustering and label-based validation. Electronics. **14**(15), 3018 (2025)
9. Boroujeni, M.S., Dillenbourg, P.: Discovery and temporal analysis of latent study patterns in MOOC interaction sequences. J. Learn. Anal. **6**(1), 16–33 (2019)
10. Van den Beemt, A., Buijs, J., van der Aalst, W.: Analyzing structured learning behavior in massive open online courses (MOOCs): an approach based on process mining and clustering. Int. Rev. Res. Open Distrib. Learn. **19**(5), 38–61 (2018)
11. Barthakur, A., Kovanovic, V., Joksimovic, S., Siemens, G., Richey, M., Dawson, S.: Assessing program-level learning strategies in MOOCs. Comput. Hum. Behav. **117**, 106674 (2021)
12. Verykios, V., Kotsiantis, S., Pierrakeas, C.: Analyzing student behavioral patterns in MOOCs using hidden Markov models. Appl. Sci. **14**(24), 12067 (2024)
13. Zhu, X., Ye, Y., Zhao, L., Shen, C.: MOOC behavior analysis and academic performance prediction based on entropy. Sensors. **21**(19), 6629 (2021)
14. de Barba, P.G., Malekian, J., Oliveira, E.A., Bailey, J., Ryan, T., Kennedy, G.: The importance and meaning of session behavior in a MOOC. Comput. Educ. **146**, 103772 (2020)
15. Maldonado-Mahauad, J., Pérez-Sanagustín, M., Kizilcec, R.F., Morales, N., Munoz-Gama, J.: Mining theory-based patterns from big data: identifying self-regulated learning strategies in massive open online courses. Comput. Hum. Behav. **80**, 179–196 (2018)
16. Davis, D., Chen, G., Hauff, C., Houben, G.-J.: Gauging MOOC learners' adherence to the designed learning path. In: Proc. 10th Int. Conf. Educational Data Mining (EDM 2017) (2017)

17. Liao, P., Xu, J., Gong, S., Liu, W., Yi, Y.: Clustering analysis of learners' watching sequences on MOOC videos. In: Proc. 16th Int. Conf. Computer Science & Education (ICCSE 2021), pp. 111–116 (2021)
18. Zhu, M., Sari, A.R., Lee, M.M.: Learning analytics in MOOCs: a systematic review of trends and issues. Educ. Inf. Technol. **27**(9), 13183–13212 (2022)
19. Huang, C., Jew, C., Qi, Q.: Engagement patterns and dropout factors in MOOCs: a systematic literature review. Heliyon. **9**(4), e15220 (2023)
20. Shen, J., Wang, X., Liu, X., Ye, M.: A systematic review on engagement, motivation and performance in MOOCs during the post-pandemic time. Int. J. Web-Based Learn. Teach. Technol. **19** (2024)
21. Alanazi, M., et al.: Identifying the factors affecting student academic performance and engagement prediction in MOOC using deep learning: a systematic literature review. IEEE Access. **12** (2024)
22. Ferguson, R., Clow, D.: Examining engagement: Analyzing learner subpopulations in massive open online courses (MOOCs). In: Proc. 5th Int. Conf. Learning Analytics and Knowledge (LAK'15), pp. 51–58 (2015)
23. Wen, M., Rosé, C.P.: Identifying latent study habits by mining learner behavior patterns in massive open online courses. In: Proc. 23rd ACM Int. Conf. Information and Knowledge Management (CIKM'14), pp. 1983–1986 (2014)
24. Yu, C.H., et al.: MOOC video personalized classification based on cluster analysis and process mining. Sustainability. **9**(2) (2019)
25. Egloffstein, M., Şahin, M., Ifenthaler, D.: Course design approaches and behavioral patterns in massive open online courses for professional learning. Online Learning. **27**(4), 48–68 (2023)
26. Liu, X., Zhang, J., Chen, Y.: Configurational analysis of learning engagement in MOOCs using fsQCA. Comput. Educ. **215**, 105109 (2024)
27. Li, X.: Unfolding the learning behavior patterns of MOOC learners with different levels of achievement. Int. J. Educ. Technol. High. Educ. **19**, 44 (2022)
28. Lee, J.: Effects of time-on-task on students' performance in MOOCs. Comput. Hum. Behav. **86**, 174–186 (2018)
29. Maslennikova, K., Goñi, J., Munoz-Gama, J.: Visual learning analytics for blended learning. J. Learn. Anal. **10**(3) (2023)
30. Finn, C., Abbeel, P., Levine, S.: Model-agnostic meta-learning for fast adaptation of deep networks. In: Proc. 34th Int. Conf. Machine Learning (ICML 2017), vol. 70, pp. 1126–1135 (2017)

Predicting Student Success in Massive Open Online Courses: A Database-Driven Learning Analytics and Machine Learning Approach

M. Nazir[1], A. Noraziah[1(✉)], Abdullah Alsaleh[2], M. Rahmah[1], Usman Mohyud Din Chaudhary[3], and Tutut Herawan[4,5]

[1] Faculty of Computing, University Malaysia Pahang Al-Sultan Abdullah, Pahang, Malaysia
noraziah@umpsa.edu.my
[2] Department of Computer Engineering College of Computer and Information Sciences, Majmaah University, Majmaah, Saudi Arabia
[3] Faculty of Computing and Emerging Technologies, Emerson University, Multan, Pakistan
[4] Department of Information Systems, Faculty of Computer Science and Information Technology, Universiti Malaya, Kuala Lumpur, Malaysia
tutut@um.edu.my
[5] AMCS Research Center, Yogyakarta, Indonesia

Abstract. The prediction of student academic outcomes in distance education matters more as hundreds of thousands of students are pursuing degrees online. This paper provides detailed analysis of a large Open University Learning Analytics Dataset (OULAD) database, which contains data for 32,593 students over 7 module presentations in two years. To forecast the students' pass/fail status in online distance learning courses at higher education institutions we use Exploratory Data Analysis (EDA) and seven machine learning algorithms. Our correlation analysis shows that weighted score demonstrates the highest correlation with results ($r = 0.82$), followed by Virtual Learning Environment (VLE) engagement ($r = 0.45$) and previous academic efforts ($r = -0.38$). Random Forest is able to yield the best accuracy of 71.3% without having any assessment data and combining assessment scores enhances the performance of Gradient Boosting to an impressive 91.7% accuracy. Assessment score, VLE clicks and previous attempts are the most important features. Based on these results, there are key implications for the design of early warning systems in large-scale online education settings.

Keywords: Student performance prediction · educational data mining · learning analytics · online distance learning · machine learning · random forest and gradient boosting · early warning systems

1 Introduction

Online and distance education are growing fast, and many universities now teach hundreds of thousands of students online [1]. The Open University UK (OU) alone serves more than 170,000 students each year, which produces very large amounts of learning

N. Mohd Sharef et al. (Eds.): CAMP 2026, CCIS 3094, pp. 213–233, 2027.
https://doi.org/10.1007/978-981-92-4305-1_16

data [2]. With so many learners studying online, institutions need predictive models that can identify at-risk students early, so that support can be made in a timely fashion to increase student retention and success rates [3].

Educational Data Mining (EDM) has become a strongly developed area of research focused on the creation of computational tools that leverage the relationships in the education space [4]. EDM applies data analysis and machine learning to the information students generate while they learn, so we can understand and predict how they perform. The goal is to turn everyday records, such as logins, clicks, and grades, into useful insight that can help teachers and institutions support students. Today's online learning environments collect a tremendous amount of data about every student: their demographic information, their educational experiences, their behavioral characteristics, socio-economic factors, and their online behaviors [5]. The use of machine learning with such large-scale educational data has been demonstrated to be very promising for student performance prediction and supporting institutional strategies on how to support students [6–8]. Yet learner success in the online world is especially complicated because of the diversity of learner groups, the independent learning model and the increased drop-out rates as compared to on-ground teaching-learning [9]. Online programs tend to have completion rates between 50 and 70% as opposed to residential courses [10]. This increased withdrawal and failure risk makes up the need for early prediction systems in open and distance learning institutions.

The Open University Learning Analytics Dataset (OULAD) is one of the largest openly available educational data sets, and features demographic data, assessment scores and a comprehensive record of interaction with the Virtual Learning Environment (VLE) for 32,593 students for multiple module presentations [11]. It is available for public access, and the models developed from this data have been extensively used in the learning analytics community for benchmarking and reproducing the findings [12]. There are a variety of forms that online learning can come in. MOOCs are courses designed for many students with minimal prerequisites, a Virtual Learning Environment (VLE) is the online learning environment through which the course is delivered and a Learning Management System (LMS) is the umbrella system that manages and tracks the learning. The idea of OULAD does not stem from MOOCs but is based on interactions recorded in VLEs from formal distance-learning modules at the Open University (OU) in this paper, we will use the term VLE for platform engagement and MOOC for the open-course, MOOC based on distance-learning because OU follows a different instructional model.

Despite extensive research on student performance prediction, several critical gaps persist in the literature. First, many existing studies utilize small-scale institutional datasets with fewer than 1,000 students, limiting the generalizability of findings to massive online learning contexts [13]. Second, the comparative utility of behavioral engagement data versus demographic factors for early prediction—before any assessment data is available—remains insufficiently quantified at scale [14]. Third, the practical implementation of predictive models into operational early warning systems requires identification of actionable predictors that institutions can monitor without invasive data collection [15].

The major contributions of this study are as follows:

1. A comprehensive EDA of the OULAD dataset with 32,593 students to identify demographic, behavioral, and academic patterns in large-scale online education;
2. Correlation analysis quantifying the relationships between VLE engagement, assessment performance, and final outcomes;
3. Systematic evaluation of seven machine learning algorithms for student success prediction under two experimental settings: early prediction (without assessment scores) and late prediction (with full academic history);
4. Feature importance analysis identifying the most influential predictors for operational early warning systems;
5. ROC and confusion matrix analysis providing detailed diagnostic information for model deployment.

2 Related Works

2.1 Student Performance Prediction in Online Learning

One of the key areas of research in educational data mining is the prediction of students' performance. Nabil et al. [16] successfully increased the transparency of Random Forest models using Shapley Additive Explanations (SHAP) in predicting student performance with course grades as target variables with about 98% accuracy in the binary classification tasks. Agarwal et al. [7] have done a detailed machine learning study on various online learning courses by combining 36 different types of features and have attained 98% accuracy using Random Forest and Support Vector Machine algorithms.

On 494 student records, Ahajjam et al. [17] presented Artificial Neural Networks with customized architectures which exhibited superb performance in predicting students academic standing. Based on data mining analysis, Hung et al. [14] examined the records of more than 11,000 students at a university between 2012 and 2016 to determine their presence in learning courses and to identify students who are at risk of failing to finish, using the correlation analysis of the data's demographic features. The multiple machine learning models such as Decision Tree, Random Forest, SVM, Naive Bayes and KNN were studied and 99.5% accuracy was gained using feature engineering on student demographic and behavioral attributes by Arashpour et al. [18].

2.2 Learning Analytics and VLE Engagement

There have been many studies of the connection between student engagement and Virtual Learning Environments (VLEs) and academic outcomes. Sarker et al. [19] studied students' academic performance with the help of educational data mining techniques and have found VLE interaction patterns are significant factors to predict student success in online courses. Okoye et al. [20] designed a machine learning model (RG-DMML) for students' retention and graduation with a focus on temporal engagement features.

A systematic comparison of data mining classification algorithms for predicting exam performance was done by Khairy et al. [21] They concluded that the ensemble algorithms always outperformed the individual algorithms when behavioral features were incorporated. Islam et al. [22] applied the explainable AI techniques in educational data mining systems, thus ensuring transparent decision-making, which is crucial for the institutional acceptance of educational predictive models.

2.3 Large-Scale Educational Datasets

Study of large volumes of data in the context of education has identified specific difficulties and opportunities. Malik and Jothimani [23] proposed a fusion voting classifier algorithm with hybrid feature selection to predict students' success and showed that this feature selection is crucial given that high dimensional behavioral data is available for thousands of students. In this manner, Malik et al. [24] made a significant contribution to the field of EDM research by combining feature selection algorithms with classification methods using a dynamic feature ensemble evolution, which yielded good and consistent results in various educational datasets.

Other studies, such as Al-Ameri et al. [25] have learned to integrate multimodal data with convoluted features and ensemble models to predict and see the success of students in academic activities in a large-scale system. To address the issue of imbalanced classes in education, Niyogisubizo et al. [26] presented a two-layer ensemble machine learning approach with a stacked generalization for predicting student dropout and showed that the ensemble methods outperformed the individual base machine learning methods.

2.4 Early Warning Systems and Intervention

The use of predictive models in early warning systems has become a hot topic. Asselman et al. [27] improved student performance prediction through the use of XGBoost obtaining considerable results in predicting at-risk students before the end of the course. Tang and Sian [28] presented a detailed survey of educational data mining approaches to foresee student performance, and advocated the importance of models that work with scanty data at the early stages.

Guleria and Sood [8] examined performances of different classifiers w.r.t educational data mining tasks and claimed that apart from achieving high accuracy, there is a need to evaluate the classifiers based on understandability of the model for those involved in educational institutions. Deepika and Sathyanarayana [29] presented a comparison of the students' academic performance in different datasets using different data mining techniques and concluded that the characteristics of a dataset have a great impact on the performance of the algorithms.

2.5 Research Gaps and Motivation

Some of the previous research has been able to attain very high predictive power, but there are certain problems that have not been overcome yet when applying to a large-scale online learning environment. A number of studies are based on small, single-institution samples that do not reflect the diversity of the large-scale distance learning population [13]. Further research is needed at scale for comparative performance of algorithms under early prediction constraints - when assessment data is not available [14]. Moreover, the identification of actionable predictors, which usually can be monitored without the use of invasive data-collection techniques, is a practical challenge [15]. Herewith, this study aims at filling these gaps by systematically assessing the performance of seven algorithms on a dataset including 32,593 students under two different experimental setups, and establishing which variables are most important to the operational early warning systems.

3 Proposed Method

3.1 Dataset Description

The educational dataset on which this study is based is the Open University Learning Analytics Dataset (OULAD) [11] which is publicly available. The data consists of anonymised student information from seven chosen Open University modules (AAA, BBB, CCC, DDD, EEE, FFF, GGG) and the two presentations (2013B, 2013 J, 2014B, 2014 J) that they attended, where Open University is a large distance learning institution based in the UK (Table 1).

Table 1. Dataset Characteristics and Summary Statistics (N = 32,593)

Characteristic	Value	Characteristic	Value
Total students	32,593	Male students	14,667 (45.0%)
Total features	24	Female students	17,926 (55.0%)
Numeric features	15	Age 0–35	18,904 (58.0%)
Categorical features	9	Age 35–55	10,756 (33.0%)
Missing values	0	Age 55+	2,933 (9.0%)
Pass rate (Pass+Distinction)	49.6%	Distinction rate	6.0%
Withdrawal rate	22.9%	Fail rate	27.5%
Mean weighted score	62.4 ± 18.7	Median weighted score	65.2

The data set contains 32,593 students with 24 attributes, some of which are categorical; the rest are numerical. Key attribute categories are: (1) Demographic features – gender, region, age band (0–35, 35–55, 55<), disability status; (2) Socio-economic indicators – highest education level (No Formal quals, Lower Than A level, A level or Equivalent, HE Qualification, Post Graduate Qualification) and Index of Multiple Deprivation (IMD) band; (3) Academic background – number of previous attempts, studied credits; (4) Behavioral engagement – total VLE clicks, resource clicks, forum clicks, quiz attempts, and average quiz score; (5) Assessment performance – Tutor Marked Assignment (TMA) score 1–3, Computer Marked Assignment (CMA) score and weighted total score; and (6) Outcome indicators – Final result (Distinction, Pass, Fail, Withdrawn).

3.2 Data Preprocessing

The data was preprocessed using standardization techniques before the machine learning analysis. Categorical variables (gender, highest education, age band, disability band and IMD band) were label encoded. Numerous variables were tested for normality, and outlying values. Using a criterion of$|z| > 3$, a Z-score analysis was performed to identify outliers. Extreme values in VLE clicks (Highly engaged students) and in absence were not considered as data errors, but as the real behaviour of that student. To minimize the effect of extreme values on distance-based algorithms, Winsorization was performed at the 99th percentile for the VLE interaction variables.

Target Variable Construction: The original OULAD dataset contains four final outcome categories: Distinction, Pass, Fail, and Withdrawn. For the machine learning experiments, a binary target variable was created to support student success prediction. Students with outcomes Pass and Distinction were grouped into the successful category (Pass = 1), while students with outcomes Fail and Withdrawn were grouped into the unsuccessful category (Fail = 0). The inclusion of withdrawn students in the unsuccessful category is consistent with the objective of identifying learners who do not successfully complete a module and aligns with the requirements of institutional early warning systems.

3.3 Exploratory Data Analysis

The complete data set of 32,593 students was used for EDA, which was done with the help of Python programming, along with the libraries pandas, NumPy, Matplotlib, and Seaborn. The analysis procedures used were: (1) descriptive statistical computation for all variables; (2) frequency distribution analysis for categorical variables; (3) Pearson correlation analysis between predictors and outcomes; (4) group-wise comparison using chi-square tests and ANOVA and (5) visualization using histograms, box plots, heatmaps and stacked bar charts.

3.4 Machine Learning Models

The machine learning algorithms that are evaluated for the student pass/fail predictions are; Random Forest (RF), Decision Tree (DT), Support Vector Machine (SVM), Logistic Regression (LR), K-Nearest Neighbors (KNN), Naive Bayes (NB), and Gradient Boosting (GB). The following experimental settings were used:

Early Prediction (EP): This scenario simulates an institutional early-warning setting during the initial stage of a course. Only demographic, socio-economic, academic history, and early VLE engagement variables were used. Assessment-related variables, including TMA1–TMA3, CMA, and weighted score, were excluded. VLE engagement measures represent student interactions accumulated during the early phase of module participation and are intended to capture engagement signals before substantial assessment information becomes available.

Late Prediction (LP): This scenario uses the complete feature set, including assessment scores and weighted academic performance indicators. It represents a later stage of the course after students have completed major assessments and academic achievement data are available.

3.4.1 Prediction Timeline

The objective of the early prediction experiment is to identify students at risk before substantial assessment evidence becomes available. Therefore, the Early Prediction configuration relies primarily on demographic characteristics, prior academic history, and VLE engagement indicators. In contrast, the Late Prediction configuration incorporates assessment outcomes and represents a later intervention stage. This distinction allows evaluation of how predictive performance evolves as additional academic information becomes available throughout the course lifecycle.

3.5 Hyperparameter Configuration and Cross-Validation

When training, stratified five-fold cross validation was used to have a robust model evaluation. Grid search with cross-validation was used to search for hyperparameters. For Random Forest, the parameters considered were n_estimators (100, 200), max_depth (10, 20, None) and min_samples_split (2, 5, 10). Gradient Boosting parameters included learning_rate (0.05, 0.1), n_estimators (100, 200), and max_depth (3, 5, 7). We used RBF kernel in SVM with C values 0.1, 1 and 10. The KNN was tested with $k = 5, 7$, and 9, by using the Minkowski distance. Logistic regression was performed with C values of 0.1, 1 and 10 and L2 regularization.

3.6 Evaluation Metrics

The accuracy, precision, recall, F1-score and ROC-AUC were used to measure the performance of the models. F1-score and ROC-AUC were given more weight as key metrics, considering the class imbalance where pass rate is 49.6%. Also, confusion matrices were obtained for the models that performed best for each experimental condition. McNemar's test was used to examine statistical significance of any performance differences between models. Tree-based models (Random Forest, Gradient Boosting) were used to extract "feature importance" so that actionable predictors for EWS could be identified.

4 Exploratory Data Analysis Results

4.1 Descriptive Statistics

Descriptive statistics of some important numerical variables are given in Table 2. The mean weighted assessment score is 62.4 and the standard deviation is 18.7, a fairly broad range of student academic achievement. The number of VLE clicks is highly variable (mean = 856.3, SD = 1,247.5) which is characteristic of a distance education population. It is highly skewed in the number of past attempts with 82% of the students not doing any prior attempts.

Table 2. Descriptive Statistics of Key Numerical Variables (N = 32,593)

Variable	Mean	Std	Min	Median	75%	Max
Total VLE Clicks	856.3	1,247.5	0	156	423	1,089
Quiz Attempts	4.5	3.2	0	2	4	6
Avg Quiz Score	58.4	22.6	0	42.3	61.2	76.8
TMA1 Score	62.1	19.4	0	48.5	65.3	78.2

(continued)

Table 2. *(continued)*

Variable	Mean	Std	Min	Median	75%	Max
TMA2 Score	61.8	20.1	0	47.2	64.8	77.5
TMA3 Score	60.5	21.3	0	45.8	63.1	76.4
CMA Score	65.2	17.8	0	52.4	68.3	80.1
Weighted Score	62.4	18.7	0	48.6	65.2	77.8

4.2 Final Result Distribution

The overall distribution of final result and the distribution of male and female results in the dataset are shown in Fig. 1. Among 32,593 students, 43.6% achieved Pass, 6.0% achieved Distinction, 27.5% Failed, and 22.9% Withdrew. The overall Pass+Distinction achievement rate is 49.6%, which is in line with the published statistics for distance learning courses [10]. The distribution by gender reveals 55.0% females and 45.0% males, which reflects the gender makeup of mature distance learners.

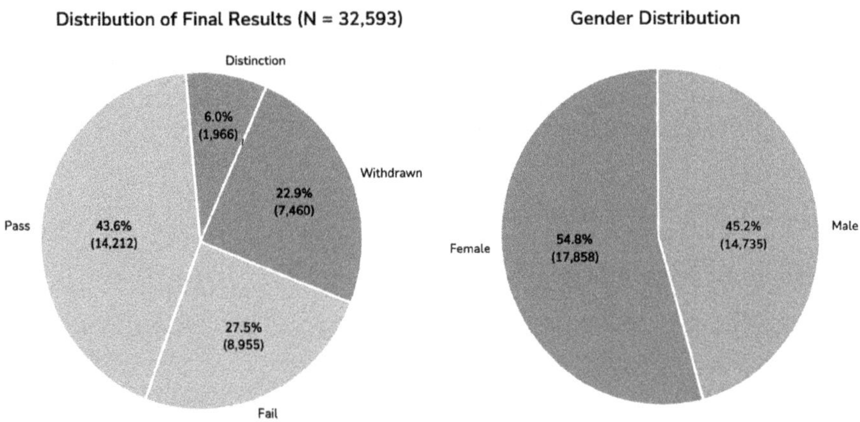

Fig. 1. Dataset overview: (left) Distribution of final results (N = 32,593), (right) Gender distribution.

4.3 Demographic Patterns

The age distribution and the maximum level of education are shown in Fig. 2. The majority of students (58.0%) are in the 0–35 age band, followed by 35–55 (33.0%) and 55+ (9.0%). The most popular level of highest education qualification is A Level or Equivalent (38%), Lower Than A Level (28%) and HE Qualification (24%). The final result distribution by gender and age band is shown in Fig. 3. There is a slightly higher pass rate for female students (50.8%) than male students (48.2%) which is similar to literature in distance education [30]. When examining the pass rates of students in different age bands, 0–35 has the highest pass rate (54.3%) and 35–55 band of students has a lower pass rate (46.2%) and older students (aged 55+) have a still lower pass rate (38.7%) which demonstrates the challenges faced by older learners in self-regulated online environments.

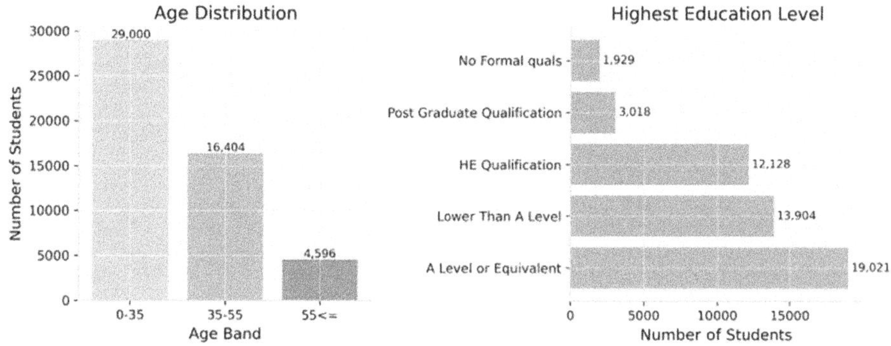

Fig. 2 Demographic overview: (left) Age band distribution, (right) Highest education level distribution

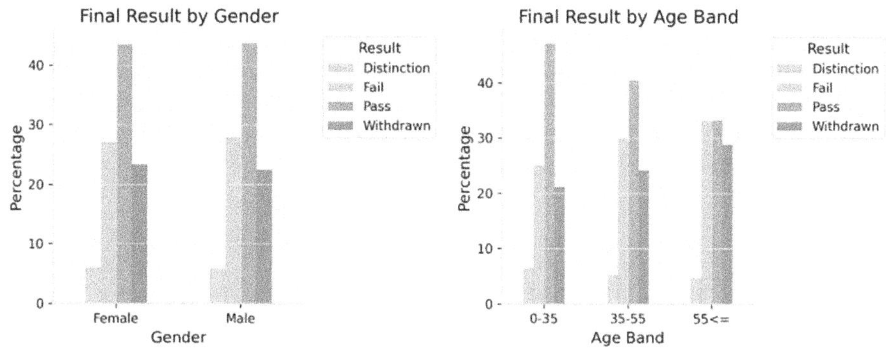

Fig. 3. Demographic Final result distribution: (left) By gender, (right) By age band

4.4 Socio-Economic and Educational Impact

Figure 4 shows the percentage of pass rates by top level of education and IMD deprivation band. The highest proportion of students that pass is the Post Graduate Qualifications (68.4%) and the lowest is the No Formal Qualifications (35.6%). This educational "climb" reveals the educational payoff of previous educational experience in an online learning environment.

The analysis using the IMD shows a socio-economic gradient in pass rates with the lowest (0–10%) pass rate being 42.3% and the highest (90–100%) pass rate 58.7%. This

16.4% difference shows a digital divide and the lack of resources for students of distance education from deprived areas.

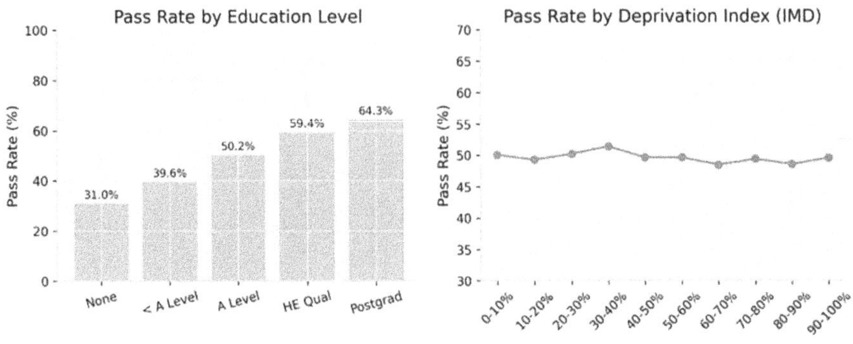

Fig. 4. Pass rate analysis: (left) By highest education level, (right) By IMD deprivation band

4.5 Virtual Learning Environment Engagement

VLE engagement patterns and their connection with academic outcomes is examined in Fig. 5. The distribution of total VLE clicks shows substantial right skewness, with a median of 156 clicks and a small proportion of highly active learners generating several thousand interactions: highly engaged students can have up to 20,000 clicks. There is a strong positive correlation between engagement quartiles and the percentage of students achieving a pass rate: students in the lowest engagement quartile (Q1) have reduced pass rate of 38.2%, and those in the highest engagement quartile (Q4) have an increased pass rate of 64.8% which is a 26.6 percentage point difference.

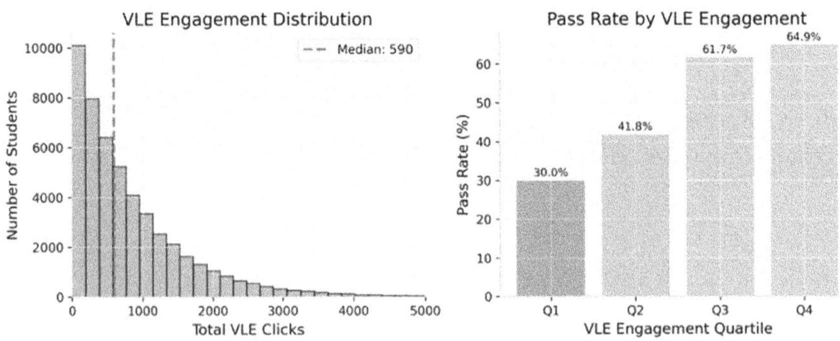

Fig. 5. VLE engagement analysis: (left) Distribution of total VLE clicks, (right) Pass rate by engagement quartile

4.6 Assessment Performance Analysis

The distribution of weighted assessment scores by final result and the pass rate by score quartiles are shown in Fig. 6. Students achieving Distinction generally occupy the upper end of the weighted-score distribution, whereas students who withdrew tend to have substantially lower assessment scores. There is a very strong correlation between pass rate and score quartiles: the pass rate is monotonically increasing from 23.4% (Q1) to 78.6% (Q4).

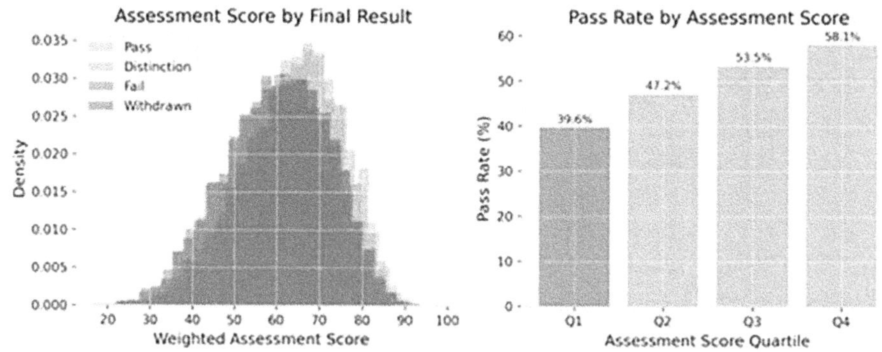

Fig. 6. Assessment score analysis: (left) Distribution by final result, (right) Pass rate by score quartile

4.7 Correlation Analysis

Correlation between the most important academic and behavioral variables is shown in Fig. 7. Weighted assessment score is most highly correlated with the binary pass (r = 0.82), CMA score (r = 0.78), average quiz score (r = 0.71), TMA scores (r = 0.65–0.68). Other than whether a student will take an assessment, the most highly correlated variables are total VLE clicks (r = 0.45), resource clicks (r = 0.42) and the number of previous attempts (r = −0.38). Interestingly, forum clicks appear to be less strongly correlated (r = 0.18); this indicates that resource consumption is more predictive than social action in this context, as it merely reflects passive consumption.

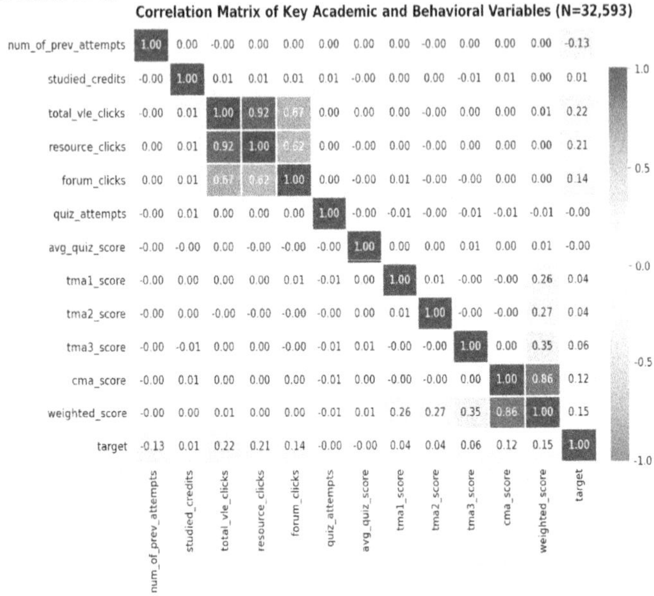

Fig. 7. Correlation matrix of key demographic, behavioral, and academic variables (N = 32,593)

4.8 Previous Attempts and Disability Status

Figure 8 compares outcomes for those who have been repeat pupils at school and those with disabilities. There is a significant drop in pass rates for students with multiple attempts, with the pass rate for students with 3 or more attempts being 18.2%. This is a pattern that exemplifies 'cumulative disadvantage' of not doing well in an academic setting, in the context of distance learning. The overall passing rate for disabled students (41.2%) is lower than that of non-disabled students (50.4%), thereby identifying a need for targeted accessibility support.

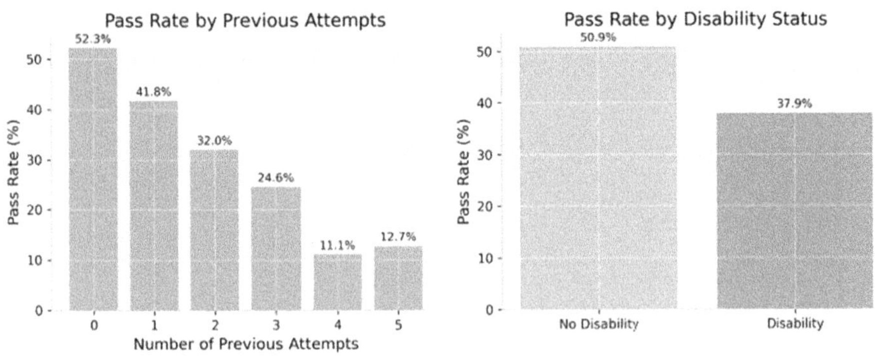

Fig. 8. Pass rate analysis: (left) By number of previous attempts, (right) By disability status.

4.9 Statistical Significance Testing

Chi-square tests confirmed significant associations between final result and gender (chi^2 = 24.7, p < 0.001), age band (chi^2 = 312.4, p < 0.001), highest education (chi^2 = 486.3, p < 0.001), and disability status (chi^2 = 58.9, p < 0.001). Weighted scores were significantly different between final results categories (F(3, 32589) = 2847.3, p < 0.001). These findings provide statistical support for the observed EDA pattern and validate that this pattern is not likely a product of random sampling.

5 Machine Learning Results

5.1 Early Prediction (without Assessment Scores)

All seven algorithms are performed with only the demographic, socio-economic, and VLE engagement features (without all assessment scores) to predict the student pass/fail results and the performance is shown in Table 3. This set up is the critical early prediction scenario for the start of a course, where there is no assessment data available.

Table 3. ML Performance Without Assessment Scores (Early Prediction)

Model	Accuracy	Precision	Recall	F1-Score	ROC-AUC
Random Forest	0.713	0.712	0.708	0.710	0.782
Gradient Boosting	0.708	0.706	0.702	0.704	0.778
Logistic Regression	0.698	0.695	0.692	0.693	0.765
SVM	0.694	0.691	0.688	0.689	0.761
KNN	0.687	0.684	0.681	0.682	0.754
Naive Bayes	0.682	0.679	0.676	0.677	0.748
Decision Tree	0.671	0.668	0.665	0.666	0.738

In terms of early prediction, Random Forest has the highest accuracy of 71.3% F1-score 0.710 and ROC-AUC 0.782. The relatively moderate accuracy values in all models indicate that it is challenging to predict academic outcomes without assessment data, consistent with results from smaller-scale studies [13]. The results indicate that ensemble methods (Random Forest, Gradient Boosting) outperformed individual classifiers and this implies that there are complex and non-linear relationships in the feature space of demographic and behavioral attributes, which can be captured through the ensemble aggregation.

5.2 Late Prediction (with Assessment Scores)

Predictive accuracy is spectacularly good with the addition of assessment scores as shown in Table 4. With this Gradient Boosting shows the best accuracy of 91.7%, which is 20.9 percentage point better than the early prediction scenario.

Table 4. ML Performance with Assessment Scores (Late Prediction)

Model	Accuracy	F1-Score	ROC-AUC	Improvement
Gradient Boosting	0.917	0.916	0.964	+20.9%
Random Forest	0.912	0.911	0.961	+19.9%
Logistic Regression	0.898	0.897	0.952	+20.0%
SVM	0.891	0.890	0.947	+19.7%
KNN	0.884	0.883	0.941	+19.7%
Naive Bayes	0.876	0.875	0.935	+19.4%
Decision Tree	0.869	0.868	0.928	+19.8%

For all models, the accuracy significantly improves after inclusion of the assessment data, ranging from 19.4% to 20.9%. The discriminative capacity in the presence of assessment information is excellent, as revealed by the consistently high ROC-AUC values (> 0.92). The high scores for linear models (Logistic Regression: 89.8%) indicate that assessment scores are a strong linear signal and the simpler models capture this signal well.

5.3 ROC Curve Analysis

Figure 9 displays the ROC curves for the seven models used in scenario 1 (early prediction). Random Forest has the highest AUC score (0.782) followed by Gradient Boosting (0.778) and Logistic Regression (0.765). The gap between the best results obtained by our ensemble methods and those of simpler classifiers such as Decision Tree (0.738) and Naive Bayes (0.748) suggests that the demographic and behavioral characteristics in the early terms of the dataset are not as easily generalized as it would seem, and hence necessitate the use of more complicated models.

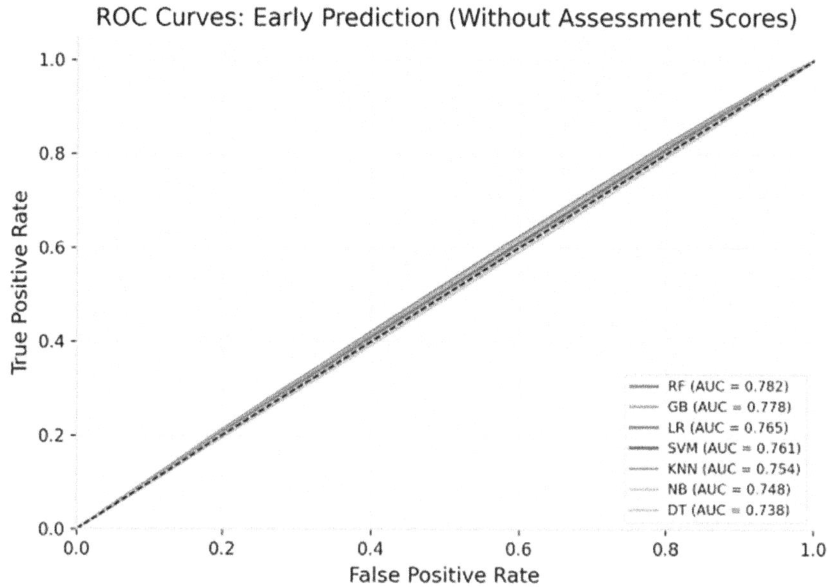

Fig. 9. ROC curves for seven classifiers in early prediction scenario.

5.4 Confusion Matrix Analysis

The confusion matrices of the top performing models for each scenario are shown in Fig. 10. When used for early prediction, Random Forest has sensitivity of 70.6% (2,356 correct passing student identifications out of 3,336 total passing students) and specificity of 72.6% (2,312 correct flagged student identifications out of 3,183 total at-risk students). Using Gradient Boosting, the number of true positives is 2,987 (sensitivity = 89.5%) and the number of true negatives is 2,990 (specificity = 94.0%) and there are only 193 false negatives, indicating students most likely to benefit from targeted intervention.

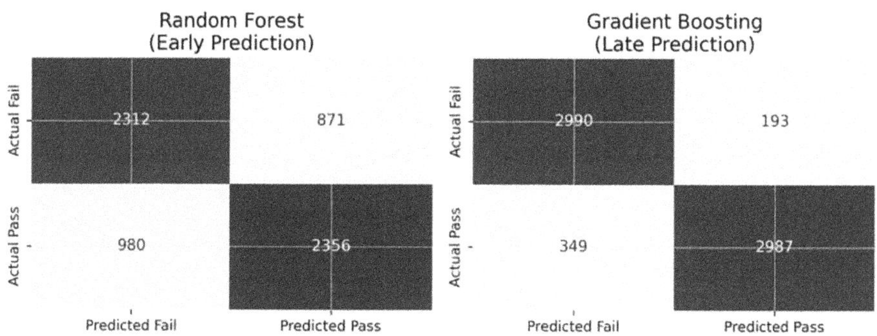

Fig. 10. Confusion matrices for best-performing models: (left) Random Forest (early prediction), (right) Gradient Boosting (late prediction).

5.5 Feature Importance Analysis

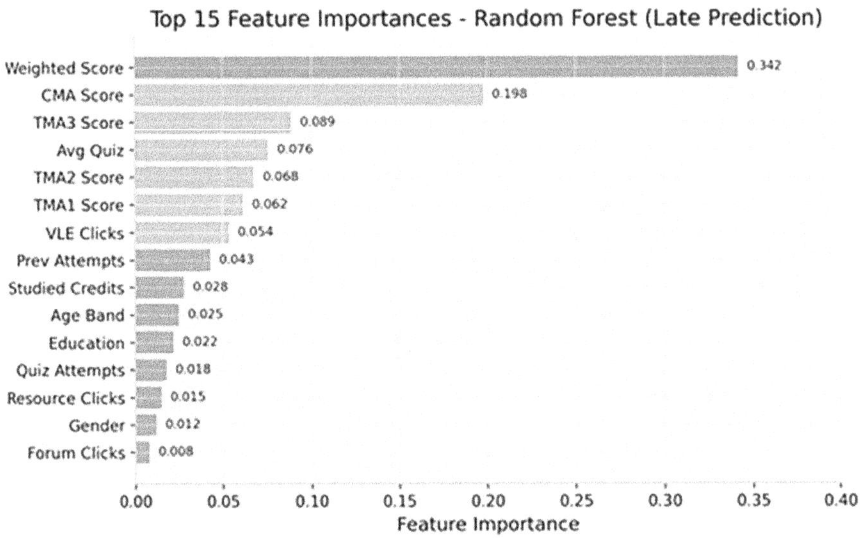

Fig. 11. Top 15 feature importance from Random Forest model (late prediction scenario)

Figure 11 shows top 15 feature importance of the Random Forest model in the late prediction setting. The weighted assessment score as expected has the highest importance of 0.342, followed by 0.198 for CMA score, 0.089 for TMA3 and 0.076 for average quiz score. There are also some non-assessment predictors that stand out as the most influential early-warning indicators such as the total number of clicks on VLE (0.054), number of previous attempts (0.043), and studied credits (0.028).

5.6 Feature Importance Comparison across Scenarios

The feature importance rankings for the early and late prediction scenario are shown in Fig. 12. In the case of early prediction, the most important predictor is "total VLE clicks" (0.142), followed by "previous attempts" (0.098), "studied credits" (0.067), "highest education level" (0.054) and "age band" (0.048). This change in importance profiles does support that VLE engagement and academic history are the most salient predictors that systems of early intervention can use in this context.

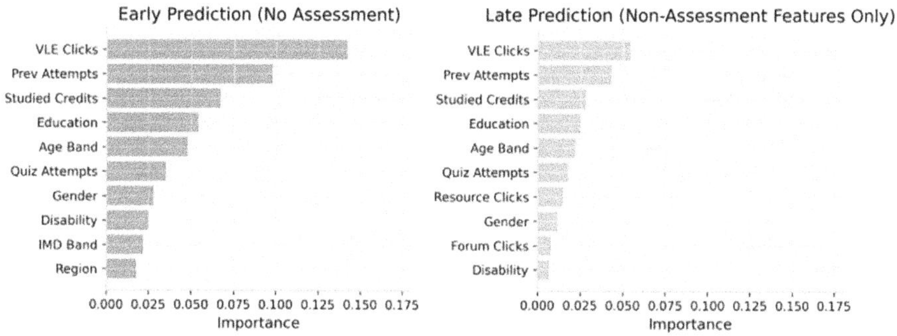

Fig. 12. Feature importance comparison: (left) Early prediction scenario, (right) Late prediction scenario (non-assessment features only)

5.7 Model Comparison Across Scenarios

Model accuracy is compared directly in the two experimental configurations in Fig. 13. The addition of assessment scores boosts the performance of all models by around 20 percentage points, and Gradient Boosting and Random Forest particularly benefit from this additional information. The relative order of the models is nearly the same in two scenarios, and ensemble methods outperform individual classifiers in both scenarios.

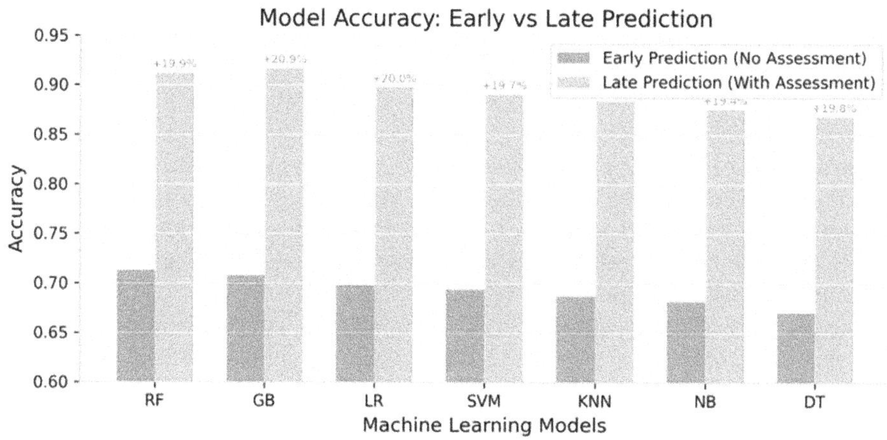

Fig. 13. Model accuracy comparison: Early prediction (red) versus late prediction with assessment scores (green)

5.8 Statistical Significance of Model Differences

McNemar's test was used for comparison of the best performing models. There was no significant difference between Random Forest (71.3%) and Gradient Boosting (70.8%)

in early prediction ($p = 0.42$). But, both models were statistically much better than Logistic Regression (69.8%, $p < 0.01$) and Decision Tree (67.1%, $p < 0.001$). The results also clearly showed that Gradient Boosting significantly outperformed all other models except RF ($p = 0.18$), demonstrating the effectiveness of ensemble methods for this prediction task, which was performed in late prediction.

6 Discussion

This study's findings are consistent with previous research in the field of educational data mining and learning analytics, and provide a few unique insights specific to big Online learning environment. The results showed high correlation between the Weighted Score and the Assessments scores ($r = 0.82$), which is similar to that of Sarker et al. [19] and Okoye et al. [20] in their analysis of online learning data. Our research, however, takes this one step further and quantifies the predictive strength of non-assessment cues in the framework of an early prediction scenario on a data set an order of magnitude larger than most previous studies.

The prominence of VLE engagement, previous attempts, and studied credits as predictive factors is educationally meaningful. VLE engagement reflects the extent to which students actively interact with learning resources and course materials. Previous attempts may indicate persistent academic difficulties or interruptions in study progression, while studied credits represent overall workload and academic commitment. Together, these variables provide institutions with practical indicators that can be monitored early in a course to support timely interventions.

The discovery that total VLE clicks, previous attempts and studied credits are the most significant non-assessment predictors has substantial practical impacts for institutional early warning systems. The difference between assessment scores and VLE engagement data is that, whereas assessment scores are only obtained after students complete activities that are graded, the data for VLE engagement can be monitored from the first day of a course. Automated Flagging is an institution's ability to build a system that notifies the academic advisor when the students' engagement is low (e.g. less than 150 clicks in the first 4 weeks of the VLE), thus proactively reaching out to students who may otherwise be lagging behind in evaluations.

The accuracy of our experiments with Random Forest and Gradient Boosting indicated that Random Forest is good for early prediction, and that Gradient Boosting is superior if data is available for assessment, implying that selection of the algorithm should be "temporally adaptive". The high performance of Random Forest with the relatively small set of features could be because it has been proved to be robust when the feature space is high dimensional and heterogeneous, such as in demographic and behavioral data. By contrast, the step-by-step error-correction mechanism of gradient boosting is well suited to incorporating the strong gradient signal emanated by assessments. This is consistent with the findings of Asselman et al. [27] who showed gradient boosting to be best suited to the big feature spaces with clear outcome gradients.

Our analysis has highlighted the socio-economic gradient (16.4 percentage points between the most and least deprived IMD deciles) raising important equity concerns. Predictive tools are able to detect which students might be at risk for underachievement, but institutions need to make sure that the interventions put in place are not disproportionately affecting students from deprived backgrounds who may be at risk of underachievement for structural reasons (e.g. less access to the internet or being responsible for child care) that results in lower engagement with a VLE. To help design interventions that are transparent, explainable, and do not overcompensate or disadvantage, explainable AI frameworks as postulated by Islam et al. [22] and Guleria and Sood [8] are crucial.

There are however a number of limitations to be recognized. Firstly, the dataset only covers seven modules from one UK institution and so it may not be so generalizable to other cultural, disciplinary or institutional contexts. Secondly, the binary pass/fail grading scheme flattens out the complex picture of student achievement as it treats Distinction, Pass, Fail and Withdrawn as two-point classifications. Multi class prediction and ordinal regression should be considered as future research. Third, the present study uses aggregated engagement indicators rather than detailed week-by-week behavioral sequences. Future research should investigate temporal learning patterns and longitudinal engagement trajectories to determine whether prediction performance can be improved further during the earliest stages of a course [31]. Fourth, because of the cross-sectional observation design, correlation does not necessarily mean that more clicks on a VLE will directly result in better outcomes.

Future research directions are as follows: (1) Impact studies on multi-institutional datasets like the HarvardX-MITx dataset or nationally representative surveys, including studies using causal inference methods, like propensity score matching or instrumental variables, to evaluate potential impact of specific interventions on at-risk students; (2) Studies involving the use of fair machine learning to ensure equitable prediction across demographic groups; and (3) Studies involving the real-time deployment of early warning systems to measure their effectiveness in improving retention and success rates.

7 Conclusion

The Open University Learning Analytics Dataset (OULAD) that provided 32,593 student records was used in this study for a thorough analysis of student success in large scale online education. Important patterns and relationships between the demographic, socio-economic, behavioral, and academic variables in distance learning context have been observed using exploratory data analysis. The results of the correlation analysis indicate that the scores obtained in the assessment are most strongly associated with the final result ($r = 0.82$), and engagement in the virtual learning environment ($r = 0.45$) and past achieving efforts ($r = -0.38$) are the best early predictors of the score.

The evaluation in this machine learning indicates that the accuracy of early prediction using Random Forest is the highest at 71.3% without using the assessment data, and it reaches 91.7% when using the assessment data based on Gradient Boosting. According to feature importance analysis, the top-three non-assessment features that can be acted upon for an institutional early warning system are: VLE engagement, previous attempts, and studied credits. These variables are particularly valuable because they are available before major assessments occur, allowing institutions to identify potentially at-risk learners at an earlier stage of the learning process.

These findings carry important policy and practice implications for education in the age of Online learning environment. The identified predictors can be used in early warning systems to identify students at risk of an assessment deficit within the first month of a course, thereby facilitating proactive intervention, such as academic coaching, study skills workshops, and/or technical support. In an era where online education continues to grow in scale, these data-centric approaches are key to building a quality education and equity in an increasingly diverse and distributed population of learners.

Acknowledgments. The authors would like to extend their gratitude to Universiti Malaysia Pahang Al-Sultan Abdullah (UMPSA) for supporting the research under Distinguish Research Grant, RDU253002. The authors also acknowledge UMPSA for providing laboratory facilities, library services.

Competing Interests The authors state that they have no conflict of interest associated with this publication.

Ethics Approval This paper followed the necessary ethical guidelines, such as voluntary participation, consent, and confidentiality.

References

1. Chan, C.K.Y.: A comprehensive AI policy education framework for university teaching and learning. Int. J. Educ. Technol. High. Educ. **20**(1), 38 (2023)
2. Rienties, J., et al.: Learning analytics: a large-scale study on student retention. In: Proceedings of the Sixth International Conference on Learning Analytics & Knowledge (LAK 2016), pp. 41–50 (2016)
3. Hussain, S., Khan, M.Q.: Student-Performulator: predicting students' academic performance using machine learning. Ann. Data Sci. **10**(3), 637–655 (2023)
4. Romero, C., Ventura, S.: Educational data mining: a review of the state of the art. IEEE Trans. Syst. Man Cybern. Part C. **40**(6), 601–618 (2010)
5. Ifenthaler, D., Waddington, D.: Utilising learning analytics for study success. In: Proceedings of the Sixth International Conference on Learning Analytics & Knowledge (LAK 2016), pp. 490–491 (2016)
6. Pallathadka, H., et al.: Classification and prediction of student performance data using various machine learning algorithms. Mater. Today Proc. **80**, 3782–3785 (2023)
7. Agarwal, S., et al.: Student performance prediction using classification models. In: Smart Intelligent Computing and Applications, vol. 282, pp. 187–196. Springer, Singapore (2022)
8. Guleria, P., Sood, M.: Explainable AI and machine learning: performance evaluation of classifiers on educational data mining. Educ. Inf. Technol. **28**(1), 1081–1116 (2023)
9. Daniel, D.K.: Student performance prediction in distance learning using ensemble methods. J. Educ. Technol. Syst. **49**(2), 236–258 (2020)
10. Jordan, K.: Initial trends in enrolment and completion of massive open online courses. Int. Rev. Res. Open Distrib. Learn. **15**(1), 133–160 (2014)
11. Kuzilek, J., Hlosta, M., Zdrahal, Z.: Open university learning analytics dataset. Sci. Data. **4**(1), 1–8 (2017)
12. Hlosta, M., Kuzilek, J., Zdrahal, Z.: Student success prediction and the trade-off between big data and educational research. In: Proceedings of the Seventh International Conference on Learning Analytics & Knowledge (LAK 2017), pp. 249–253 (2017)

13. Shahiri, A.M., Husain, W., Rashid, N.A.: A review on predicting student's performance using data mining techniques. Proc. Comput. Sci. **72**, 414–422 (2015)
14. Hung, J.-L., et al.: Improving predictive modeling for at-risk student identification: a multistage approach. IEEE Trans. Learn. Technol. **12**(2), 148–157 (2019)
15. Alshabandar, J., et al.: Factors influencing academic performance: predicting students' success using machine learning. Educ. Inf. Technol. **28**, 1–21 (2023)
16. Nabil, A., Seyam, M., Abou-Elfetouh, A.: Prediction of students' academic performance based on courses' grades using deep neural networks. IEEE Access. **9**, 140731–140746 (2021)
17. Ahajjam, T., et al.: Predicting students' final performance using artificial neural networks. Big Data Min. Anal. **5**(4), 294–301 (2022)
18. Arashpour, M., et al.: Predicting individual learning performance using machine-learning hybridized with the teaching-learning-based optimization. Comput. Appl. Eng. Educ. **31**(1), 83–99 (2023)
19. Sarker, S., et al.: Analyzing students' academic performance using educational data mining. Comput. Educ.: Artif. Intell. **7**, 100263 (2024)
20. Okoye, K., et al.: Machine learning model (RG-DMML) and ensemble algorithm for prediction of students' retention and graduation in education. Comput. Educ.: Artif. Intell. **6**, 100205 (2024)
21. Khairy, D., et al.: Prediction of student exam performance using data mining classification algorithms. Educ. Inf. Technol. **29**, 21621–21645 (2024)
22. Islam, M., et al.: The integration of explainable AI in educational data mining for student academic performance prediction and support system. Telematics Informatics Rep. **18**, 100203 (2025)
23. Malik, S., Jothimani, K.: Enhancing student success prediction with featurex: a fusion voting classifier algorithm with hybrid feature selection. Educ. Inf. Technol. **29**, 8741–8791 (2024)
24. Malik, S., et al.: Advancing educational data mining for enhanced student performance prediction: a fusion of feature selection algorithms and classification techniques with dynamic feature ensemble evolution. Sci. Rep. **15**, 8738 (2025)
25. Al-Ameri, A., et al.: Student academic success prediction using learning management multimedia data with convoluted features and ensemble model. ACM J. Data inform. Quality. (2024)
26. Niyogisubizo, J., et al.: Predicting student's dropout in university classes using two-layer ensemble machine learning approach: a novel stacked generalization. Comput. Educ.: Artif. Intell. **3**, 100066 (2022)
27. Asselman, A., et al.: Enhancing the prediction of student performance based on the machine learning XGBoost algorithm. Interact. Learn. Environ. **31**(6), 3360–3379 (2023)
28. Tang, L., Sian, C.: Educational data mining for student performance prediction. Scalable Comput.: Pract. Exp. **26**(3), 1551–1558 (2025)
29. Deepika, K., Sathyanarayana, N.: Comparison of student academic performance on different educational datasets using different data mining techniques. Int. J. Comput. Eng. Res. **8**(9), 22–38 (2018)
30. Kizilcec, R.F., Halawa, S.: Attrition and achievement gaps in online learning. In: Proc. 2nd ACM Conf. Learning @ Scale (L@S 2015), pp. 57–66 (2015)
31. Hlosta, M., et al.: Visual learning analytics of an online course. In: Proceedings of the Eighth International Conference on Learning Analytics & Knowledge (LAK 2018), pp. 151–155 (2018)

Digital Innovation and Transformation

A Multisource Data-Driven Model for Personalized Exercise Prescription for Diabetes Remission

Xiongji Liu[1](✉), Suepphong Chernbumroong[1], Pradorn Sureephong[1], and Jun Wu[2]

[1] College of Arts, Media and Technology, Chiang Mai University, Chiang Mai, Thailand
xiongji_liu@cmu.ac.th

[2] Guangdong CAS Cogniser Information Technology Co., Ltd, Guangzhou, China

Abstract. This study focuses on constructing a multisource data-driven large model for personalized exercise prescription tailored for patients in diabetes remission. By integrating wearable physiological monitoring data, in-hospital clinical examination data, and real-world exercise records, we develop a Transformer-based large model to generate safe, standardized, and personalized exercise prescriptions. A prospective comparison study will be conducted to compare AI-generated prescriptions with clinicians' exercise prescriptions to evaluate the feasibility and safety of the proposed system. This research aims to provide an effective and scalable artificial intelligence solution for diabetes chronic disease management.

Keywords: Diabetes Remission · Exercise Prescription · Transformer · Multisource Data · Wearable Device · IoT · Gymbot · Digital Health

1 Introduction

Type 2 diabetes mellitus (T2DM) remains a significant public health burden in Thailand. Recent diabetes care has increasingly emphasized remission-oriented management, defined as maintaining blood glucose levels below the diagnostic threshold for diabetes without glucose-lowering medication. Diabetes remission may improve quality of life, reduce diabetes-related complications, and decrease long-term healthcare costs (Division of Noncommunicable Diseases, Department of Disease Control, 2023). In Thailand, intensive lifestyle modification has been recommended as a key strategy for achieving T2DM remission. This approach integrates dietary modification, increased physical activity, and regular exercise over 3–6 months, with a target of approximately 10–15% weight reduction (Royal College of Family Physicians of Thailand, 2022). Exercise is particularly important because it improves insulin sensitivity, supports weight reduction, enhances glycemic control, and helps preserve skeletal muscle mass during weight loss (Colberg et al., 2016). Physical activity is also recognized as a core diabetes self-care behavior that supports sustained disease management (Association of Diabetes Care & Education Specialists, n.d.).

N. Mohd Sharef et al. (Eds.): CAMP 2026, CCIS 3094, pp. 237–245, 2027.
https://doi.org/10.1007/978-981-92-4305-1_17

Despite its importance, exercise monitoring in remission services remains limited. Existing practices commonly rely on fragmented tools, including glucometers, pedometers, smartwatches, LINE Official Account, Google Forms, and spreadsheets. These approaches often depend on self-reported data, manual verification, and spreadsheet-based tracking, while unequal technology access and limited digital literacy further constrain implementation (Mongkolsucharitkul et al., 2025). Consequently, delayed feedback may hinder timely exercise adjustment, reduce adherence, and increase the risk of dropout. To address these limitations, this study proposes GymBot, an edge-computing platform for automated real-time exercise monitoring in diabetes remission care. This study proposes GymBot as the platform which integrates AI-enabled camera-based motion analysis for posture detection, exercise movement assessment, and interactive feedback, with real-time glucometer data recording. By integrating exercise and glucose data into a unified platform, GymBot aims to support personalized exercise guidance, reduce manual follow-up burden, and enhance primary care support for T2DM remission.

2 Literature Review

2.1 Diabetes Management and Digital Transformation

Diabetes mellitus is a prevalent chronic metabolic disease characterized by persistent hyperglycemia and abnormal insulin secretion. Unregulated long-term glycemic control commonly triggers severe complications, including cardiovascular disorders, renal failure, retinopathy and peripheral neuropathy, imposing heavy medical burdens on global public health systems (Rhee et al., 2020). To alleviate the pressure of diabetes treatment, modern healthcare has gradually shifted from hospital-dominated diagnosis to integrated home-community-hospital management modes.

Driven by mobile Internet, IoT and big data technologies, digital health has become an indispensable auxiliary tool for diabetes management. Digital platforms and portable monitoring devices support continuous physiological tracking, real-time data recording and personalized health guidance, which effectively optimize self-management awareness of diabetic patients and reduce operational costs during long-term intervention (Cho et al., 2006). Numerous studies have verified that digital interventions can stabilize blood glucose levels and improve patients' treatment adherence.

Nevertheless, current digital diabetes solutions still have obvious limitations. Most platforms only focus on basic data storage and passive blood glucose monitoring, while neglecting intelligent and closed-loop exercise intervention services (Chu et al., 2023). The lack of integrated functions covering automatic data collection, adaptive exercise prescription generation and real-time safety monitoring restricts further improvements of digital rehabilitation systems, which creates research space for this study's dynamic exercise prescription framework.

2.2 IoT and Wearable Technology in Diabetes Care

The rapid development of IoT and wearable sensing techniques has revolutionized chronic disease management by supporting continuous, real-time monitoring for type 2 diabetes mellitus (T2DM) (Didyuk et al., 2021). Intelligent wearable devices can collect multi-source data covering daily exercise behaviors, calorie consumption and physiological indicators, including body weight, blood pressure, blood glucose and HbA1c. Compared with traditional hospital periodic detection, IoT-based monitoring effectively captures longitudinal behavioral and physiological changes, providing reliable data support for personalized diabetes intervention (Jafleh et al., 2024).

Although existing wearable-assisted solutions have proven effective for auxiliary glycemic regulation, most studies only focus on static data tracking and simple health alerts. Current intervention strategies still rely on fixed, standardized exercise prescriptions, which fail to adjust training intensity according to users' daily fatigue level and prior exercise status (Kanaley et al., 2022). Such rigid regimens frequently cause poor exercise adherence, unsatisfactory training effects and potential glycemic fluctuations.

Recently, deep learning-based adaptive strategies have been widely discussed in intelligent healthcare to compensate for the limitations of static prescriptions. However, few studies combine sequential exercise data and clinical indicators to generate dynamic prescriptions via Transformer architecture. Therefore, this study adopts the GymBot IoT system to collect synchronized exercise and clinical data, and develops a Transformer-based framework to generate adaptive daily exercise prescriptions, aiming to improve patients' exercise adherence and achieve safer and more efficient T2DM lifestyle intervention.

2.3 Artificial Intelligence in Exercise Prescription and Healthcare

Artificial intelligence (AI) has emerged as a core enabling technology for modern intelligent healthcare, exhibiting powerful capabilities in processing multivariate time-series data, mining potential physiological correlations and realizing individualized health intervention. In the field of chronic disease rehabilitation, AI algorithms are extensively utilized to analyze user activity habits, vital signs and historical rehabilitation records, assisting clinicians in formulating quantitative and targeted exercise plans instead of empirical and generalized schemes (Zhang et al., 2024).

Conventional AI-based exercise prescription studies primarily adopt machine learning algorithms and shallow recurrent neural networks to classify exercise types or predict fixed training loads. These methods struggle to capture long-range temporal dependencies between fluctuating physiological indicators and cumulative exercise fatigue, resulting in insufficient personalized adjustment capability for daily dynamic scenarios. With the advancement of sequential modeling techniques, the Transformer architecture has gradually replaced traditional recurrent models due to its superior performance in multivariate time-series learning. It effectively integrates multi-day historical data to quantify the interactive relationship between exercise load and physiological responses.

Despite these advancements, current AI-driven exercise prescription systems for T2DM patients still possess noticeable defects. Most existing models ignore prior-day exercise feedback and fail to embed clinical safety constraints into the inference process, thereby lacking closed-loop adjustment mechanisms oriented to diabetes-specific physiological characteristics (Xiang et al., 2026). To address such research gaps, this study employs a Transformer-based time-series model to learn dynamic patterns from wearable exercise data and clinical measurements, generating daily updated exercise prescriptions that balance exercise effectiveness, user fatigue tolerance and diabetes safety requirements.

3 Conceptual Framework

This work develops a closed-loop framework to create customized and safe exercise interventions for patients with type 2 diabetes (T2DM). As shown in Fig. 1, the framework comprises five interconnected modules to streamline the whole process from data collection to adaptive exercise adjustment. The system integrates two core data sources: daily workout data recorded by the GymBot device and routine clinical indicators. The combined data effectively reflects participants' exercise behaviors and real-time health status. The integrated data was preprocessed to train a time-series Transformer model. The model was utilized to capture intrinsic correlations between users' exercise activities and physiological responses for personalized prescription development. The trained model supports online inference and generates preliminary exercise plans, including workout type, duration and intensity based on each patient's latest health and exercise data. To eliminate potential exercise risks for T2DM patients, all preliminary prescriptions were validated and adjusted via a rule-based module. Guided by ACSM guidelines and clinical thresholds, this step ensures the safety and practicality of final prescriptions. Revised prescriptions were delivered to patients through the GymBot platform. Post-workout data including updated physiological indicators were collected and added to the dataset. This closed feedback loop allows the model to be continuously optimized, so as to adapt exercise prescriptions to the dynamic physical conditions of individual patients.

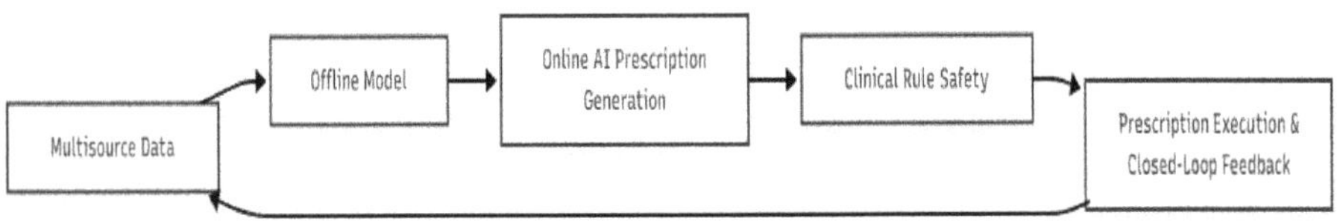

Fig. 1 Simplified conceptual framework of the closed-loop personalized exercise prescription system. The framework encompasses five core components covering data collection, model training, intelligent prediction, safety constraint and iterative feedback for T2DM patients.

4 Methodology

4.1 Participants

This study will recruit participants diagnosed with type 2 diabetes mellitus (T2DM) who use, or are able to use, the GymBot intelligent fitness system for regular physical training. Informed consent will be obtained from all participants before data collection.

The planned exclusion criteria will include severe cardiovascular complications, physical movement disorders, and unstable acute metabolic conditions that restrict exercise. Recruitment and data collection will begin only after the required ethical approval has been obtained. Enrolled participants are expected to contribute longitudinal workout records and routine clinical indicators, which will be used for model development and subsequent evaluation.

4.2 Model Development & Intervention Workflow

The proposed study will integrate two categories of heterogeneous data and develop a Transformer-based time-series framework equipped with dual-gating weighting mechanisms (Fig. 2). The planned workflow includes offline data preprocessing, gated feature weighting, online prescription inference, rule-based safety checking, and potential closed-loop refinement. The framework is intended to generate adaptive preliminary exercise prescriptions for T2DM patients, subject to empirical validation and clinical review.

4.2.1 Multisource Data Collection

Behavioral exercise data and clinical physiological data will be collected and integrated into a unified longitudinal dataset using user ID and acquisition date as a composite key.

1) GymBot Exercise Data
   The GymBot system will be configured to record multidimensional daily exercise data, including cumulative exercise days, daily exercise duration, real-time exercise curves, estimated calorie consumption, completed workout videos, historical training records, AI-generated exercise scores, and summary grades. These candidate indicators will be evaluated for their ability to represent daily exercise load, preferences, and movement execution quality.
2) Clinical Physiological Data
   Core clinical indicators associated with diabetes management and exercise safety will be collected, including body weight (BW), glycated hemoglobin (HbA1c), fasting blood glucose (FBG), postprandial blood glucose (PPG), and blood pressure (BP). Relevant medical history will also be collected as static baseline information, subject to ethical approval and data-protection requirements. These variables are intended to serve as candidate constraint features for individualized prescription generation and the proposed static-gating calculation.

4.2.2 Proposed Transformer Model Development and Exercise Prescription Generation

1) Offline Data Processing and Dual-Gating Feature Weighting
   Raw data will be preprocessed using prespecified procedures for missing values and outlier handling. The planned feature engineering will include prior-day exercise features based on a t-1 window and 3-day and 7-day rolling summaries of exercise and blood glucose trends. The study proposes a dual-gating weighting mechanism

consisting of static and dynamic components to examine differences in feature contributions across personalized T2DM scenarios. The static component will use relatively stable demographic and clinical baseline variables, including age, T2DM duration, complications, and relevant chronic medical history. Its weighting rules will be specified before model evaluation and will be treated as candidate constraints on exercise intensity and exercise-type selection. The dynamic component will use time-varying exercise and physiological variables, such as exercise intensity, estimated calorie consumption, movement completion quality, blood glucose, and blood pressure. It will be designed to assign greater attention to potentially abnormal physiological or exercise-load conditions, but the appropriateness of these weights will require validation. Continuous variables will undergo Z-score normalization, and categorical variables will be encoded for model input. Target exercise prescriptions, including duration, intensity, and exercise type, will be formulated with reference to ACSM guidance and input from qualified clinical or exercise professionals. The longitudinal dataset is planned to be divided chronologically into 70% training, 15% validation, and 15% test subsets to reduce time-series leakage. A time-series Transformer model will then be trained to estimate continuous and categorical prescription targets. Early stopping will be considered to reduce overfitting, and all model-development choices will be documented for subsequent evaluation.

2) Online Inference and Rule-Based Safety Correction

During the planned online-inference stage, input features will be organized using the same static and dynamic gating rules specified for offline processing. The Transformer model is expected to generate preliminary personalized prescriptions, which will then be checked by a rule-based safety module using prespecified clinical thresholds. The module will be designed to reduce exercise intensity or duration when selected indicators suggest hyperglycemia, hypertension, or excessive exercise fatigue. This rule-based process is intended to mitigate identifiable risks rather than guarantee safety, and AI-generated prescriptions will require clinical review before participant use.

3) Closed-Loop Iterative Mechanism

Following clinical review, eligible prescriptions may be delivered through the GymBot terminal. Exercise and clinical data collected after training sessions may then be added to the longitudinal dataset. Subject to data quality, safety review, and validation, the expanded dataset may support periodic model retraining or refinement. The proposed feedback process will be evaluated to determine whether it improves prescription personalization or adaptability.

5 Discussion

This work proposes a Transformer framework paired with the GymBot platform to design tailored exercise plans for patients with type 2 diabetes in remission. The closed-loop system adopts multi-source exercise and clinical data alongside a dual-gating weighting design, aiming to produce flexible and safe exercise arrangements and make up for the shortcomings of scattered monitoring tools and uniform workout plans used in routine diabetes care. This study also has several shortcomings. The recruited subjects do not

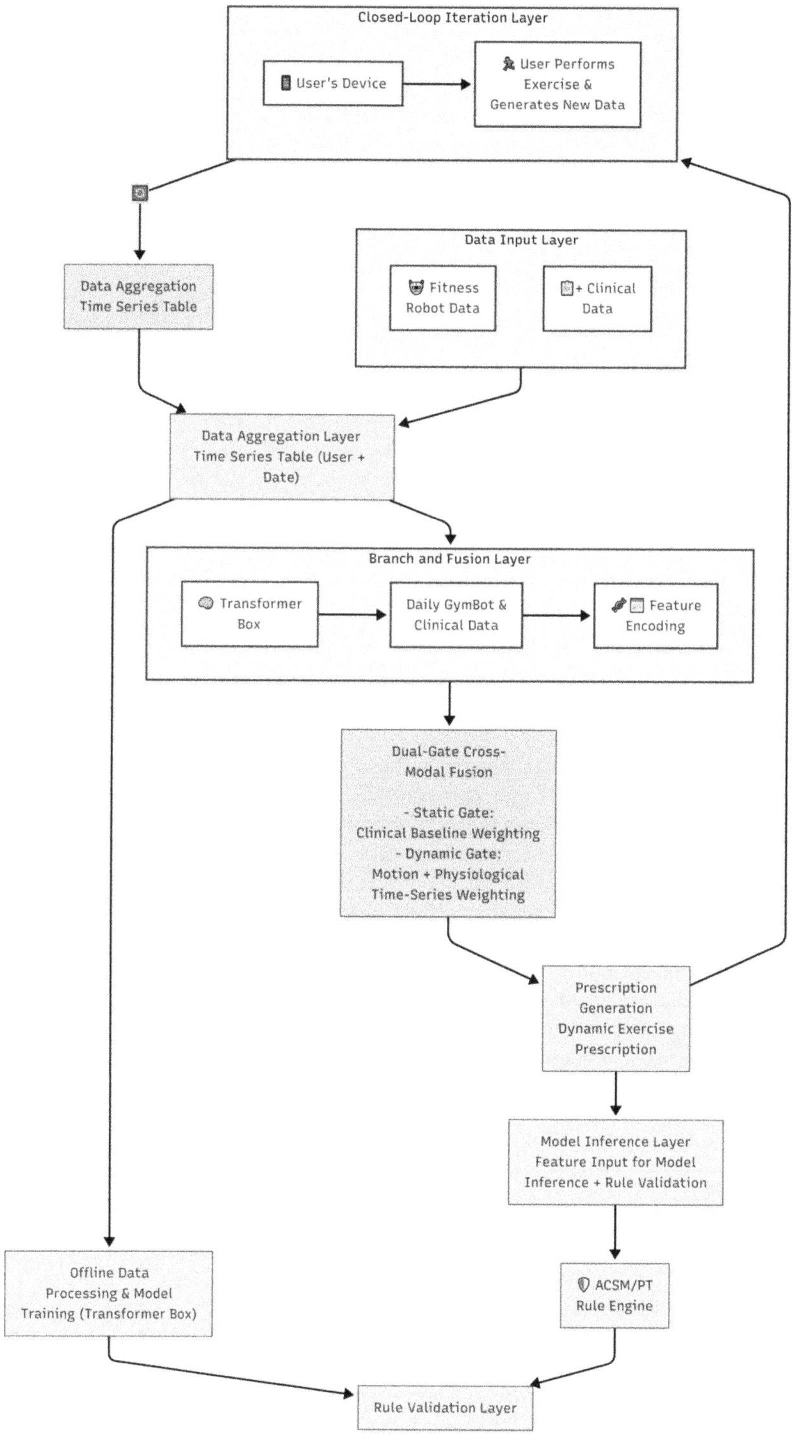

Fig. 2 Conceptual Framework of the Transformer-Based Dual-Gate Cross-Modal Closed-Loop Exercise Prescription Model for Remission-Stage Type 2 Diabetes Mellitus Patients

include patients with serious complications or physical mobility issues, and daily factors such as diet and sleep are not included in the research data. In addition, safety control depends on established rules, and the evaluation is limited to short-term indicators, so the long-term effect and applicability among different groups remain unknown.

Follow-up research may recruit more types of participants and add more lifestyle data to make the research more comprehensive. The algorithm design can also be improved, and longer observation periods can be arranged to study sustained intervention outcomes. In real-world use, connecting the platform with primary care services will help popularize this digital tool for diabetes management.

6 Conclusion

This research intends to build a multi-source data based model and a matching closed-loop system to work out personalized exercise prescriptions for patients with diabetes remission. Combining IoT technologies, wearable devices and the Transformer structure, the proposed framework is expected to achieve automatic data collection, smart plan formulation and real-time safety monitoring, which can effectively improve existing lifestyle intervention for diabetes. Since the model and supporting system are still under development, follow-up tests and verification are needed to check their practical effects. On the whole, this research offers a new direction for digital diabetes management. Subsequent work will focus on finishing model development, carrying out comparative trials and optimizing the whole system to support chronic disease treatment and care.

Acknowledgement. This work is supported by the European Union's Horizon Europe programme through the Marie Skłodowska-Curie Actions (MSCA), under the projects ACMod (Affective Computing Models: from Facial Expression to Mind-Reading, Grant Agreement No. 101130271) and HarmonicAI (Human-guided Collaborative Multi-Objective Design of Explainable, Fair and Privacy-Preserving AI for Digital Health, Grant Agreement No. 101131117). The authors would like to express their sincere appreciation to the College of Arts, Media and Technology, Chiang Mai University, for its academic and institutional support. Special thanks are extended to the Lamphun Provincial Public Health Office for its generous collaboration, valuable field insights, facilitation of local coordination, and continuous support throughout the development of this study. Their practical knowledge and close engagement with primary healthcare services in Lamphun have greatly contributed to shaping the research direction and ensuring its relevance to real-world diabetes remission care.

References

Association of Diabetes Care & Education Specialists: ADCES7 self-care behaviors. (n.d.). https://www.adces.org/diabetes-education-dsmes/adces7-self-care-behaviors

Colberg, S.R. et al.: Physical activity/exercise and diabetes: a position statement of the American Diabetes Association. Diabetes Care. **39**(11), 2065–2079 (2016). https://doi.org/10.2337/dc16-1728

Division of Noncommunicable Diseases, Department of Disease Control: Guidelines for Implementing Diabetes Remission Care Services (Remission Service). Ministry of Public Health, Department of Disease Control (2023)

Mongkolsucharitkul, P. et al.: Insights into diabetes remission services: perspectives from general practitioners, family physicians and multidisciplinary teams. Family medicine and community. Health. **13**, e003631 (2025). https://doi.org/10.1136/fmch-2025-003631

Royal College of Family Physicians of Thailand: Diabetes Remission in Type 2 Diabetes with Intensive Lifestyle Intervention Guide for Healthcare Providers. Royal College of Family Physicians of Thailand (2022)

Rhee, S.Y., Kim, C., Shin, D.W., Steinhubl, S.R.: Present and future of digital health in diabetes and metabolic disease. Diabetes Metab. J. **44**(6), 819–827 (2020)

Cho, J.H., et al.: Long-term effect of the internet-based glucose monitoring system on HbA1c reduction and glucose stability: a 30-month follow-up study for diabetes management with a ubiquitous medical care system. Diabetes Care. **29**, 2625–2631 (2006)

Chu, Y., Li, S., Tang, J., Wu, H.: The potential of the medical digital twin in diabetes management: a review. Front. Med. **10**, 1178912 (2023)

Didyuk, O., et al.: Continuous glucose monitoring devices: past, present, and future focus on the history and evolution of technological innovation. J. Diabetes Sci. Technol. **15**(3), 676–683 (2021)

Jafleh, E.A., et al.: The role of wearable devices in chronic disease monitoring and patient care: a comprehensive review. Cureus. **16**, e68921 (2024)

Kanaley, J.A., et al.: Exercise/physical activity in individuals with type 2 diabetes: a consensus statement from the American College of Sports Medicine. Med. Sci. Sports Exerc. **54**, 353–368 (2022)

Zhang, L., et al.: T2DFitTailor: a tool for type 2 diabetes patients to tailor exercise plan. medRxiv (2024)

Xiang, Y., et al.: Progress on exercise therapy in type 2 diabetes mellitus with cognitive impairment. Front. Sport. Act. Liv. **8**, 1742195 (2026)

A PRISMA-Informed Structured Narrative Review of Narrative Frameworks in AI-Powered Chatbot Counselling: Toward a Thai-Language Narrative Therapy Model

Ungkana Kerttongmee Warnett(✉), Pradorn Sureephong, Suepphong Chernbumroong, and Tama Duangnamol

Faculty of Digital Technology Management, Chiang Mai University, Chiang Mai, Thailand
ungkana_k@ymail.com

Abstract. Mental-health need among young adults is a growing global concern, and is especially acute where structural and cultural barriers limit access to care [3, 18]. In Thailand, limited service capacity and stigma-related constraints widen the treatment gap, and help-seeking is shaped by face-saving norms and relational understandings of the self [4, 21]. AI-powered chatbots have emerged as scalable tools that can support self-reflection and emotional engagement [16, 25], yet most validated systems implement Cognitive Behavioural Therapy (CBT), whose individualistic, skill-and-homework structure aligns poorly with relational, collectivist contexts [14, 22]. Narrative Therapy (NT) — which foregrounds storytelling, meaning-making, and non-pathologising language [24] — offers a potentially more culturally compatible alternative. This paper presents a PRISMA-informed structured narrative review of six empirically implemented narrative-based chatbot systems. We identify five categories of narrative approach and synthesise cross-cutting findings: narrative frameworks are technically feasible but clinically unvalidated; structured conversational design improves engagement and coherence; long-term narrative memory remains a critical technical limitation; cultural adaptation is largely underdeveloped; and an emerging ethical concern is that users may adopt AI-generated interpretations without critical reflection [17]. Building on these findings, we propose a conceptual model for a Thai-language NT chatbot integrating a narrative-framework layer, structured dialogue, retrieval-based memory, narrative-sensitive evaluation, and a cultural-adaptation layer, with cross-cutting ethical safeguards. The work contributes to the emerging field of culturally responsive therapeutic AI by consolidating fragmented evidence and setting out concrete design and research priorities.

Keywords: Narrative therapy · AI chatbot · Mental health · PRISMA-informed review · Cultural adaptation · Thailand · Large language models

N. Mohd Sharef et al. (Eds.): CAMP 2026, CCIS 3094, pp. 246–257, 2027.
https://doi.org/10.1007/978-981-92-4305-1_18

1 Introduction

Mental-health challenges among young adults represent a growing global public-health concern, and longitudinal evidence points to a marked decline in the mental well-being of the young over the past two decades [3]. Analyses of the Global Burden of Disease data show that mental disorders impose a substantial and rising burden among those under twenty-four, with projections indicating continued growth without effective intervention [18]. In Thailand, this challenge is amplified by structural limitations in mental-health service capacity and by culturally influenced barriers to help-seeking — stigma, concern about social judgment, and norms surrounding emotional restraint and the preservation of social harmony [4, 21].

AI-powered chatbot systems have emerged as a scalable, low-cost, and stigma-reducing complement to traditional services. Existing research shows that conversational agents can support self-reflection and emotional engagement and can produce measurable symptom benefit [1, 25], although high attrition and a fragile digital therapeutic alliance remain persistent limitations [2, 15]. Critically, most validated systems are grounded in Cognitive Behavioural Therapy (CBT), a highly structured framework rooted in Western, individualistic assumptions [14]. Its reliance on identifying and correcting maladaptive thoughts and on between-session homework presumes a self that is willing to be examined and corrected — an assumption in tension with contexts where identity and well-being are understood relationally, as in Thailand [22]. Evidence further indicates that interventions adapted to a specific cultural group are substantially more effective than generic, predominantly Western programmes [13], yet such adaptation is rarely implemented in chatbot systems.

Narrative Therapy (NT) offers a potentially more culturally compatible framework. It emphasises storytelling, meaning-making, and non-pathologising language, positioning individuals as the authors of their own experience rather than as bearers of a disorder to be corrected [23, 24]. Its externalising stance — separating the person from the problem — resonates with Thai and Buddhist attributions of suffering to circumstance, and because it does not require the client to accept a diagnosis, it does not activate the shame that blocks help-seeking. These characteristics also align naturally with the open-ended, co-constructive capacities of modern large language model (LLM)-based systems.

The case for a digital, narrative-based approach in Thailand is reinforced by the structure of the treatment gap. Specialist mental-health capacity is concentrated in urban centres, leaving rural and lower-income young adults underserved, and the cost and visibility of attending a clinic interact with stigma to deter help-seeking before it begins. A conversational system delivered on a familiar, private messaging platform lowers several of these barriers at once — cost, distance, and exposure — while a narrative stance sidesteps the diagnostic framing that itself carries stigma. The promise is therefore not merely scalability but a closer fit between the mode of support and the cultural conditions under which Thai young adults are willing to seek it.

Despite this conceptual alignment, the application of narrative frameworks in AI-powered chatbot counselling remains underexplored and fragmented across disciplines including social work, human–computer interaction, and computational linguistics. There is limited synthesis of how such systems are designed, implemented, and evaluated, particularly for culturally specific contexts. This study addresses that gap through a PRISMA-informed structured narrative review of empirically implemented narrative-based chatbot systems.

Research Question What narrative-framework methodologies and design principles from the existing AI chatbot counselling literature are adoptable for a Thai-language Narrative Therapy chatbot system for young adults?

2 Methodology

2.1 Review Design

This study adopts a PRISMA-informed structured narrative review, combining systematic search and screening procedures with qualitative, comparative synthesis. A narrative (meta-narrative) synthesis is appropriate for an emerging field in which the available studies are few, methodologically heterogeneous, and drawn from different research traditions, making statistical aggregation neither feasible nor meaningful [12]. The review is therefore intended as a targeted synthesis of empirically implemented narrative chatbot systems rather than an exhaustive systematic review of the broader literature, while still applying explicit, reproducible search and selection procedures.

2.2 Search Strategy

A structured search was conducted across major academic databases — PubMed, PsycINFO, ACM Digital Library, IEEE Xplore, Web of Science, and Google Scholar — covering publications from 2021 to 2026 and supplemented by hand-searching of reference lists. Search terms combined three concept blocks: (i) narrative frameworks (e.g. "narrative therapy", "storytelling", "re-authoring", "reminiscence"); (ii) chatbot and conversational technologies (e.g. "chatbot", "conversational agent", "large language model", "LLM"); and (iii) mental health and counselling (e.g. "counselling", "well-being", "mental health").

2.3 Inclusion Criteria

Studies were included if they (i) reported original empirical research; (ii) implemented an AI-based conversational system; (iii) applied a narrative-based or self-reflective framework; and (iv) provided sufficient detail on system design and evaluation to support comparative analysis. Purely conceptual papers, non-narrative systems, and works lacking any system implementation were excluded.

2.4 Selection Logic

Final selection rested on three criteria: relevance to narrative-based therapeutic or reflective frameworks; implementation of an AI conversational system; and the availability of evaluable design and outcome parameters. This two-tier logic distinguished a broad contextual literature — used to situate the field — from a focused analytical corpus of directly relevant systems analysed in depth, ensuring that the studies compared were chosen for relevance rather than convenience.

2.5 Study Selection Summary

The review followed PRISMA-informed procedures: an initial database search, removal of duplicates, screening of titles and abstracts, and full-text assessment of potentially eligible records. A final corpus of six studies was identified [6–9, 16, 17]. The selection flow is shown in Fig. 1. The small size of the focused corpus reflects the early stage of research on narrative chatbot systems rather than the absence of broader related literature, and is itself a substantive finding of the review.

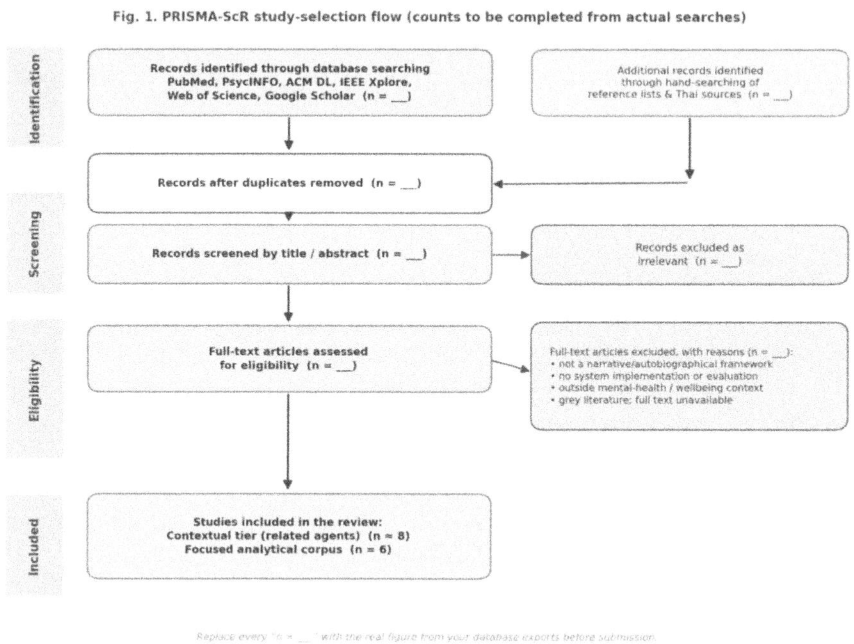

Fig. 1. PRISMA-informed study-selection flow. The record counts are placeholders to be completed from the actual database searches; the contextual tier situates the field and the focused corpus ($n = 6$) is analysed in depth.

3 Results

3.1 Overview of Included Studies

The six studies represent a small but diverse set of narrative-based chatbot implementations, spanning clinical practice, training, and observational research, and drawn from Chinese-, English-, and Norwegian-speaking contexts. Five categories of narrative approach were identified: (A) clinical narrative-therapy systems that operationalise NT directly [6, 7]; (B) training and simulation platforms that use LLMs to train or assess practitioners [8]; (C) autobiographical and reminiscence agents that structure storytelling for well-being [16]; (D) general-purpose LLM self-formation, where users employ everyday chatbots for autobiographical reasoning [17]; and (E) structured computational NT architectures with formal therapeutic-stage modelling [9].

Across these categories, the corpus moves along a spectrum from broad self-reflection to formal clinical practice. At one end, general-purpose models are appropriated by users for autobiographical reasoning with no therapeutic design at all [17]; in the middle, reminiscence and training systems impose structure on storytelling for well-being or skill development [8, 16]; and at the other end, purpose-built systems operationalise NT directly, culminating in a fully staged computational architecture with therapy-specific evaluation [6, 7, 9]. This range is useful precisely because it lets the review compare what structure, memory, and evaluation each design choice buys — and what each omits. Table 1 summarises the corpus, and Table 2 maps the systems against eight design dimensions used in the synthesis that follows.

Table 1. Five narrative categories are represented across six systems, with evaluation ranging from single-case autoethnography to a randomised controlled trial.

Study [ref]	Narrative framework	Method & sample	Platform / system
Chan (2025) [6]	NT — not-knowing stance, externalisation	Exploratory qualitative; $n = 23$ social-work students; China	WhatsApp + GPT-4 ("Nico")
Chan, Zhao & Zhao (2024) [7]	NT — practitioner-configured	Autoethnographic case study; $n = 1$ practitioner	WhatsApp + OpenAI Assistant API
Chan, Adnan & Wu (2026) [8]	NT training — AI-simulated client + fidelity scoring	Pre–post single group; $n = 35$; HK / China / Pakistan	Web platform + LLM-simulated client
Jiang et al. (2025) [16]	Autobiographical reminiscence — SFAM five-phase	Mixed-methods RCT; $n = 48$ (24 dyads)	Remini (PACM HCI / CSCW)
Kvietkute & Ness (2026) [17]	Narrative self-formation — Ricœur + Foucault	Qualitative interviews; $n = 16$; adults 20–35; Norway	ChatGPT (general-purpose)
Feng et al. (2025) [9]	INT — four-stage NT + IMA evaluation	Experimental; simulated + human raters (18–30)	GPT-4o-based INT

From Table 2, no system addresses cultural adaptation or transparency, and clinical validation is essentially absent — the principal gaps the proposed model targets.

Table 2. Design dimensions across the corpus (● present · ◑ partial · — absent).

Dimension	[6]	[7]	[8]	[16]	[17]	[9]
Formal NT framework	◑	●	●	–	–	●
Structured dialogue stages	–	◑	●	●	–	●
Long-term narrative memory	–	–	–	◑	–	◑
Narrative-sensitive evaluation	–	–	◑	◑	–	●
Clinical outcome validation	–	–	–	◑	–	–
Cultural adaptation	–	–	–	–	–	–
Transparency / explainability	–	–	–	–	–	–
Over-reliance safeguard	–	–	–	–	◑	–

4 Synthesis of Findings

Rather than describing each study in turn, this section synthesises cross-cutting patterns across the corpus.

4.1 Narrative Frameworks

Narrative approaches vary substantially in their depth of alignment with clinical NT. Some systems implement core NT principles such as externalisation, the not-knowing stance, and re-authoring [6, 7], and Feng et al.'s Interactive Narrative Therapist (INT) formalises the full therapeutic arc into four stages with explicit reflection-level planning [9]. Others adopt broader storytelling or self-reflection mechanisms — structured reminiscence [16] or unstructured autobiographical reasoning with a general-purpose model [17] — without formal therapeutic grounding. This spread confirms both that narrative frameworks are implementable in conversational AI and that few systems operationalise NT with clinical fidelity.

4.2 Chatbot Architecture

Two architectural approaches emerge. Unstructured conversational systems prioritise flexibility but lack therapeutic guidance, whereas structured architectures embed staged therapeutic flows. The evidence favours structure: in the only randomised comparison, a five-phase reminiscence design produced significantly greater positive affect, connectedness, and narrative co-construction than minimally guided prompting [16], and the four-stage INT outperformed generic LLM role-play across narrative-therapy dimensions [9]. Embedded therapeutic scaffolding, not raw model capability, appears to drive coherence and engagement. The lesson for design is that the conversational model is necessary but not sufficient: it is the explicit representation of where the user is in the therapeutic process — and what kind of move is appropriate next — that distinguishes a therapeutic interaction from an open-ended chat. Structure also makes a system auditable, since a staged flow exposes points at which safety checks, progress measurement, and cultural adaptation can be inserted.

4.3 Memory and Narrative Continuity

A consistent limitation is the inability to maintain long-term narrative continuity. Practitioner-configured systems lost the thread of a developing story once it exceeded the context window, breaking the cumulative re-authoring that NT requires [7]; retrieval-augmented generation partially addresses this in only one system [9]. This matters more for NT than for symptom-focused approaches: re-authoring is inherently longitudinal, depending on the system remembering which unique outcomes have already surfaced, which alternative storylines are being built, and how the preferred narrative has evolved across sessions. Without durable, structured memory of these elements, each session restarts the story, and the therapeutic arc cannot accumulate. Persistent narrative memory is therefore an architectural necessity rather than an enhancement, and remains the most pervasive technical constraint in the corpus.

4.4 Evaluation Methods

Evaluation remains underdeveloped. Most studies rely on qualitative analysis or engagement metrics, and clinical outcomes measured with validated instruments are largely absent — the most rigorous study assessed therapeutic quality rather than symptom change, and the single RCT measured affect and connectedness rather than clinical symptoms [9, 16]. The most promising direction is therapy-specific measurement: the Innovative Moment Assessment tracks genuine narrative transformation rather than surface engagement [9, 11], but lacks cross-cultural validation. This distinction is consequential, because engagement metrics can rise even when no therapeutic change occurs — indeed, an over-engaged user deferring to the model would register as a success on conventional measures while failing on the only criterion NT cares about. Process measures that detect genuine narrative shifts are therefore not a refinement but a necessity, and adapting them to Thai linguistic and cultural expression is a precondition for any credible effectiveness claim. Effectiveness in a mental-health population thus remains unverified.

4.5 Cultural Adaptation

Cultural adaptation is minimally addressed. The corpus is developed predominantly in Western or East-Asian-diaspora contexts and does not explicitly model the linguistic or cultural variation that NT delivery in Thai would require. This matters because help-seeking among East and Southeast Asian populations is shaped by self-reliance, public and self-stigma, and loss-of-face concerns, with a marked preference for informal or alternative sources of support [20, 22], and because culturally targeted interventions are markedly more effective than generic ones [13]. The absence of any Thai or wider Southeast Asian implementation is the clearest gap for the present work.

Adaptation here means more than translation. A Thai NT system would need to encode indirectness and face-preservation in how it asks questions, draw on Buddhist and relational idioms when externalising distress, and recognise that the family and community often figure in the preferred story as much as the individual does. Surface localisation that leaves a Western conversational stance intact risks the same mismatch that limits CBT chatbots; genuine adaptation treats culture as a design parameter running through dialogue, memory, and evaluation alike.

4.6 Ethical Considerations

An emerging ethical concern is that users may passively adopt AI-generated interpretations of their own lives instead of authoring them — a form of narrative deference directly at odds with NT's purpose of restoring authorship [17]. No system in the corpus detects, measures, or mitigates this risk, and none incorporates a transparency or explainability mechanism. Designs must therefore actively reinforce user agency rather than encourage passive acceptance.

4.7 Convergences, Divergences, and Research Gaps

Three convergences hold across the corpus: narrative frameworks are technically implementable in conversational AI; relational, structured designs outperform unstructured ones on engagement and coherence; and therapy-specific process measurement is both feasible and more informative than generic engagement metrics. The systems diverge sharply, however, in purpose (direct care, practitioner training, or observational study), in evaluation rigour (from single-case autoethnography to a randomised trial), and in cultural context — with no Thai or Southeast Asian work at all. Reading these patterns against Table 2 surfaces five shared gaps that any future system in this space must address:

- **G1 — Clinical validation is absent.** No study evaluates therapeutic effectiveness in a mental-health population using validated instruments; the strongest designs measured therapeutic quality or affect rather than symptom change [9, 16].
- **G2 — Cross-session narrative memory is fragile.** Context-window loss breaks the cumulative re-authoring NT depends on, and retrieval-based memory is only partially implemented in a single system [7, 9].
- **G3 — Narrative deference is unmonitored.** The risk that users cede authorship to the model — the failure mode most corrosive to NT — is neither detected nor mitigated anywhere in the corpus [17].
- **G4 — Cultural adaptation is missing.** No system models the language, face-saving norms, or relational idioms required for a non-Western context, despite strong evidence that culturally targeted interventions are more effective [13, 20, 22].
- **G5 — Transparency is absent.** No system makes its reasoning visible or contestable, even though this is the most direct design-level response to the narrative-deference risk in G3.

5 Proposed Model for a Thai Narrative Chatbot

Drawing the synthesis together, we propose a conceptual model for a Thai-language NT chatbot (TH-NTC) that composes the field's demonstrated strengths and supplies what it lacks. The model comprises five layers and a cross-cutting ethical component, delivered through LINE — the messaging platform most Thai young adults already use — to minimise the access barrier. Its overall architecture is shown in Fig. 2.

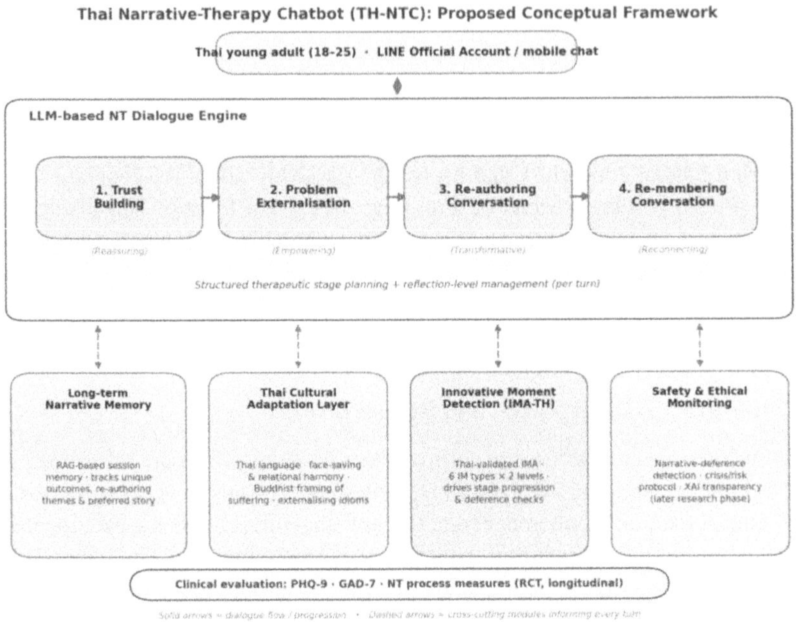

Fig. 2. The proposed TH-NTC conceptual framework: a staged NT dialogue engine with cross-cutting modules for memory, culture, evaluation, and ethical safeguards.

5.1 Narrative Framework Layer

Implements core NT processes — externalisation, re-authoring, and reflective questioning from a not-knowing stance — adopting the staged INT architecture shown to outperform generic role-play [9] and grounded in the foundational NT literature [23, 24].

5.2 Dialogue Structure Layer

Provides a multi-stage conversational flow — trust building, problem externalisation, re-authoring, and re-membering — that guides users through structured reflection while managing reflection level at each turn, answering the evidence that structured design improves coherence and engagement [16].

5.3 Memory Layer

Supports long-term narrative tracking through Retrieval-Augmented Generation (RAG), maintaining which unique outcomes have surfaced, which re-authoring themes are emerging, and how the preferred story is developing across sessions — directly addressing the continuity failure observed across the corpus [7, 9].

5.4 Evaluation Layer

Combines interaction indicators with a Thai-validated adaptation of narrative-change measurement (the Innovative Moment Assessment) to track genuine narrative transformation and, crucially, to distinguish authentic self-authoring from narrative deference [9, 11]. Translating and validating this coding for Thai pragmatics is a prerequisite for any future clinical trial using validated symptom instruments.

5.5 Cultural Adaptation Layer

Localises language and pragmatics and encodes face-saving sensitivity, relational framing, and Buddhist, externalising idioms for distress, so that engagement does not activate shame and the system aligns with Thai help-seeking norms [4, 20–22].

- Cross-cutting component — Ethical safeguards

  A monitoring component encourages active reflection and detects signs of over-reliance, prompting the user back into an authoring role rather than a receiving one, and keeps the system's reasoning visible and open to challenge. This directly targets the narrative-deference and transparency gaps identified in the synthesis [17].

  Together these layers provide a foundation for developing culturally aligned narrative AI systems in Thailand, and a concrete agenda for their empirical evaluation.
- Implementation and evaluation roadmap.

  Development would proceed in three phases. The first establishes the cultural-adaptation and dialogue layers and validates the Thai Innovative Moment coding scheme against expert ratings, since reliable process measurement is a prerequisite for every later claim. The second integrates the retrieval-based memory layer and conducts usability and engagement testing with Thai young adults, comparing structured against unstructured delivery as the corpus evidence would predict. The third pursues clinical validation — a randomised, longitudinal design using validated instruments such as the PHQ-9 and GAD-7 alongside narrative-process measures — while the ethical-safeguard component is monitored throughout for signs of narrative deference. This sequence ensures that effectiveness claims rest on measurement that has itself been validated for the Thai context, directly answering gaps G1 through G5.

6 Discussion and Limitations

The synthesis carries a clear implication for design: the value of a narrative chatbot lies less in raw conversational fluency than in the therapeutic scaffolding around it — staged dialogue, persistent narrative memory, process-level measurement, and cultural fit. The proposed TH-NTC model is deliberately composed from components already demonstrated in isolation across the corpus, lowering implementation risk, while the genuinely novel work — Thai cultural adaptation, narrative-deference monitoring, and clinical validation — is precisely where the field is currently silent.

Several limitations should be acknowledged. As a structured narrative review, this study maps breadth and identifies gaps but does not perform a formal quality appraisal or meta-analysis; pooled effect estimates await a more mature literature. The focused corpus is small ($n = 6$), heterogeneous in design and population, and drawn entirely from non-Thai, predominantly English- and Chinese-speaking contexts, which limits direct inference to the Thai setting — though that scarcity is itself a substantive finding. The included systems are recent (2024–2026) and lack long-term follow-up, and the proposed model is conceptual and, as yet, empirically untested. Finally, because narrative chatbot research is fast-moving and dispersed across disciplines, some relevant work may sit outside the searched databases; the review should therefore be read as a targeted synthesis rather than an exhaustive account.

7 Conclusion

This study has presented a PRISMA-informed structured narrative review of narrative frameworks in AI-powered chatbot counselling. The findings indicate that narrative approaches are technically feasible and conceptually well matched to conversational AI, but remain early-stage, with limited clinical validation and inconsistent system design. Four insights recur: narrative frameworks can be implemented in chatbot systems; structured dialogue improves engagement and coherence; long-term memory remains a critical constraint; and cultural adaptation and ethical design are underdeveloped.

Contributions. This paper makes three contributions. **First**, it provides a structured categorisation of narrative chatbot research that consolidates work previously fragmented across social work, human–computer interaction, and computational linguistics. **Second**, it synthesises the field's key design parameters, technical constraints, and research gaps into a comparable framework. **Third**, it proposes a conceptual model for a Thai-language Narrative Therapy chatbot that integrates narrative principles, system architecture, memory, evaluation, and cultural adaptation, together with a concrete empirical agenda. A Thai-language NT chatbot that tracks therapeutic progress over time, identifies innovative moments with cultural and linguistic accuracy, guards against narrative deference, and is validated against clinical outcomes would be a genuinely novel contribution to both digital mental health and the emerging field of culturally responsive therapeutic AI.

References

1. Abd-Alrazaq, A.A., Rababeh, A., Alajlani, M., Bewick, B.M., Househ, M.: Effectiveness and safety of using chatbots to improve mental health: systematic review and meta-analysis. J. Med. Internet Res. **22**(7), e16021 (2020). https://doi.org/10.2196/16021
2. Berger, T.: Rethinking the therapeutic alliance in digital mental health interventions. World Psychiatry. **24**(3), 345–346 (2025). https://doi.org/10.1002/wps.21343
3. Blanchflower, D.G., Bryson, A.: The global decline in mental health of the young. In: NBER Working Paper No. 33688. National Bureau of Economic Research (2025)
4. Burnard, P., Naiyapatana, W.: Culture and communication in Thai nursing: a report of an ethnographic study. Int. J. Nurs. Stud. **43**(6), 755–765 (2006)
5. Cabrera Lozoya, D., Conway, M., De Duro, E.S., D'Alfonso, S.: Leveraging large language models for simulated psychotherapy client interactions: development and usability study of Client101. JMIR Med. Educ. **11**, e68056 (2025). https://doi.org/10.2196/68056
6. Chan, C.: An exploratory study of a narrative therapy chatbot with social work students. J. Technol. Hum. Serv. (2025). https://doi.org/10.1080/15228835.2025.2559342

7. Chan, C., Zhao, J., Zhao, Y.: Our journey with an AI assistant offering narrative therapy on WhatsApp. J. Soc. Work. Pract. **39**(2), 165–189 (2024). https://doi.org/10.1080/02650533.2024.2420314
8. Chan, C., Adnan, M., Wu, J.: Feasibility of an AI-enhanced platform for training beginners in narrative therapy. J. Technol. Hum. Serv. (2026). https://doi.org/10.1080/15228835.2026.2663553
9. Feng, Y. et al.: Reframe your life story: interactive narrative therapist and innovative moment assessment with large language models. In: Proceedings of EMNLP 2025, pp. 24484–24509. ACL (2025)
10. Fitzpatrick, K.K., Darcy, A., Vierhile, M.: Delivering cognitive behavior therapy to young adults with symptoms of depression and anxiety using a fully automated conversational agent (Woebot): a randomized controlled trial. JMIR Ment. Health. **4**(2), e7785 (2017). https://doi.org/10.2196/mental.7785
11. Gonçalves, M.M., Ribeiro, A.P., Mendes, I., Matos, M., Santos, A.: Innovative moments as markers of meaningful change in psychotherapy. J. Constr. Psychol. **37**(1), 1–16 (2024)
12. Greenhalgh, T., Robert, G., Macfarlane, F., Bate, P., Kyriakidou, O., Peacock, R.: Storylines of research in diffusion of innovation: a meta-narrative approach to systematic review. Soc. Sci. Med. **61**(2), 417–430 (2005). https://doi.org/10.1016/j.socscimed.2004.12.001
13. Griner, D., Smith, T.B.: Culturally adapted mental health intervention: a meta-analytic review. Psychotherapy. **43**(4), 531–548 (2006). https://doi.org/10.1037/0033-3204.43.4.531
14. Im, C.-H., Woo, M.: Clinical efficacy, therapeutic mechanisms, and implementation features of cognitive behavioural therapy–based chatbots for depression and anxiety: narrative review. JMIR Ment. Health. **12**, e78340 (2025). https://doi.org/10.2196/78340
15. Jabir, A.I., Lin, X., Martinengo, L., Sharp, G., Theng, Y.L., Car, L.T.: Attrition in conversational agent-delivered mental health interventions: systematic review and meta-analysis. J. Med. Internet Res. **26**, e48168 (2024). https://doi.org/10.2196/48168
16. Jiang, Z., Yeo, S., Seow, W.X.D., Perrault, S.T.: Remini: leveraging chatbot-mediated mutual reminiscence for promoting positive affect and feeling of connectedness among loved ones. Proc. ACM Hum. Comput. Interact. **9**(7) (2025). https://doi.org/10.1145/3757650
17. Kvietkute, D., Ness, I.J.: Encountering generative AI: narrative self-formation and technologies of the self among young adults. Societies. **16**(1), 26 (2026). https://doi.org/10.3390/soc16010026
18. Liu, W. et al.: Global burden and trends of major mental disorders in individuals under 24 years of age from 1990 to 2021, with projections to 2050: insights from the global burden of disease study 2021. Front. Public Health. **13**, 1635801 (2025). https://doi.org/10.3389/fpubh.2025.1635801
19. National Alliance on Mental Illness: Mental health by the numbers (2025). https://www.nami.org/mhstats
20. Park, J., Lee, S., Park, G., Woodward, S.: Mental health help-seeking behaviours of east Asian immigrants: a scoping review. Eur. J. Psychotraumatol. **16**(1), 2514327 (2025). https://doi.org/10.1080/20008066.2025.2514327
21. Phadsri, S., Shioji, R., Tanimura, A.: Mental health stigma in Thai society: cultural barriers and psychological implications. J. Explor. Interdiscip. Methodol. **1**(6) (2024)
22. Shi, W., Shen, Z., Wang, S., Hall, B.J.: Barriers to professional mental health help-seeking among Chinese adults: a systematic review. Front. Psych. **11**, 442 (2020). https://doi.org/10.3389/fpsyt.2020.00442
23. White, M.: Maps of Narrative Practice. W. W. Norton, New York (2007)
24. White, M., Epston, D.: Narrative Means to Therapeutic Ends. W. W. Norton, New York (1990)
25. Zhang, Q., Zhang, R., Xiong, Y., Sui, Y., Tong, C., Lin, F.-H.: Generative AI mental health chatbots as therapeutic tools: systematic review and meta-analysis. J. Med. Internet Res. **27**, e78238 (2025). https://doi.org/10.2196/78238

AI-Assisted User-Centered Storyboarding for Serious Simulation Game Integrating Human Factors Engineering, Design Thinking, and AI Automation

Mya Thu Zar Aung(✉), Pradorn Sureephong, Suepphong Chernbumroong, and Tama Duangnamol

Faculty of Digital Technology Management, College of Arts, Media and Technology, Chiang Mai University, Chiang Mai, Thailand
{myathuzar_aung,pradorn.s,suepphong.c,Tama.d}@cmu.ac.th

Abstract. Serious simulation games are widely used in healthcare, aviation, military, and corporate training to support decision-making and skill development within realistic training environments. However, their development is often costly and time-consuming, mainly because of the complex storyboard design. Storyboard development is a critical backbone of the game production where traditional text-based storyboards cause communication difficulties, production delays, and limited user-centered design. This study proposes a conceptual workflow to achieve user-centered storyboard with AI-Automation for serious simulation design. That integrates with combination of Human Factors Engineering (HFE), serious simulation Design Thinking and AI-automation. HFE identifies human abilities and limitations, such as memory constraints, physical capabilities, and environmental stressors, while Design Thinking translates these insights into user-centered design decisions. AI is then applied to automate storyboard generation and visualize simulation scenarios. The proposed approach aims to improve design efficiency, reduce development costs, enhance stakeholder communication, and support the creation of more effective and learner-centered serious simulation games.

Keywords: Serious games · Simulation · AI storyboarding · Human Factors Engineering · Design Thinking · User-centered design · Generative AI · Digital game-based learning · UX/UI Design

1 Introduction

Learning and practicing surgery without touching a real patient or practicing how to handle a plane crash without any real danger would be questionable. But serious simulation games make this possible. They combine the entertainment parts of serious games such as points, rewards, story and feedback with simulations technique (Mitgutsch and Alvarado 2012; Vlachopoulos and Makri 2017). Healthcare, aviation, defense, and corporate training use them to let people practice safely and efficiently (Vlachopoulos and

N. Mohd Sharef et al. (Eds.): CAMP 2026, CCIS 3094, pp. 258–272, 2027.
https://doi.org/10.1007/978-981-92-4305-1_19

Makri 2017; Gómez et al. 2025). These games market is growing fast. Digital health game-based learning market alone was worth USD 4.65 billion in 2024 and is expected to reach USD 15.95 billion by 2030 (Grand View Research 2024). 33.54% of the total serious games market is from healthcare sector (IMARC Group 2024; Mordor Intelligence 2025). Research shows these games help to reduce the risk of clinical mistakes (Diaz-Navarro et al. 2024) and that 85% of medical educators believe that they improve student engagement (Grand View Research 2024).

Despite all this success, there is a big problem hiding at the start of pre-production called, storyboard. A storyboard is like a blueprint that shows every scene, learning goal, and decision point before the game is built. If the storyboard is wrong or unclear, the whole game suffers with scope creep, poor learning outcomes, unclear stakeholder communications, wasted project investment, production delays and in severe cases, project abandonment or contract termination (Mitgutsch and Alvarado 2012; Aleem et al. 2016; Ben Amara et al. 2024; Bunt et al. 2024). One small healthcare serious game took 14 months and USD 10,000 to build, and still needed a complete rebuild because the pre-production planning was not good enough (Miyosawa et al. 2025). Mixed of domain content, pedagogical theory and game design concept cause knowledge paradox as the 'chocolate-covered broccoli' that incoherence between entertainment and educational goals (Mitgutsch and Alvarado 2012).The core problem is that manual storyboards are mostly text based, time consuming to build and most narrative stories are without thinking and defining carefully about what real players can and cannot do. As a human, game players have their abilities and limitations such as physical abilities, memory limits and environmental stress where they are working.

This paper address both human-related and process-related problems of storyboarding inefficiency in serious simulation game design. The proposed framework aims to replace manual, text-heavy workflows with an AI automated pipeline capable of generating scene visualizations, scenario flow diagrams, competency maps, and UX/UI prepared data from a creative template scenario brief. The solution finds in combining three approaches. First, Human Factors Engineering (HFE) tells us exactly what human abilities and limitations to design for. Second, Design Thinking gives a user-centered process to turn those abilities and limitations into good design decisions of serious simulation game techniques. Third, AI automates the creation of storyboard once the user needs are clear. Together, these components form the proposed AI-Assisted User-Centered Storyboard (AI-AUCS) framework for serious simulation game design. In this paper, Section 2 reviews the related work, Section 3 explains the methodology, Section 4 presents the framework and then Section 5 concludes.

2 Related Work

2.1 The Storyboard Problem in Serious Simulation Games

Storyboards are the main communication tool between game designers, educators, developers and stakeholders before a game is built (Bunt et al. 2024). Three well-known problems make manual storyboards fail. First, they are full of text, which is hard for non-designers like doctors or teachers to read and check (Ben Amara et al. 2024). Second, they take a long time to make, especially for games with many branching scenarios.

Third, they do not show how each scene links to a learning goal, so it is impossible to check whether the game teaches what it is supposed to teach before production starts (Ben Amara et al. 2024). Research confirms that a poor storyboard or game design document directly causes rework and financial losses in production (Aleem et al. 2016). Carlier et al. (2023) showed that since there is no standard process, serious game designers must restart the whole design process from scratch every single time when needed to edit that occurs wasting time and money. None of the 47 existing serious game design frameworks reviewed by Maxim and Arnedo-Moreno (2025) address storyboard automation or AI-driven pre-production framework yet.

2.2 Human Factors Engineering: Designing for What Humans Can and Cannot Do

A critical question is why serious simulation games sometimes fail to achieve effective training outcomes even containing the good content. One major reason is that interactive game was designed without thinking about player abilities and limitations (ISO 2019). Human Factors Engineering (HFE) is the science of designing systems that fit real human abilities and limits. It bridges the gap between human capabilities, limitations and technology design. It covers three areas that are highly relevant to storyboard UX/UI design to ensures system design are safe, intuitive and efficient by aligning technology with natural human nature.

The first is Cognitive Load, how much mental effort player can take needs to think in design. Human working memory can only hold about seven things at one time (Sevcenko et al. 2021). Sevcenko et al. (2021) proved this directly in a serious simulation game study, players perform best at a medium level of cognitive load. Human brain has limitation, if scenarios give too many instructions at once, players become overwhelmed and stop learning. If the game is too easy, they get bored. Excitement or playful enough to practice in the game repeatedly is also the objective of serious simulation games. Thus, designer needs to balance the cognitive overload or underload. HFE-based storyboard design can identify both problems by carefully controlling how much information appears in each scene. The second is Physical Ergonomics, what movements and interactions the human body can realistically perform. A storyboard that requires a simulated surgeon to do ten things simultaneously ignores the fact that a real surgeon does them one step at a time in the real world. The third is Environmental Stressors such as noise, time pressure, and uncertainty that real professionals face on the job (ISO 2019). Unlike entertainment games, serious simulation games such as medical emergency games, flight simulators, combat zone trainings need to use these stressors to replicate real-world pressures. Designers intentionally add and balance them to train players situational awareness avoid critical errors under pressure. HFE tells designers to embed these stressors authentically into every scene. When designers know these three constraints from the beginning, they can build storyboards to produce training that works in the real world.

2.3 Design Thinking: Turning HFE + Serious Simulation Design Decisions

Knowing human abilities and limits are not enough, story designer also needs a process to turn that knowledge into a good serious simulation design. Existing frameworks for serious game design such as Serious Game Design Assessment (SGDA) Framework (Mitgutsch and Alvarado 2012), the stakeholder-centered enterprise architecture model (Bunt et al. 2024), and the iterative evaluation approach SGDA-IE (Ben Amara et al. 2024), have meaningfully structured the broader design process. To combine those frameworks with HFE, Design Thinking provides this process through five phases as mentioned in Empathize, Define, Ideate, Prototype, and Test (Schoormann et al. 2024). The Empathic Design Thinking Framework (EDTF), which integrates co-design sessions and user storyboards into the empathy phase of educational game design, confirms that user-centered storyboarding significantly improves alignment between design intentions and learner experience (Aslan et al. 2025). In standard Design Thinking, the Empathize phase collects information about users through interviews and observation. The proposal of this research is to infuse HFE into this phase. Instead of just asking users what they feel, HFE-infused empathy measures what they can cognitively handle, what physical steps they can realistically perform, and what environmental stress they face. This turns soft user feelings into hard design requirements and constraints (Brauner et al. 2022). These abilities and constraints then combine it in the Define and Ideate phases of design thinking. So, every storyboard decision of serious simulation technique such as UX/UI integration, pedagogical features, competency mapping, how many choices appear in a scene, how tasks are sequenced, whether alarms or time pressure needed in the game is grounded in real user data rather than designer guesswork.

2.4 AI for Storyboard Automation

Once user needs and HFE abilities constraints are clearly defined and prepared storyboard data, AI can automate much of the storyboard generation process once user requirements and HFE constraints have been clearly defined. Large Language Models (LLMs) can read a scenario script and extract structured scene data, generate descriptions for each scene, and produce branching decision flows (Gallotta et al. 2024). Zhang et al. (2024) showed that LLMs can generate complete workflow models from plain text descriptions. Chen and Jhala (2025) demonstrated a pipeline that turns narrative prompts into visual game scenes. Generative image AI can produce mock-up visuals for each scene (Feuerriegel et al. 2024; Chen and Jhala 2025). These three capabilities together such as text parsing, workflow generation, and image synthesis can replace the most time-consuming parts of manual storyboarding. The AI in Game Development Survey found that over 85% of game studio executives already use AI in their work (a16zGames 2024), showing the industry is ready for this shift.

2.5 Research Gap

Existing serious game design frameworks (Mitgutsch and Alvarado 2012; Bunt et al. 2024; Ben Amara et al. 2024) do not address pre-production storyboard automation.

AI tools can automate scene generation but do not consider human limitations or user-centered design (Chen and Jhala 2025; Feuerriegel et al. 2024). HFE and Design Thinking address user needs but have not been combined with AI into a storyboard-specific framework for serious simulation games. I will design into this research proposes to integrate HFE, Design Thinking of Serious Simulation and AI Automation.

3 Research Methodology

The overall research follows the Design Science Research (DSR) paradigm. (Peffers et al. 2007), which is appropriate when the goal is to design, build, and evaluate a new artefact that solves a clearly identified practical problem. Design Science Research (DSR) paradigm requires three iterative cycles of identifying the problem and defining requirements, designing and demonstrating the artefact, and evaluating its utility against those requirements (Peffers et al. 2007).

This paper presents to identify problems from the first phase of an ongoing research project. The upcoming phases such as define requirements, design and demonstrate the artefact, evaluation and validation are planned as subsequent phases.

3.1 Phase1: Identify Problem and Define Requirements

3.1.1 Identify Problem

Venable et al. (2017) analyzed, DSR starts with finding problems since initial state based on (Peffers et al. 2007). Thus, this paper contributed the problem identification with related literature synthesis, gap analysis and conceptual framework design as follow.

- Conducted a systematic literature review in five areas such as serious simulation game design frameworks, storyboarding practices, Human Factors Engineering (HFE), Design Thinking and generative AI.
- Identified a research gap and found no existing storyboard framework integrates HFE, Design Thinking, and AI automation into a pre-production workflow for serious simulation games.
- Developed a preliminary conceptual framework based on the problems and solution through related works analysis.

This phase onwards, the ongoing processes planned to operate phase by phase as the following structure.

3.1.2 Define Requirements

Planned to gather user evidence on storyboard challenges and their needs through semi-structured interviews. In this phase, three participant groups of specific domain expertise related to the proposed framework will be recruited. Group 1, serious game designers/instructional designers. Group 2, healthcare simulation educators and subject-matter experts. Group 3, development team members such as UX/UI designers, developers and testers with experience in educational technology. Estimated 3 to 6 person per focus group, giving a total of approximately 10 to 18 participants. Interviews will be structured using Design Thinking Empathise phase, deliberately infused with HFE principles to move beyond subjective user preferences and capture three measurable human limitation constraints: Cognitive Load, Physical Ergonomics, and Environmental Stressors (ISO 2019). These HFE points illustrate connection between each human limitation and its corresponding storyboard design decision. Interview data will be analysed using Design Thinking Define process. To define and realize whether solution and conceptual framework in 3.1.1 is a suitable solution based on the active user problems and requirements of in this industry.

3.2 Phase 2: Design and Demonstrate the Artefact (Prototype)

This phase intends to design and develop a working prototype. Hence, planned to use Design Thinking Ideate and Prototype processes to prepare all necessary design knowledge based on phase1. Below is the list to operate.

- Design workflow concept
- Design necessary serious simulation game variables
- Design AI-automation
- UX/UI simulation variables, auto generated scene images (LLMs) text-to-image AI technologies
- Branching scenario HFE-based competency-to-scene alignment maps
- Branching scenario flow diagram

3.3 Phase 3: Evaluation and Validation

Evaluation phase will assess both framework and prototype.

The framework will be reviewed by a panel of 8 to10 domain experts such as serious game designers, subject matter experts, players. By using a structured review instrument based on four criteria:

- Relevance to the identified storyboard problems
- User-Centered effectiveness via combination of HFE, Serious Simulation Design Thinking and AI
- Usability for practitioners
- Pedagogical alignment with serious simulation game design theory.

Each criterion will be rated using a 5-point Likert scale. Scores and qualitative feedback will be analyzed to identify framework strengths and areas for improvement.

Prototype quality will be assed using an expert review approach involving stakeholders, serious simulations designers, subject matter experts, UX/UI designers, developers and testers. Evaluation will conduct through usability testing, user experience assessment, storyboard quality evaluation, and comparative analysis against human-created storyboard. A structured evaluation will be developed based on serious simulation game design principles, Human Factors Engineering (HFE), and instructional design requirements as follow:

- User input scenario narrative coherence and logical flow
- Relevance to learning objectives

- Accuracy of competency alignment
- Integration of HFE considerations
- Visual clarity and communication effectiveness
- Decision-point and branching quality

Each criterion will be rated on a five-point scale, with a mean score of 4.0 or above used as the acceptance threshold. The prototype will be evaluated with approximately 25 end-users using the System Usability Scale, SUS (Brooke 1996) targeting a score above 70 to measured quality of the prototype, and the User Experience Questionnaire, UEQ (Laugwitz et al. 2008) will be access for broader user experiences such as attractiveness, efficacy, easy to follow, dependability. These measures will provide insights into user-centeredness of this prototype.

Since the manual storyboard creation is one of the problems stated in this research, compare AI-generated storyboards against human-created storyboards is one of the matters. A comparative evaluation will be conducted using equivalent scenario briefs. For each scenario, both an AI-generated storyboard and a human-created storyboard will be produced. Those 25 participants and expert reviewers mentioned earlier, will evaluate both versions using the storyboard quality. The comparison will focus on:

- Story quality and scenario realism
- Learning objective alignment
- HFE integration
- Creativity and innovation
- Visual communication quality
- Development Time (Efficiency)
- Overall usefulness

Quantitative scores and qualitative feedback will be analysed to identify similarities, differences, and potential advantages of the AI-assisted approach.

Success Metrics The proposed overall research will be considered successful if it achieves the following criteria:

- SUS score greater than 70, indicating acceptable usability.
- Positive UEQ benchmark ratings across key user experience dimensions, storyboard quality ratings of 4.0 or above on a 5-point scale.
- Demonstrated alignment between storyboard outputs and intended learning objectives.
- Evidence of HFE integration addressing cognitive load, physical ergonomics, and environmental stressors.
- Reduced storyboard development time compared with traditional manual approaches.
- Positive qualitative feedback from designers, educators, and subject-matter experts regarding usefulness, practicality, and adoption potential.

Together, these measures will provide comprehensive evidence of the usability, quality, effectiveness, and user-centered solution of the proposed AI-assisted storyboard framework for serious simulation games.

4 Preliminary Conceptual Framework

To develop a preliminary conceptual framework, this research follows based on the problems and requirements as in sects. 3 DSR method. Through related work literature reviews in sect. 2, research gap analyzed as there is no existing storyboard framework integrates with HFE, Design Thinking, and AI automation for storyboarding phase of serious simulation games.

Identify Problem

Two main problem categories identified. Process-related and human-related problems. Under the category of human-related problem, cause of ignorance or incomplete player center concept in game design is concerning. When storyboard designers ignore player's practical human natures such as players abilities and limitations, it will not contribute a good design to motivate while playing the game. Another problem is also crucial in the game design stage. It is process-related problem manual workflows, text-based narrative story without visualization, slow iteration and high cost due to impact of those processes.

Solution

In this paper, solution identified based on the research gap and problems analyzed in related work literatures.

Table 1. Link of problems and proposed solution.

Problem Category	Problem	Proposed Solution
Human-Related Problems	Designers ignore player's HFE factors, no user-centered process	Pillar 1 and 2: Creative template prepares with HFE + Design Thinking+ Serious Simulation game design theory to ensure every design decision. It reflects purposely user-centered design within the serious games and simulation techniques
Process-Related Problems	Manual workflows, no visualization, slow iteration, high cost	Pillar 3: AI automates scene generation, flow diagrams, and competency mapping based on pillar 1 and 2

Based on Table 1, combination of three pillars integrates for the proposed solution. This paper contributes the conceptual frameworks to develop and demonstrate the AI-assisted storyboarding prototype. By addressing technical details of the AI pipeline, HFE carriables and competency-mapping logic as in following pillars.

Pillar 1: Creative Data Preparation through HFE

This is the Human-centered engine of the framework. It prepares important data based on scenario story, game objective and learner's profile input. Three types of HFE human abilities and limitation, Cognitive Load limits, Physical Ergonomics requirements, and Environmental Stressor profiles (ISO 2019; Brauner et al. 2022) become the rules that shape every storyboard scene. Each HFE constraint identified during the Empathize phase is not meant to treat as a design impression, but intended to capture as a structured, machine-readable variable that can later be retrieved and reasoned over by the AI pipeline in Pillar 3.

Defining the HFE Variables

HFE three constraint categories will be operationalized as the following variables:

HFE Constraint	Variable Name	Variable Data type	Scenario Complex Decisions Data	Source
Cognitive Load	HFE_cog_load_index	Limit (1–5) (Data will input by storyboard designer)	1 = minimal 2 = easy 3 = moderate 4 = high 5 = complex	Derived from number of simultaneous decision points per scene, informed by working memory limits (Miller 1956 cited in Sevcenko et al. 2021)
Physical Ergonomics	HFE_task_sequence_type	Categorical (Data will input by storyboard designer)	sequential / parallel / interrupted	Follow sect. 3, Interview transcripts describing realistic clinical task flow
Environmental Stressors	HFE_stressor_profile	label	noise, time_pressure, uncertainty, distraction	Coded from expert descriptions of real-world working conditions (ISO 2019)

These three variables aimed to store as structured metadata attached to every scene node in the storyboard. This distinguishes the proposed framework from existing approaches by grounding purposeful user-centered storyboard design decisions in measurable human capabilities, limitations and contextual requirements.

Competency-Mapping Logic
Competency mapping links each scene to a specific learning objective to make sure and see interconnected of each logic. The mapping concept will be structured as below:

Competency Mapping Concept	Example
Scene_ID → Competency_ID → Assessment_Criterion	Scene_07 → Competency:"Airway Management Under Pressure" → Assessment: "Learner identifies correct intervention within 90 s under HFE_stressor_profile = {time_pressure, noise}"

This automatic structure easily allows user to see linkage and verify whether every scene in a storyboard serves a defined learning objective and directly addressing the competency misalignment problem identified in Sect. 2. Each Competency_ID is drawn from a controlled vocabulary (e.g., a healthcare simulation competency classification) related to the input story document, allowing the system to check coverage across an entire storyboard and flag any learning objective that has no corresponding scene.

Pillar 2: Serious Simulation Game Design Theory + Design Thinking
Combine pillar1 HFE data into every UX/UI ensures align serious simulation game design principles. SGDA Framework (Mitgutsch and Alvarado 2012) guides that each scene serves the game's learning purpose. Integrate Design Thinking with SGDA framework achieve user-centered serious simulation design. This will contribute as in stakeholder-centered design (Bunt et al. 2024) ensures that educators, clinicians, and developers all clear vision to the storyboard from the start.

Pillar 3: AI Generative Technologies
This is an important AI-assisted intelligent decision-support-layer of the framework. It retrieves data and automates the storyboard production based on pillar 1 and 2. An LLM parses the scenario brief and extracts structured scene data (Zhang et al. 2024; Gallotta et al. 2024). A generative image system produces visual mock-ups for each scene (Feuerriegel et al. 2024). A workflow engine maps scenario branching logic and links each scene to its learning objective. The output replaces the manual text-heavy storyboard with a visual, structured, competency-aligned, user-centered pre-production reference that all stakeholders can understand.

Technical Pipeline
The prototype pipeline consists of four connected stages:

Stage 1: Scenario Parsing (LLM-based NLP) a Large Language Model parses the data input from storyboard designer text-based scenario brief and extracts structured scene variables such as setting, characters, decision points, and learning contents. This is consistent with demonstrated LLM capability to extract structured data from narrative text (Zhang et al. 2024; Gallotta et al. 2024).
Stage 2: Knowledge Retrieval (Retrieval-Augmented Generation) aim of this stage is to make sure **retrieval of pillar 1, 2 and 3 design knowledge during generation**. Before generating any scene content, the system performs a **retrieval step** against a

structured knowledge base containing (a) prior HFE constraint variables from Pillar 1, (b) the competency taxonomy, and (c) a library of validated serious simulation game design patterns (Mitgutsch and Alvarado 2012; Bunt et al. 2024). Combination of this retrieval ensures that scene generation is grounded in prior expert knowledge rather than relying solely on the LLM's general training data.

Stage 3: Constraint-Conditioned Generation LLM generates each scene descriptions and the generative image model produce the visual mock-up, both conditioned on the retrieved HFE variables (HFE_cog_load_index, HFE_task_sequence_type, HFE_stressor_profile) and the assigned competency triple mapping. For example, a scene with HFE_cog_load_index = 5 is automatically generated with a single clear instruction and minimal on-screen choices, rather than the multiple simultaneous prompts a generic LLM might otherwise produce.

Stage 4: Decision Support and Validation

The system cross-checks the generated storyboard against the competency-mapping logic from Pillar 1, flagging any scene with a missing competency link or any learning objective without a corresponding scene. This functions as an **intelligent decision-support tool** for the designer, rather than simply producing content, the system actively recommends corrections, surfaces design gaps, and ranks candidate scene variations against the HFE constraint profile.

Selected Components Combined Solution

The reason of each component was selected to combine as a solution for this paper, and no other components are selected because it solves a specific problem identified in Sect. 2. No single existing framework addresses all four problems simultaneously, because each one was developed to solve only one category of problem. HFE alone has no game design theory. Design Thinking alone has no specific game design grounding domain. SGDA alone has no process for gathering user data or automating production. Generative AI alone has no grounding in human limitations or pedagogical theory. This proposed combination produces one capability that does not exist in any single component. The ability to automatically generate a storyboard scene that is simultaneously constrained by human ergonomics validated against a specific learning objective and aligned with serious game design purpose without requiring a human designer to manually check all three conditions for every scene.

To achieve more clarification for the process workflow of these components are shown as in the following Fig. 2.(AI-AUCS) Technical Workflow. The workflow process designed based on the combination of pillar 1,2 and 3 to improve the dual problems of human-related and process-related. Aimed to handle by technologies and purposeful and creative design formulation. To achieve clarity of the process, illustrated Fig. 2 (AI-AUCS) Technical Workflow as the reference for the prototyping stage.

Preliminary Conceptual Framework Development

The preliminary conceptual framework is the primary contribution of this paper. Combination of HFE, Serious Simulation Design Thinking and AI technology intended to get a user-centered storyboard for serious simulation design. This concept will process in four steps input, user-centered integrations, AI-automation and then result output will produce as in a storyboard.

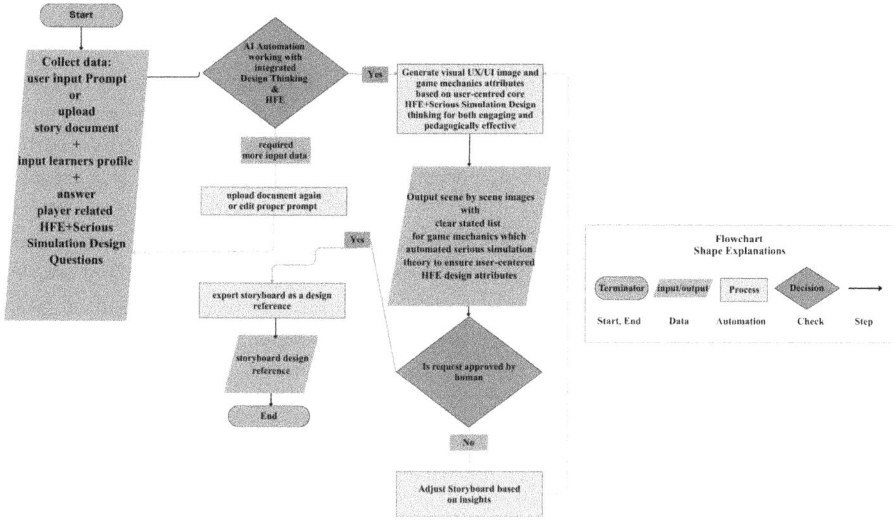

Fig. 1 Preliminary conceptual framework

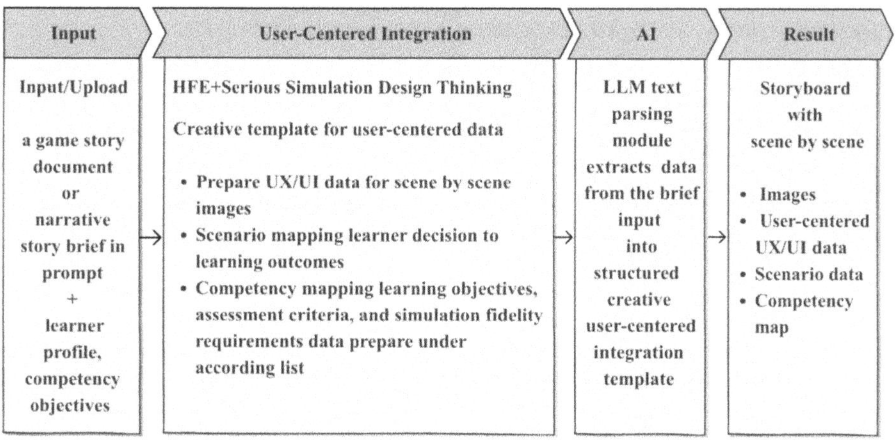

Fig. 2. (AI-AUCS) Technical Workflow

The workflow described as in Fig. 1, when storyboard designer or educator input the text-based story document or narrative story prompt together with learner profile and competency objectives then AI-automation will process based on creative user-centered integration and result output as a storyboard efficiently. This proposed framework aimed to prototype a functional AI-automated storyboarding tool for user-centered serious simulation games in the future.

5 Conclusion

Serious simulation games are evolving into a modern learning and training industry. Design development of serious teaching and learning game is crucial. This research provides a timely investigation into the use of AI automation for user-centered storyboarding in serious simulation game design. This paper presented the first phase of that ongoing research project. AI may help efficient in process-related problems such as manual workflow, visualization, and automation. Meanwhile, Design Thinking and Human Factors Engineering may help address human-related problems by making the design process more user-centered. Proposed concept is expected to reduce production time, improve stakeholder communication, and support more effective serious simulation game design that train people well. The outcomes of overall research will also benefit practitioners across healthcare simulation, education, emergency response, and corporate training, while advancing the scholarly understanding of AI-assisted design processes in serious simulation game development.

Acknowledgments. This work is supported by the European Union's Horizon Europe programme through the Marie Skłodowska-Curie Actions (MSCA), under the projects ACMod (Affective Computing Models: from Facial Expression to Mind-Reading, Grant Agreement No. 101130271) and HarmonicAI (Human-guided Collaborative Multi-Objective Design of Explainable, Fair and Privacy-Preserving AI for Digital Health, Grant Agreement No. 101131117).

References

Aleem, S., Capretz, L.F., Ahmed, F.: Game development software engineering process life cycle: a systematic review. J. Softw. Eng. Res. Dev. **4**, 6 (2016). https://doi.org/10.1186/s40411-016-0032-7

a16zGames AI in game development survey 2024. (2024). https://a16z.com/ai-in-game-development-survey. Accessed 20 Jan 2025

Aslan, S., Genc Kumtepe, E., Kumtepe, A.T.: EDTF: a user-centered approach to digital educational games design and development. Information. **16**, 794 (2025). https://doi.org/10.3390/info16090794

Ben Amara, B., Mhiri Sellami, H., Ben Said, L.: An approach for serious game design and development based on iterative evaluation. J. Softw. Evol. Process. (2024). https://doi.org/10.1002/smr.2680

Brauner, P. et al.: Beyond playful learning — serious games for the human-centric digital transformation of production and a design process model. Soc. Sci. **7**, 100580 (2022). https://doi.org/10.1016/j.ssaho.2022.100580

Brooke, J.: SUS: a quick and dirty usability scale. In: Jordan, P.W., Thomas, B., Weerdmeester, B.A., McClelland, I.L. (eds.) Usability Evaluation in Industry, pp. 189–194. Taylor & Francis (1996)

Bunt, L., Greeff, J., Taylor, E.: Enhancing serious game design: expert-reviewed, stakeholder-centered framework. JMIR Serious Games. **12**, e48099 (2024). https://doi.org/10.2196/48099

Carlier, S., Naessens, V., De Backere, F., De Turck, F.: A software engineering framework for reusable design of personalized serious games for health: development study. JMIR Serious Games. **11**, e40054 (2023). https://doi.org/10.2196/40054

Chen YC, Jhala A (2025) Narrative-to-scene generation: an LLM-driven pipeline for 2D game environments. arXiv preprint https://arxiv.org/abs/2509.04481

Diaz-Navarro, C., et al.: Global consensus statement on simulation-based practice in healthcare. Adv. Simul. **9**, 19 (2024). https://doi.org/10.1186/s41077-024-00288-1

Feuerriegel, S., Hartmann, J., Janiesch, C., Zschech, P.: Generative AI. Bus. Inf. Syst. Eng. **66**, 111–126 (2024). https://doi.org/10.1007/s12599-024-00884-5

Gallotta, R. et al.: Large language models and games: a survey and roadmap. IEEE Trans. Games. **1–18** (2024). https://doi.org/10.1109/TG.2024.3418618

Gómez, C., Vallejo, P., Aguilar, J.: A systematic literature review on serious games methodologies for training in the mining sector. Information. **16**, 389 (2025). https://doi.org/10.3390/info16050389

Grand View Research: Healthcare Gamification Market Size, Share and Trends Analysis Report, 2025–2030. Grand View Research, San Francisco (2024) https://www.grandviewresearch.com/industry-analysis/healthcare-gamification-market-report. Accessed 20 Jan 2025

IMARC Group: Serious Games Market Report by Gaming Platform, Application, Industry Vertical, and Region, 2025–2033. IMARC Group, New York (2024) https://www.imarcgroup.com/serious-games-market. Accessed 20 Jan 2025

ISO: Ergonomics of Human-System Interaction — Part 210: Human-Centered Design for Interactive Systems (ISO 9241-210:2019). International Organization for Standardization, Geneva (2019)

Laugwitz, B., Held, T., Schrepp, M.: Construction and evaluation of a user experience questionnaire. In: Symposium of the Austrian HCI and Usability Engineering Group, pp. 63–76. Springer Berlin Heidelberg, Berlin, Heidelberg (2008)

Maxim, R.I., Arnedo-Moreno, J.: Identifying key principles and commonalities in digital serious game design frameworks: scoping review. JMIR Serious Games. **13**, e54075 (2025). https://doi.org/10.2196/54075

Mitgutsch, K., Alvarado, N.: Purposeful by design? A serious game design assessment framework. In: Proceedings FDG'12, pp. 121–128. ACM, New York (2012). https://doi.org/10.1145/2282338.2282364

Miyosawa, Y., et al.: Development of a serious game to simulate NICU experiences: collaborative quasi-experimental study. JMIR Serious Games. **13**, e73009 (2025). https://doi.org/10.2196/73009

Intelligence, M.: Serious Games Market — Size and Companies, 2026–2031. Mordor Intelligence, Hyderabad (2025) https://www.mordorintelligence.com/industry-reports/serious-games-market. Accessed 20 Jan 2025

Peffers, K., Tuunanen, T., Rothenberger, M.A., Chatterjee, S.: A design science research methodology for information systems research. J. Manag. Inf. Syst. **24**(3), 45–77 (2007). https://doi.org/10.2753/MIS0742-1222240302

Schoormann, T., Behrens, D., Kundisch, D., Möller, F.: Integrating learning and serious game design: the development of the LGI framework. In: Proc 18th European Conference on Games Based Learning. Academic Conferences International (2024)

Sevcenko, N., Ninaus, M., Wortha, F., Moeller, K., Gerjets, P.: Measuring cognitive load using in-game metrics of a serious simulation game. Front. Psychol. **12**, 572437 (2021). https://doi.org/10.3389/fpsyg.2021.572437

Vlachopoulos, D., Makri, A.: The effect of games and simulations on higher education: a systematic literature review. Int. J. Educ. Technol. High. Educ. **14**, 22 (2017). https://doi.org/10.1186/s41239-017-0062-1

Venable, J.R., Pries-Heje, J., Baskerville, R.L.: Choosing a design science research methodology. In: ACIS 2017 Proceedings, p. 112 (2017). https://aisel.aisnet.org/acis2017/112

Zhang, J., Zhao, C., Zhao, Y., Yu, Z., He, M., Fan, J.: LLM4Workflow: an LLM-based automated workflow model generation tool. In: Proc 39th IEEE/ACM International Conference on Automated Software Engineering (ASE). ACM, Sacramento, CA (2024)

Aspect-Based Multilabel Classification and Emotional Polarity Analysis of E-Commerce Reviews for Marketing Strategies

Aji Akbar Firdaus[1], Moch. Nafkhan Alzamzami[2], Tesa Eranti Putri[1], Sri Devi Ravana[3], and Nasa Zata Dina[1,3](✉)

[1] Department of Engineering, Faculty of Vocational Studies, Universitas Airlangga, Surabaya, Indonesia
{aa.firdaus,tesaep,nasazatadina}@vokasi.unair.ac.id
[2] Department of Informatics, Faculty of Intelligent Electrical and Informatics Technology, Institut Teknologi Sepuluh Nopember, Surabaya, Indonesia
nafkhan@its.ac.id
[3] Department of Information Systems, Faculty of Computer Science and Information Technology, Universiti Malaya, Surabaya, Malaysia
sdevi@um.edu.my

Abstract. Customer reviews on e-commerce platforms offer significant insights into product quality and consumer satisfaction. Nonetheless, a singular review frequently addresses various prod- cut attributes, making aspect identification a multilabel classification challenge. This study presents an aspect-based multilabel classification and emotion analysis framework to identify product aspects and evaluate customer emotions. Four aspects were examined: finishing, materials, construction, and color. Three embedding techniques (Word2Vec, SBERT, and e5- large-v2) were integrated with two multilabel problem transformations (Classifier Chains and Label Powerset) and three classifiers (Logistic Regression, Random Forest, and Stochastic Gradient Descent). Performance was evaluated with Accuracy, F1-score, and Hamming Loss. The findings indicated that transformer-based embeddings surpassed Word2Vec in performance. The e5-large-v2 utilizing Label Powerset and SGD attained the highest F1-score (0.8228), whilst e5-large-v2 employing Label Powerset and Random Forest reached the highest accuracy (0.8946). Emotional analysis indicated that trust, joy, and anticipation were the predominant emotions. These findings provide valuable insights for product improvement and marketing strategies.

Keywords: customer behavior; natural language processing · multilabel classification · customer review analysis · product quality improvement · consumer feedback · sustainable consumption · e-commerce analytics

1 Introduction

Online customer reviews have emerged as a significant resource for comprehending consumer opinions around products. These reviews provide essential insights that might assist organizations in enhancing product quality and consumer satisfaction [1, 2].

N. Mohd Sharef et al. (Eds.): CAMP 2026, CCIS 3094, pp. 273–286, 2027.
https://doi.org/10.1007/978-981-92-4305-1_20

Nonetheless, a singular review sometimes addresses many product aspects concurrently, complicating analysis through traditional classification methods.

Aspect-based analysis enables the categorization of customers opinions based on distinct product aspects [3]. In the context of clothing products, customers often cite finishing, materials, construction, and color when assessing product quality. Given that a review may pertain to multiple aspects, multilabel classification is essential for accurately identifying all relevant aspects.

Recent advancements in natural language processing have created several word embedding techniques, including Word2Vec, SBERT, and e5-large-v2, which can represent textual information more efficiently. Moreover, multilabel classification problem transformation, like Classifier Chains (CC) [4] and Label Powerset (LP) [5, 6], have been extensively employed to tackle multilabel classification challenges. Nonetheless, their efficacy in aspect classification inside customer reviews necessitates additional evaluation.

In addition to recognizing product aspects, comprehending the emotions conveyed by customers can yield profound insights into their perspectives and experiences. Emotion analysis with the NRC Emotion Lexicon and Plutchik's emotion model can determine whether buyers exhibit positive emotions, such as trust and joy, or negative emotions, such as sadness and anger, regarding specific product aspects. Customer reviews on e-commerce platforms provide significant insights into customer opinions and experiences regarding products [10]. Examining these reviews might assist organizations in comprehending customer requirements and enhancing product quality [11]. Customer reviews frequently address various product aspects within a single text, complicating the accurate identification and analysis of aspect-specific feedback.

Prior research has predominantly utilized sentiment analysis on customer reviews, although many concentrate on assessing overall sentiment rather than concurrently finding numerous product aspects [12–16]. Consequently, critical information pertaining to specific product aspects, including finishing, materials, construction, and color, may not be sufficiently recorded. Despite the growing interest in aspect-based sentiment analysis, research on aspect-based multilabel classification is still rather scarce, especially concerning clothing product reviews.

Furthermore, recent advancements in natural language processing have yielded robust embedding approaches like SBERT and e5-large-v2, which offer contextualized representations of text. Nevertheless, the majority of current research continues to depend on conventional embeddings like Word2Vec, and there is limited comparative evidence concerning the efficacy of these contemporary embedding models for aspect-based multilabel classification tasks. The efficacy of several multilabel classification methods, including Classifier Chains (CC) and Label Powerset (LP), has not been comprehensively examined in this application domain.

Moreover, whereas sentiment analysis is frequently utilized to categorize reviews as positive, negative, or neutral, there is limited number of studies investigating emotion analysis at the aspect level [17]. Comprehending distinct emotions like trust, joy, anticipation, anger, or sadness might yield more profound insights into customer perceptions than sentiment polarity alone [18]. An integration of aspect-based multilabel classification with emotion analysis is yet inadequately investigated in customer review research.

Consequently, a framework is required that recognizes various product aspects from customer reviews and analyzes the emotions linked to those a. This study addresses these gaps by comparing word embedding techniques and multilabel classification methods for aspect identification, while also analyzing customer emotions regarding the aspects of finishing, materials, construction, and color through the NRC Emotion Lexicon and Plutchik's emotion model.

Therefore, this study aims to develop and evaluate an aspect-based multilabel emotion analysis framework for customer reviews. Specifically, the study compares different word embedding techniques and multilabel classification approaches to identify the most effective model for aspect classification. Furthermore, it analyzes customer emotions associated with the aspects of finishing, materials, construction, and color to better understand customer perceptions of the products. The study addresses the following research questions:

- How can aspect-based multilabel classification be used to identify product-related aspects in customer reviews?
- Which combination of word embedding and multilabel classification method achieves the best performance for aspect classification?
- What emotions do customers express toward the aspects of finishing, materials, construction, and color, and how are these emotions distributed across the aspects?

2 Materials and Method

Understanding customer responses to products is important for gaining insights into their emotions and opinions. Therefore, this study developed two main analyses:

(1) aspect-based multilabel classification, which identifies multiple product aspects mentioned in customer reviews, and (2) emotion analysis, which examines the emotions expressed toward those aspects. The relationship between product aspects and customer emotions was then analyzed to better understand customer perceptions. The overall research framework is presented in Fig. 1.

2.1 Data Sources and Collection

Online customer reviews (OCRs) of clothing products were collected from Amazon (amazon.com), one of the largest B2C e-commerce platforms worldwide [19]. Amazon was chosen due to its extensive customer review system and the large volume of available reviews, which provide valuable insights into customer opinions and experiences. The collected data consisted of product names, ratings, and review texts. These data were obtained through web crawling and stored in Microsoft Excel for subsequent analysis. The dataset comprises 54,036 customer reviews from 23 clothing categories, including dresses, blouses, and knits. The review distribution by category is shown in Fig. 2. Examples of the collected reviews are presented in Fig. 3.

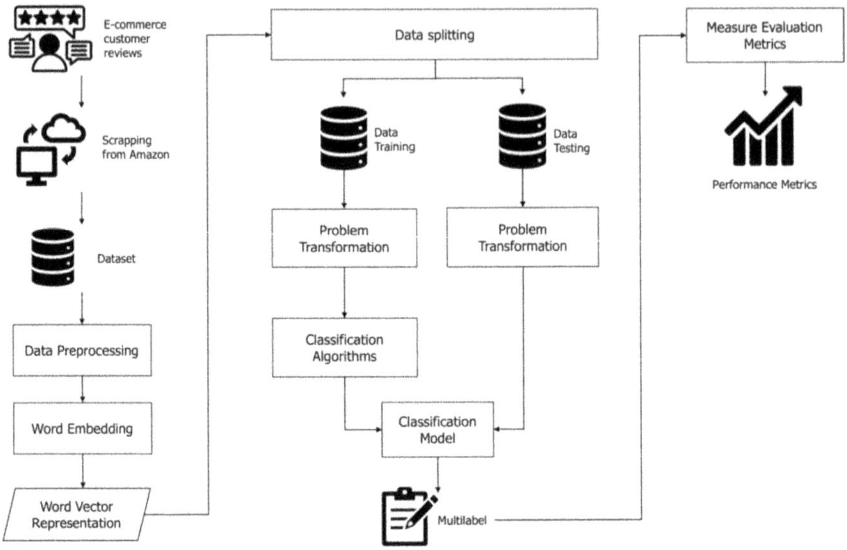

Fig. 1. Aspect-based Multilabel Classification

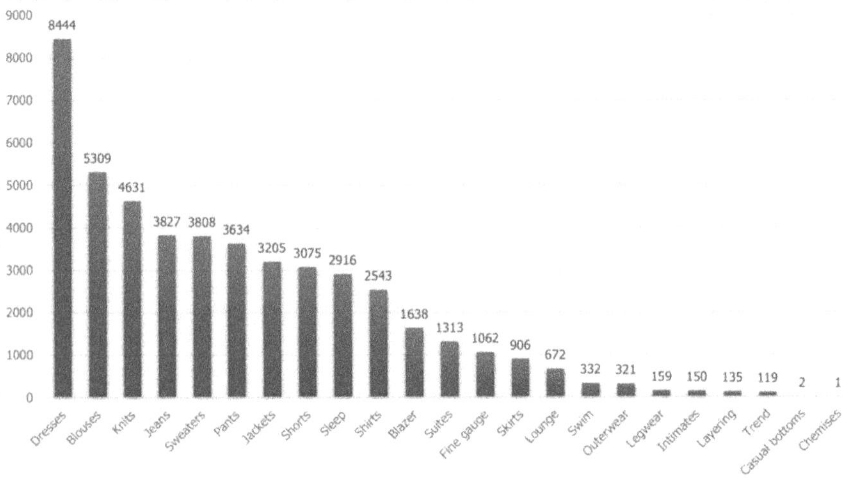

Fig. 2. Aspect-based Multilabel Classification

2.2 Text Preprocessing

The textual data preprocessing technique comprised six primary stages [20]. Initially, tokenization was executed to divide sentences into discrete words or tokens. Secondly, data cleansing was conducted to eliminate irrelevant components, such as punctuation marks, stop words, URLs, special characters, and numerical values. Third, Part-of-Speech (POS) tagging was performed to allocate grammatical categories, including nouns, verbs, and adjectives, to each token. Fourth, stemming was employed to change words to their root or base forms. Fifth, lowercasing was implemented to transform all text into lowercase letters, ensuring uniformity in text representation. A Term-Document Matrix (TDM) was ultimately created to illustrate the frequency of each distinct token inside the documents. The resulting TDM was the basic of subsequent feature selection and emotion classification procedures.

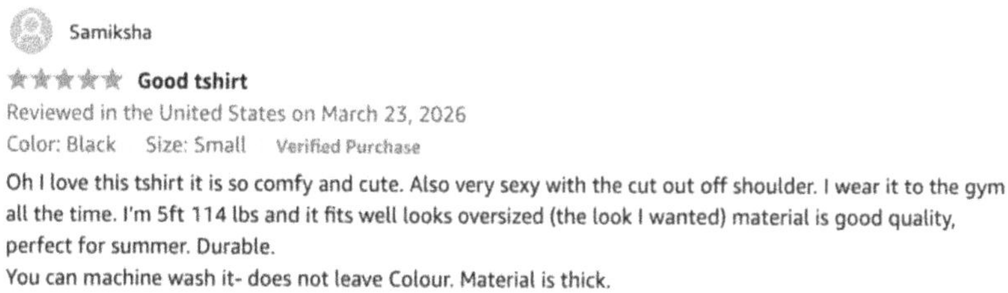

Fig. 3. Sample E-commerce Customer Review

2.3 Data Balancing

Class imbalance occurs when one class possesses a significantly greater number of instances than another class [21]. This may diminish the efficacy of machine learning models, as they often prioritize the dominant class and may inadequately recognize occurrences of the minority class. To address this issue, random over- sampling was employed by replicating samples from the minority class [22]. This augments the quantity of minority-class cases and helps in balancing the dataset. Consequently, the model may more effectively learns the characteristic of underrepresented classes, hence enhancing classification performance, particularly in terms of recall. Excessive oversampling may result in overfitting, as the replicated samples do not contribute more information to the dataset.

2.4 Word Embedding

This study used three text embedding methods, Word2Vec, Sentence-BERT (SBERT), and E5-large-v2, to convert text into numerical vectors for machine learning analysis. Word2Vec is a neural network-based method that learns word meanings from their surrounding context and represents words as dense vectors [24]. As a result, words with similar meanings are placed close to each other in the vector space. SBERT is an extension of the BERT model that generates embeddings for entire sentences rather than individual words [25]. It is designed to capture the semantic meaning of sentences and is commonly used for tasks such as semantic similarity measurement, text clustering, and semantic search. E5-large-v2 is a transformer-based embedding model that produces rich contextual representations of text [26]. With 24 transformer layers and 1,024-dimensional embeddings, it is effective for capturing semantic relationships and is widely used in tasks such as semantic similarity analysis and information retrieval.

Table 1. Aspects and description

Aspects	Description
Finishing	Finishing refers to the final processes applied to a garment to improve its appearance, quality, and durability. This includes hemming, trimming loose threads, ironing, and adding details that give the clothing a neat and polished look.
Materials	Materials are the fabrics or textiles used to make the garment, such as cotton, polyester, wool, or silk. This aspect determines the clothing's comfort, texture, durability, and suitability for different conditions.
Construction	Construction refers to how the garment is put together, including stitching, seam types, pattern cutting, and assembly techniques. It affects the fit, strength, and overall quality of the clothing.
Color	Color refers to the hue or shade of the garment, which con- tributes to its aesthetic appeal. It includes aspects like color consistency, dye quality, and resistance to fading over time.

2.5 Problem Transformation

This study used two multilabel classification methods: Classifier Chains (CC) and Label Powerset (LP). Classifier Chains predicts multiple labels by linking several binary classifiers in a sequence. Each classifier predicts one label, and the prediction is used as additional information for the next classifier. This allows CC to capture relationships between labels, although its performance can be affected by the order of the labels [27]. Label Powerset takes a different approach by treating each unique combination of labels as a separate class [28]. Instead of predicting labels individually, it predicts a single class that represents a set of labels. This method can effectively capture label relationships, but it may become less efficient when there are many different label combinations.

2.6 Classification Algorithm

The study employed three machine learning algorithms for classification: Logistic Regression (LR), Random Forest (RF), and Stochastic Gradient Descent (SGD). Logistic Regression is a linear classification method that calculates the likelihood of class membership by employing a sigmoid function to represent the connection between predictor variables and the target class, rendering it appropriate for both binary and multiclass classification applications. Stochastic Gradient Descent is an optimization method that incrementally adjusts model parameters using individual samples or small data batches, facilitating effective training on extensive datasets. Random Forest is an ensemble learning technique that builds numerous decision trees and consolidates their predictions via a voting mechanism, thereby improving classification accuracy and mitigating the risk of overfitting relative to a singular decision tree model.

2.7 Emotional Polarity Analysis

This study utilized emotion analysis to discern emotions conveyed in customer reviews. The NRC Word-Emotion Association Lexicon was employed to identify emotional polarity inside the text. The lexicon categorizes words into eight fundamental emotions according to Plutchik's emotion model: anger, anticipation, disgust, fear, joy, sadness, surprise, and trust. This approach categorizes joy, surprise, anticipation, and trust as positive emotions, whereas anger, fear, sadness, and disgust are classified as negative emotions. The NRC Emotion Lexicon has 14,182 English words annotated by emotion categories. Plutchik's paradigm was chosen for its simple and comprehensive foundation for understanding and interpreting human emotions. Figure 4 illustrates Plutchik's eight-emotion paradigm, whereas Table 2 delineates the emotions along with the corresponding number of words linked with each emotion in the NRC Emotion Lexicon.

Table 2. Words related to each emotion [7]

Emotion	Words
anger	rage, annoyance, hostility, frustration
anticipation	expectation, hope, interest, excitement
disgust	dislike, revulsion, contempt, aversion
fear	worry, anxiety, terror, concern
joy	happiness, pleasure, delight, satisfaction
sadness	grief, disappointment, sorrow, regret
surprise	amazement, astonishment, wonder
trust	confidence, reliability, faith, assurance

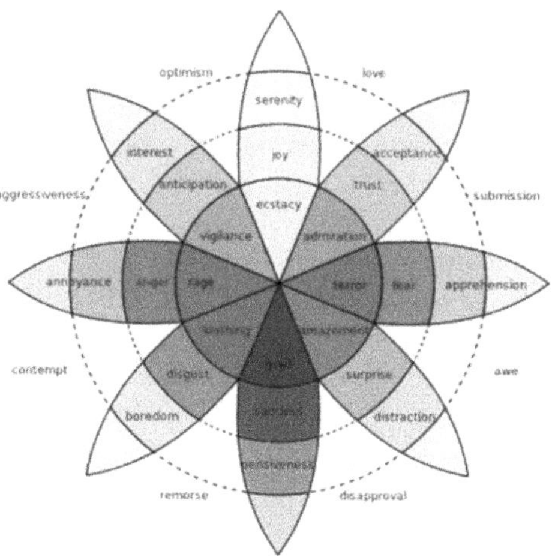

Fig. 4. Emotion Wheel [8]

2.8 Performance Metrics

A confusion matrix was utilized to evaluate the classification performance of the proposed model. The confusion matrix encapsulates prediction results by com- paring predicted labels with actual labels, comprising four elements: true positives (TP), true negatives (TN), false positives (FP), and false negatives (FN). Several evaluation metrics were derived from these values. Accuracy quantifies the ratio of successfully classified occurrences to the total number of predictions. Precision de- notes the proportion of predicted positive instances that are genuinely positive, while recall signifies the model's capacity to recognize all actual positive instances. The F1-score offers a balanced assessment of precision and recall through the computation of their harmonic mean. Moreover, Hamming loss quantifies the proportion of inaccurately predicted labels, with diminished values signifying superior classification efficacy. These metrics provide a thorough evaluation of the model's prediction performance.

3 Result and Discussion

This section reports the experimental results and provides a comprehensive discussion of the findings obtained through the application of the proposed methods.

3.1 Aspect-Based Multilabel Classification

The examined aspect-based multilabel has four dimensions: (a) finishing; (b) materials; (c) construction; (d) color. The detailed explanation of each aspect is shown in Table 1. This study employs the word embeddings E5-large-v2, SBERT, and Word2Vec. Additionally, two problem transformations have been implemented. The Classifier Chain (CC) and Label Powerset (LP). The outcomes are il- lustrated in Fig. 5.

Figure 5 illustrates the performance comparison of three word embedding techniques (E5-large-v2, SBERT, and Word2Vec) in conjunction with three machine learning classifiers (CC-LR, CC-RF, CC-SGD, LP-LR, LP-RF, and LP-SGD). The models were evaluated using Accuracy, F1-score, and Hamming Loss, with elevated Accuracy and F1-score signifying superior classification performance, whilst diminished Hamming Loss indicates fewer prediction errors.

The results indicate that the E5-large-v2 embedding consistently attained superior performance across the majority of evaluation metrics. The combination of E5- large-v2 and LP-SGD achieved the greatest F1-score (0.8228), indicating an exceptional balance between precision and recall. Regarding classification accuracy, E5- large-v2 with LP-RF attained the highest score (0.8946), closely succeeded by E5- large-v2 with CC-RF (0.8942). Moreover, E5-large-v2 with CC-SGD yielded the minimal Hamming Loss (0.0463), indicating the highest precision in label assignments across all evaluated setups.

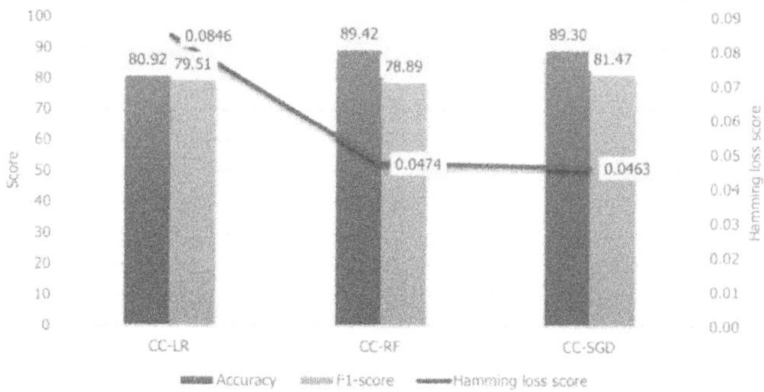

(a)

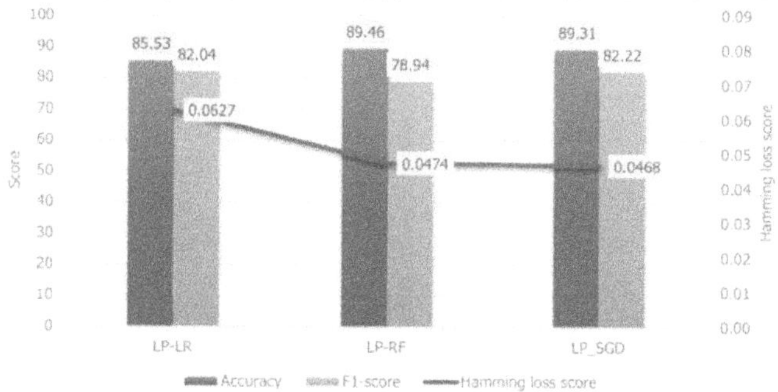

(b)

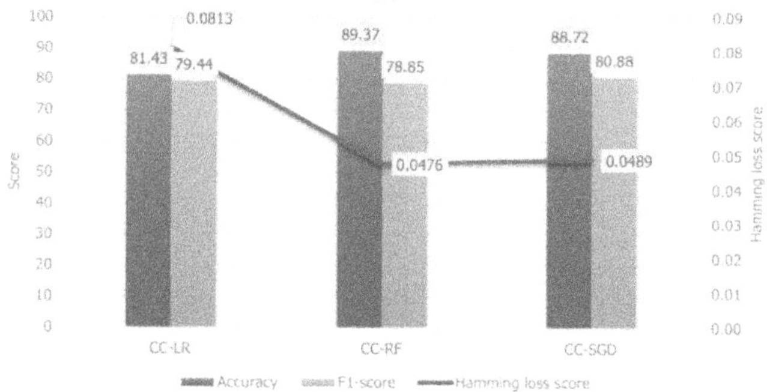

(c)

Fig. 5. Aspect-based multilabel classification results using CC and LP with E5-large-v2, SBERT, and Word2Vec embeddings: (a) CC–E5-large-v2; (b) LP–E5-large-v2; (c) CC–SBERT; (d) LP–SBERT; (e) CC–Word2Vec; (f) LP–Word2Vec.

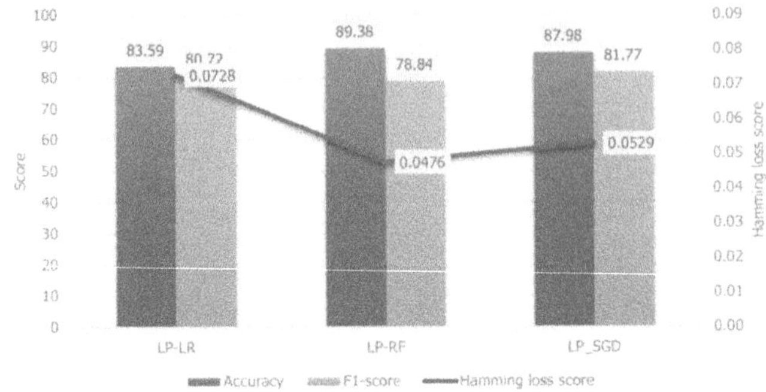

(d)

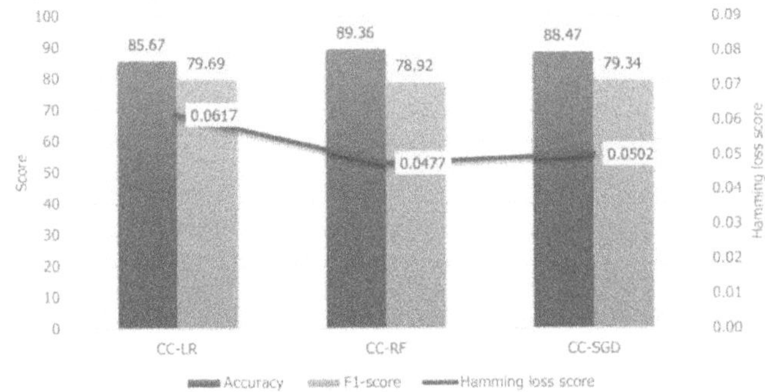

(e)

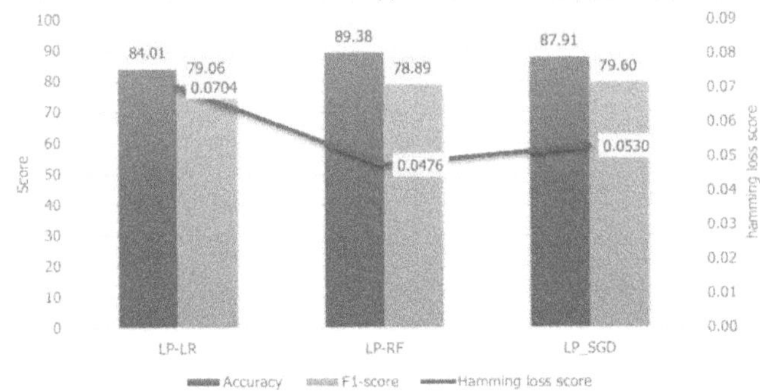

(f)

Fig. 5. (*continued*)

The SBERT embedding yielded optimal performance with LP-SGD, which attrained an F1-score of 0.8177, although LP-RF earned the maximum accuracy of 0.8938 and a comparatively low Hamming Loss of 0.0476. The results demonstrate that SBERT is effective for the classification task, however its performance is often marginally worse to that of E5-large-v2.

Conversely, Word2Vec produced relatively inferior outcomes across the as- sassed measures. The optimal Word2Vec configuration was CC-SGD, attaining an F1-score of 0.7954, but CC-RF and LP-RF yielded the greatest accuracy values of 0.8936 inside the Word2Vec framework. Nonetheless, the aggregate F1-scores and Hamming Loss metrics indicate that Word2Vec was inferior to the transformer- based embeddings.

In the comparison of classification methodologies, the SGD-based models, spacivically LP-SGD and CC-SGD, consistently attained the greatest F1-scores, signifying robust predictive efficacy for the classification task. The Label Powerset (LP) method typically surpassed the Classifier Chains (CC) approach regarding F1- score, particularly when integrated with E5-large-v2 and SBERT embeddings.

The findings demonstrate that transformer-based embeddings, specifically E5-large-v2 and SBERT, surpass the conventional Word2Vec representation, with the E5-large-v2 and LP-SGD combination yielding the highest classification performance overall. These findings underscore the significance of contextualized word representations in enhancing classification efficacy.

3.2 Emotional Polarity

Figure 6 illustrates the distribution of customers emotions about four products aspects: finishing, materials, construction, and color, as derived from the NRC Emotion Lexicon and Plutchik's emotion model. Trust emerged as the predominant emotion across all dimensions, with 38% of emotional expressions for finishing, 35% for materials, and 34% for both construction and color. This was succeeded by joy and anticipation, which consistently represented significant percentages of the emotional responses. For the finishing aspect, joy and anticipation accounted for 28% and 19%, respectively, resulting in a combined positive emotion rate of 85%. The materials aspect exhibited 24% anticipation and 22% joy, with positive feelings constituting 81% of the overall responses. The construction and color elements demonstrated roughly equivalent emotional responses, with 34% trust, 28% joy, and 25% anticipation, resulting in a total of 87% positive emotions.

Conversely, negative emotions were comparatively constrained across all product aspects. Sadness emerged as the predominant negative emotion, fluctuating be- tween 5% and 8%, with the highest proportion observed in the materials aspect at 8%. Other negative emotions, such as anger, fear, and disgust, typically accounted 1–2% of the responses. The emotion of surprise was minimal, fluctuating between 2% and 5%. Overall, the findings indicate that customers hold overwhelmingly positive perceptions of the product's finishing, materials, construction, and color. The predominance of trust, joy, and anticipation suggests strong customer confidence, satisfaction, and favorable expectations toward the product, while the low prevalence of negative emotions reflects minimal dissatisfaction with these aspects.

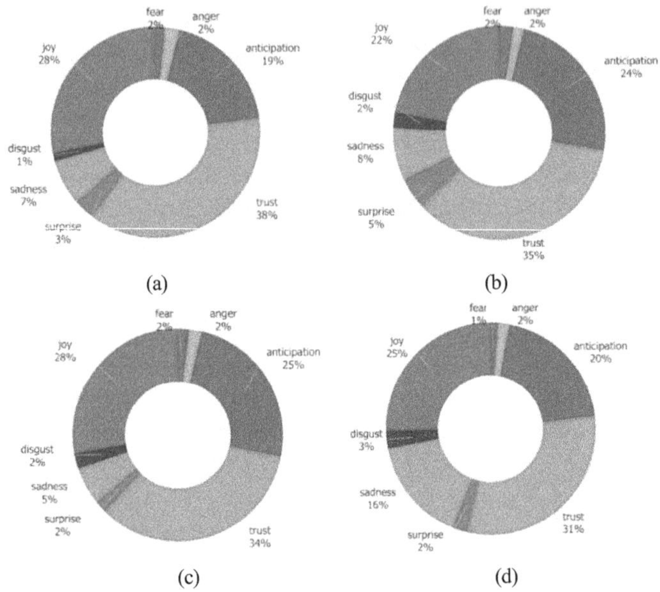

Fig. 6. Customers' emotional polarity of each aspect: (a) finishing; (b) materials; (c) construction; (d) color.

4 Conclusion

This study examined aspect-based multilabel classification and emotion analysis of customer reviews focusing on four product aspects: finishing, materials, construction, and color. The results showed that transformer-based embeddings per- formed better than the traditional Word2Vec model. Among the tested models, E5- large-v2 with Label Powerset-SGD (achieved the highest F1-score (0.8228), E5- large-v2 with LP-RF achieved the highest accuracy (0.8946), and E5-large-v2 with CC-SGD obtained the lowest Hamming loss score (0.0463). These results demonstrate that transformer-based embeddings are effective at identifying various product aspects in customer reviews. The sentiment research revealed that customer opinions were predominantly favorable, with trust emerging as the most prevalent emotion, succeeded by joy and anticipation. Positive emotions accounted for 81–87% of all identified emotions, indicating high customer satisfaction and confidence in the products.

This study has certain drawbacks. The dataset was obtained from a singular product category and e-commerce platform, thus constraining the applicability of the findings to other domains. The investigation concentrated on four product aspects and a limited number of embedding and classification algorithms. Furthermore, the NRC Emotion Lexicon may inadequately encompass contextual meanings, sarcasm, or implicit emotions in customer reviews.

Acknowledgement. This research was supported by Universitas Airlangga under Penelitian Dasar Unggulan-Airlangga Research Fund (ARF) 2024 (Contract No. 2463/UN3.FV /PT.01.03/2024).

References

1. Davoodi, L., Mezei, J., Nikou, S., Espinosa-Leal, L.: Automating customer feedback analysis in E-commerce: a multi-model approach. Expert Syst. Appl. **306**, 130865 (2026). https://doi.org/10.1016/j.eswa.2025.130865
2. Dina, N.Z., Triwastuti, R., Silfiani, M.: TF-IDF decision matrix to measure customers' satisfaction of ride hailing mobile application services: multi-criteria decision-making approach. Int. J. Interact. Mobile Technol. **15**(17), 104–118 (2021). https://doi.org/10.3991/ijim.v15i17.22729
3. Maghsoudi, M., Mohammadi, N., Bakhtiari, M.: A novel approach to customer segmentation for product development on social media data: integrating aspect-based sentiment analysis and text mining. Knowl.-Based Syst. **328**, 114269 (2025). https://doi.org/10.1016/j.knosys.2025.114269
4. Rocha, V.F., Varejão, F.M., Segatto, M.E.V.: Ensemble of classifier chains and decision templates for multi-label classification. Knowl. Inf. Syst. **64**, 1–21 (2022)
5. Read, J., Pfahringer, B., Holmes, G., Frank, E.: Classifier chains: a review and perspectives. J. Artif. Intell. Res. **70**, 683–718 (2021)
6. Read, J., Puurula, A., Bifet, A.: Multi-label classification with meta-labels. In: Proceedings of IEEE International Conference on Data Mining (ICDM), Shenzhen, China, pp. 941–946 (2014)
7. Mohammad, S.M., Turney, P.D.: Crowdsourcing a word-emotion association lexicon. Comput. Intell. **29**(3), 436–465 (2013)
8. He, J. et al.: MoWER: multimodal emotion recognition in conversation with emotion wheel-enhanced representation. Knowl.-Based Syst. **345**, 116044 (2026). https://doi.org/10.1016/j.knosys.2026.116044
9. Le, H.-S. et al.: Predictive model for customer satisfaction analytics in E-commerce sector using machine learning and deep learning. Int. J. Inf. Manage. Data Insights. **4**(2), 100295 (2024). https://doi.org/10.1016/j.jjimei.2024.100295
10. Ben Youssef, K., Leicht, T., Pellicelli, A.C.: The effect of customer education on brand associations and brand preference in business-to-business relationships. J. Bus. Ind. Mark. **41**(3), 307–317 (2026). https://doi.org/10.1108/JBIM-04-2025-0292
11. Prova, N.N.I., Ravi, V., Singh, M.P., Srivastava, V.K., Chippagiri, S., Singh, A.P.: Multilingual sentiment analysis in e-commerce customer reviews using GPT and deep learning-based weighted- ensemble model. Int. J. Cogn. Comput. Eng. **7**, 268–286 (2026). https://doi.org/10.1016/j.ijcce.2025.10.003
12. Liu, Y., Huang, J., Huang, F., Zhu, Z., Ma, L.: Deep learning-driven sentiment analysis for customer satisfaction optimization in airline reviews. Ind. Manage. Data Syst. **126**(3), 833–864 (2025). https://doi.org/10.1108/IMDS-08-2024-0836
13. Laghbi, Y.A., Al Dhoayan, M.: Examining how customers perceive community pharmacies based on Google maps reviews: multivariable and sentiment analysis. Explor. Res. Clin. Soc. Pharm. **15**, 100498 (2024). https://doi.org/10.1016/j.rcsop.2024.100498
14. Bian, Y., Ye, R., Zhang, J., Yan, X.: Customer preference identification from hotel online reviews: a neural network-based fine-grained sentiment analysis. Comput. Ind. Eng. **172**, 108648 (2022). https://doi.org/10.1016/j.cie.2022.108648

15. Dina, N.Z., Juniarta, N.: Deriving customers' preferences for hotels from unstructured data. GeoJ. Tour. Geosites. **43**(3), 872–877 (2022). https://doi.org/10.30892/gtg.43305-899
16. Cam, H., Cam, A.V., Demirel, U., Ahmed, S.: Sentiment analysis of financial twitter posts on twitter with the machine learning classifiers. Heliyon. **10**(1), e23784 (2024). https://doi.org/10.1016/j.heliyon.2023.e23784
17. Sun, P., Li, L., Hossain, M.S., Ray, S., Law, K.A.: Predicting and explaining customer satisfaction: a deep learning and sentiment analysis of emotional impacts. Acta Psychol. **260**, 105597 (2025). https://doi.org/10.1016/j.actpsy.2025.105597
18. Ahmad, H., Gillani, S.M.A.H., Ghufran, H., Ramakrishnan, S.: Digital platform growth and overconsumption: a comparative study of Amazon through a sustainability lens in the United States. In: Qureshi, M.I., Liu, J., Saritas, O., Khan, N. (eds.) Digital Platforms for Sustainable Business, pp. 427–438. Elsevier, Amsterdam (2026). https://doi.org/10.1016/B978-0-443-29293-4.00002-4
19. Al-Milli, N., Elsoud, E.A., Al-Shaikh, A., Omar, H., AlNajjar, A.: Emotion detection and sentiment analysis of textual data. Egypt. Inform. J. **34**, 100976 (2026). https://doi.org/10.1016/j.eij.2026.100976
20. van Zyl, J.-P., Engelbrecht, A.P.: Analysis of classification metric behaviour under class imbalance. Egypt. Inform. J. **31**, 100711 (2025). https://doi.org/10.1016/j.eij.2025.100711
21. García-Pedrajas, N.: Partial random under/oversampling for multilabel problems. Knowl.-Based Syst. **302**, 112355 (2024). https://doi.org/10.1016/j.knosys.2024.112355
22. Sağlam, F., Cengiz, M.A.: Uncertainty guided oversampling: a classifier-specific approach for effective handling of class imbalance in machine learning. Knowl.-Based Syst. **342**, 115886 (2026). https://doi.org/10.1016/j.knosys.2026.115886
23. Zhou, J., Ye, Z., Zhang, S., Geng, Z., Han, N., Yang, T.: Investigating response behavior through TF-IDF and Word2Vec text analysis: a case study of PISA 2012 problem-solving process data. Heliyon. **10**(16), e35945 (2024). https://doi.org/10.1016/j.heliyon.2024.e35945
24. Wagay, F.A., Jahiruddin: Classification of mental illnesses from reddit posts using sentence-BERT embeddings and neural networks. Procedia Comput. Sci. **258**, 1669–1676 (2025). https://doi.org/10.1016/j.procs.2025.04.398
25. L. Wang et al. and : Text embeddings by weakly-supervised contrastive pre-training. arXiv preprint https://arxiv.org/abs/2212.03533, (2022)
26. Boutell, M.E., Luo, J., Shen, X., Brown, C.M.: Learning multi-label scene classification. Pattern Recogn. **37**(9), 1757–1771 (2004). https://doi.org/10.1016/j.patcog.2004.03.009
27. Li, J., Zhu, X., Wang, J.: AdaBoost.C2: boosting classifier chains for multi-label classification. In: Proceedings of AAAI Conference Artificial Intelligence (AAAI), Washington, DC, USA, pp. 8580–8587 (2023)
28. Plutchik, R.: The Psychology and Biology of Emotion. HarperCollins College Publishers (1994)

Conceptual Framework for Multimodal AI-Augmented Depression Screening in Type 2 Diabetes Remission Care

Yootthapong Tongpaeng[1(✉)], Suepphong Chernbumroong[2], Warinda Sricharoen[3], and Pokin Sakarinkhul[4]

[1] School of Applied Digital Technology, Mae Fah Luang University, Chiang Rai, Thailand
yootthapong.ton@mfu.ac.th
[2] College of Arts, Media and Technology, Chiang Mai University, Chiang Mai, Thailand
[3] Faculty of Pharmacy, Chiang Mai University, Chiang Mai, Thailand
[4] Lamphun Provincial Health Office, Lamphun, Thailand

Abstract. Type 2 diabetes mellitus (T2DM) remission requires sustained lifestyle modification, which may be associated with psychological burden, thereby increasing the need for timely depression screening. Although the Patient Health Questionnaire-9 (PHQ-9) is widely used for depression screening, it primarily relies on self-reported symptoms and may not fully capture verbal, non-verbal, and physiological cues expressed during assessment. This study proposes a Multimodal Depression Screening Framework to augment PHQ-9-based screening among patients undergoing T2DM remission care. The framework consists of three integrated components: Multimodal Indicator Acquisition, Multimodal Large Language Model-Based Interpretation, and Visualization and Hyper-Realistic Interaction. It captures visual, acoustic, linguistic, heart rate, and breathing rate indicators through webcam, microphone, automatic speech recognition, and non-contact mmWave sensing. These indicators are transformed into structured patient-level summaries and interpreted through an MLLM-based pipeline supported by vision-language modelling, speech-to-text processing, physiological signal interpretation, large language model-based reasoning, and retrieval-augmented generation. The resulting evidence-grounded output is delivered through a hyper-realistic virtual nurse interface that provides real-time dialogue, voice rendering, and emotionally aligned visual feedback. The proposed framework is intended to support, rather than replace, clinical judgment by providing additional interpretable evidence on how patients present during PHQ-9-based assessment.

Keywords: type 2 diabetes remission · depression screening · multimodal large language models · retrieval-augmented generation · hyper-realistic avatar

1 Introduction

Type 2 diabetes mellitus (T2DM) remission has become an important goal in diabetes care, as it allows patients to maintain glycemic levels below the diagnostic threshold without glucose-lowering medication. In Thailand, intensive lifestyle modification has

N. Mohd Sharef et al. (Eds.): CAMP 2026, CCIS 3094, pp. 287–296, 2027.
https://doi.org/10.1007/978-981-92-4305-1_21

been promoted as a key pathway to remission, focusing on dietary control, physical activity, and sustained exercise over 3–6 months, with a target weight loss of approximately 10–15% (Division of Noncommunicable Diseases, Department of Disease Control, 2023; Royal College of Family Physicians of Thailand, 2022). However, not all patients achieve remission, and some may experience relapse after entering remission (Royal College of Family Physicians of Thailand, 2022). The expectations and pressures associated with intensive behavioral change may contribute to psychological distress, including depression.

Depression is common among individuals with T2DM and may negatively affect self-care, treatment adherence, and quality of life (Ali et al., 2006; Anderson et al., 2001; Jafari et al., 2024). Evidence also suggests a bidirectional relationship between T2DM and depression, highlighting the importance of psychosocial assessment as part of comprehensive diabetes care (American Diabetes Association, 2021; Fang et al., 2022; Lyra e Silva et al., 2019). Early identification of depressive symptoms may therefore support more effective follow-up and prevent worsening psychological conditions in patients undergoing lifestyle modification.

The Patient Health Questionnaire-9 (PHQ-9) is a widely used and validated self-report instrument for depression screening and has demonstrated acceptable screening performance among individuals with T2DM (Kroenke et al., 2001; Janssen et al., 2016; van Steenbergen-Weijenburg et al., 2010). The Thai version has also shown acceptable validity for screening major depression in general practice (Lotrakul et al., 2008). However, the interpretation of PHQ-9 scores in patients with T2DM requires careful consideration because optimal cut-off points may vary across populations, care settings, and validation methods. Previous studies in T2DM populations have reported different suggested cut-offs, ranging from lower thresholds for first-stage screening to higher thresholds in specialized diabetes clinics (Cichoń et al., 2020; Janssen et al., 2016; van Steenbergen-Weijenburg et al., 2010). This variation may partly reflect the overlap between depressive symptoms and diabetes-related somatic symptoms, such as sleep disturbance, fatigue, and appetite change (Twist et al., 2013). Therefore, PHQ-9 scores should be treated not only as a fixed binary classification but also as a continuous or threshold-sensitive screening indicator that requires clinical interpretation and follow-up assessment. Nevertheless, PHQ-9-based screening mainly captures episodic self-reported symptoms and may not fully reflect behavioral or affective cues expressed during assessment. In clinical practice, depression assessment often considers broader verbal and non-verbal presentation, including speech, affect, behavior, and motor activity (Voss & Das, 2024). This creates an opportunity to augment PHQ-9-based screening with multimodal indicators that capture facial expression, voice, linguistic content, heart rate, and breathing rate during the assessment process.

Therefore, this study proposes a Multimodal Depression Screening Framework as part of the ACMod project, Affective Computing Models: from Facial Expression to Mind-Reading. The proposed framework is organized into three main components. First, Multimodal Indicator Acquisition captures visual, acoustic, linguistic, heart rate, and breathing rate indicators of patient presentation during PHQ-9-based assessment. Second, Multimodal Large Language Model-Based Interpretation integrates these indicators through VLM, STT/ASR, mmWave-based physiological processing, LLM, and retrieval-augmented generation to support interpretable and evidence-grounded screening. Third, Visualization and Hyper-Realistic Interaction present interpreted feedback through a virtual nurse interface using real-time dialogue, voice rendering, and emotionally aligned visual responses. The framework aims to support, rather than replace, clinical judgment by providing additional interpretable evidence for identifying patients who may require further psychological assessment or follow-up.

2 Conceptual Framework

The proposed framework illustrates how multimodal artificial intelligence can be used to augment PHQ-9-based depression screening among patients undergoing type 2 diabetes remission care (see Fig. 1). It integrates self-reported responses with verbal and non-verbal indicators of patient presentation, including facial expression, voice, linguistic content, heart rate, and breathing rate.

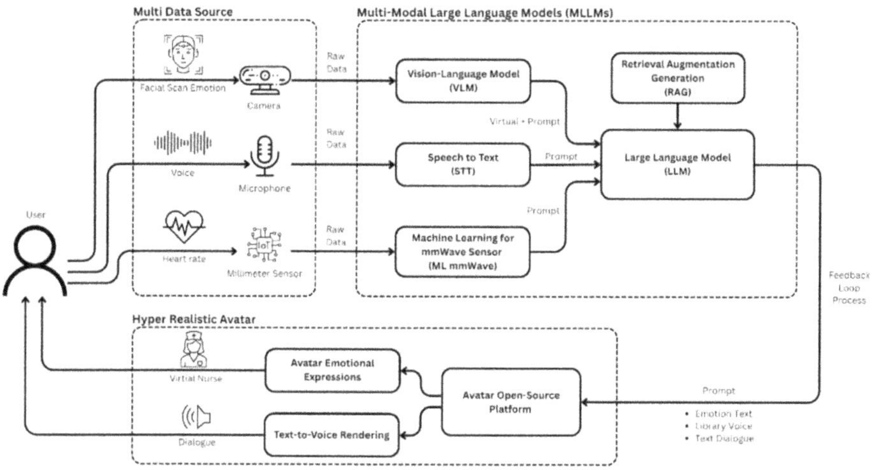

Fig. 1. Multimodal depression screening framework

These indicators are processed through an MLLM-based pipeline supported by VLM, STT/ASR, mmWave-based physiological processing, and RAG, before being translated into patient-facing feedback through a hyper-realistic avatar interface. The framework is organized into three main components: Multimodal Indicator Acquisition, Multimodal Large Language Model-Based Interpretation, and Visualization and Hyper-Realistic Interaction.

3 Multimodal Indicator Acquisition

In this study, visual, acoustic, linguistic, heart rate (HR), and breathing rate (BR) features are conceptualized as multimodal indicators of verbal and non-verbal patient presentation expressed during PHQ-9-based self-report assessment. Data acquisition is performed only after explicit user consent and combines standard web-based sensing with non-contact physiological monitoring. Visual input is captured using a standard webcam stream at approximately 640 × 480 pixels and 30 frames per second, and processed using MediaPipe FaceLandmarker, which extracts 468 facial mesh landmarks and 52 facial blend shape coefficients per frame. From these outputs, the system derives facial expressivity and blunted affect from the overall activation and temporal variance of expressive blend shapes, blink dynamics from eye-blink blend shape activation, head-motion energy from inter-frame displacement of the facial centroid, and smile genuineness from the co-activation of mouth-smile and cheek/eye-squint blend shapes. These indicators may provide supportive evidence related to reduced affective expression, psychomotor slowing, and possible social withdrawal, while reducing the risk of misinterpreting a culturally shaped polite smile as genuine positive affect. A forehead region of interest is also localized from facial landmarks to support physiological signal extraction, while only one low-resolution facial snapshot per PHQ-9 response is retained for clinician review and raw video is not stored.

Acoustic input is captured through the microphone waveform and analyzed using the Web Audio API to extract prosodic and temporal speech features, including fundamental frequency (F0), F0 variability, root mean square vocal energy, pause ratio, speaking rate, and response latency. These features are used to represent monotone speech, reduced vocal energy, delayed response, and slowed speech patterns that may reflect psychomotor retardation or depressive presentation. In parallel, spoken responses are transcribed using automatic speech recognition through the Web Speech API in Thai or English, and lightweight linguistic analysis is applied to detect first-person singular pronoun use, negative-affect vocabulary, and safety-related phrases requiring clinical escalation. For physiological input, HR and BR are estimated using a 60 GHz frequency-modulated continuous-wave radar sensor, specifically the MicRadar R60ABD1, which detects subtle chest-wall displacement associated with breathing and heartbeat without requiring electrodes, straps, microphones, or wearable devices (Marty et al., 2023). The sensor provides non-contact indicators of breathing activity, heartbeat, presence, and subtle motion, which are incorporated as supportive physiological indicators of patient arousal and presentation during the assessment. To maintain signal quality, privacy, and patient comfort, the radar sensor should be positioned at an appropriate non-contact distance from the participant, preferably within approximately 0.4–1.5 m between the participant's chest area and the radar antenna surface, following the manufacturer's recommended detection range for breathing and heartbeat monitoring (MicRadar Technology, 2023). Together, these visual, acoustic, linguistic, and physiological indicators extend PHQ-9 self-report screening by providing objective and interpretable evidence of how patients express symptoms during assessment, while limiting unnecessary storage of sensitive raw data.

4 Multimodal Large Language Model-Based Interpretation

The objective of this part is to transform the multimodal indicators acquired from the previous part into an interpretable, evidence-grounded clinical summary for PHQ-9-based depression screening in T2DM remission. In this study, multimodal large language models (MLLMs) are used as the central interpretation layer that integrates visual, acoustic, linguistic, and physiological indicators into a structured representation of patient presentation. During preprocessing, visual indicators extracted from facial landmarks and blend shape features are interpreted through a vision-language model (VLM), which enables facial-expression-related observations, such as blunted affect, reduced expressivity, head-motion reduction, blink dynamics, and smile genuineness, to be translated into clinically meaningful textual descriptors. Acoustic and linguistic indicators are processed through automatic speech recognition (ASR), which converts spoken PHQ-9 responses into machine-readable text and supports subsequent analysis of speech timing, response latency, pronoun use, negative-affect vocabulary, and safety-related phrases. Physiological indicators derived from the 60 GHz mmWave radar, including HR, BR, presence, and subtle-motion information, are converted into structured numerical and textual summaries before being integrated with the visual, acoustic, and linguistic outputs.

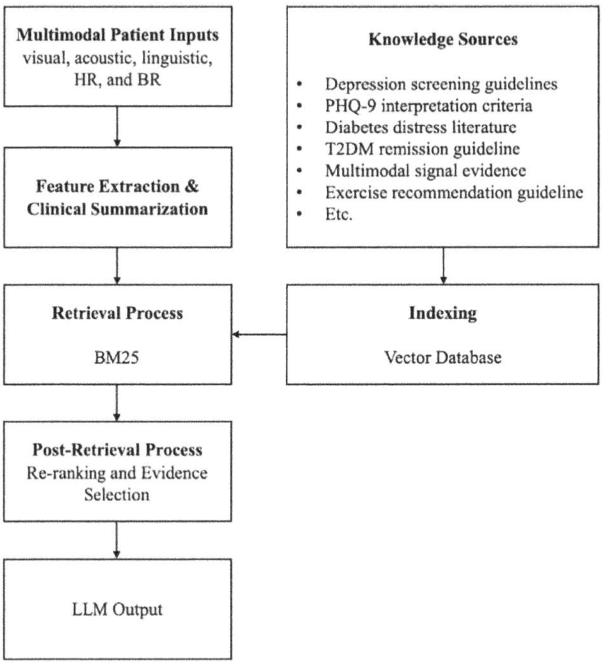

Fig. 2. Retrieval-augmented generation pipeline for evidence-grounded depression screening

These modality-specific summaries are then combined into a unified patient-level clinical summary through rule-guided summarization, allowing the system to represent not only what the patient reports, but also how the patient presents during the assessment. In addition, the framework should include a PHQ-9 threshold calibration and statistical modelling layer to reduce the risk of relying on a single fixed cut-off point. Instead of treating PHQ-9 scores only as binary labels, the system can model PHQ-9 as a continuous severity score, an ordinal severity category, and a set of alternative cut-off thresholds. Correlations between multimodal indicators and PHQ-9 outcomes can then be examined using appropriate statistical models, such as Pearson or Spearman correlation, logistic regression, ordinal regression, receiver operating characteristic analysis, and sensitivity analysis across different PHQ-9 cut-off points. This approach allows the framework to evaluate whether multimodal AI outputs remain stable across clinically plausible thresholds and whether diabetes-specific cut-offs improve screening performance.

To improve factuality and reduce hallucination, retrieval-augmented generation (RAG) is used as an evidence-grounding layer that retrieves relevant knowledge from curated sources, including depression screening guidelines, PHQ-9 interpretation criteria, diabetes-specific PHQ-9 cut-off evidence, diabetes distress literature, T2DM remission guidelines, behavioral-signal evidence, and exercise recommendation guidelines (Lewis et al., 2020; Gao et al., 2023). These knowledge sources are preprocessed, segmented into meaningful chunks, enriched with clinical metadata, and indexed as searchable knowledge representations. During inference, BM25-based sparse retrieval is applied to identify relevant evidence, followed by re-ranking and evidence selection before the retrieved information is passed to the LLM for evidence-grounded output generation (see Fig. 2). The final output consists of an interpretable depression-screening summary, evidence-supported reasoning, and a real-time feedback package for patient interaction, including emotion text, text dialogue, and voice-instruction prompts. This feedback loop enables the system to respond to the patient during PHQ-9 assessment while maintaining clinical interpretability, evidence grounding, and human reviewability.

5 Visualization and Hyper-Realistic Interaction

After the multimodal indicators have been interpreted and evidence-grounded by the MLLM-RAG layer, the system generates a real-time feedback package consisting of emotion text, text dialogue, and voice-instruction prompts. This output is delivered to the patient through a hyper-realistic virtual nurse interface designed to support natural interaction, user engagement, and self-disclosure during PHQ-9-based assessment. The avatar platform maps the generated emotion text to facial animation using Facial Action Coding System (FACS)-based blend shapes, allowing emotional expressions to be represented through anatomically meaningful facial muscle movements. The expressive design is based on ten emotional states: anger, disgust, fear, happiness, sadness, surprise, interested, bored, ashamed, and neutral. The first six emotions correspond to Ekman's universal basic emotions, which provide a well-established theoretical foundation for biologically grounded facial expressions (Ekman, 1992). Interested, bored, and ashamed extend the avatar's capacity to represent socially relevant affective states that are important in clinical and human–computer interaction contexts, where both basic and complex emotional expressions may influence user interpretation and engagement (Fridenson-Hayo et al., 2016; Dechsling et al., 2023). Neutral is used as a technical reference state, providing a baseline for expression transitions and reducing abrupt or unnatural facial movement. However, the avatar should be clearly disclosed to users as an AI-based virtual interface, not a human clinician.

Text-to-voice rendering is implemented through a neural text-to-speech pipeline that converts the validated dialogue into natural and intelligible speech, with voice characteristics selected to match the avatar's appearance and supportive clinical role. Neural TTS is suitable for this function because it enables more natural and expressive speech synthesis than traditional rule-based approaches (Tan et al., 2021). The vocal output is synchronized with speech-driven facial animation so that phonetic sequences are mapped to mouth and facial movements in real time. Prior work on interactive facial animation shows that deep neural networks and hybrid facial representations can support realistic facial expression generation through deformable 3D geometry and dynamic texture modelling (Paier et al., 2020). This cross-modal alignment between facial expression, voice, and dialogue is treated as a functional requirement because mismatched visual and vocal cues may reduce trust and increase perceptual discomfort. In this study, the hyper-realistic interaction layer does not provide diagnostic judgment; rather, it serves as a patient-facing communication interface that presents supportive feedback, encourages continued engagement, and maintains consistency between the system's affective interpretation and its verbal and non-verbal response.

6 Discussion

Although the proposed framework demonstrates the potential of multimodal artificial intelligence to augment PHQ-9-based depression screening in T2DM remission care, several practical considerations should be acknowledged. First, the framework consists of multiple computational components, including visual feature extraction, acoustic and linguistic processing, mmWave-based physiological interpretation, retrieval-augmented generation, large language model reasoning, and hyper-realistic avatar rendering. Processing heterogeneous data streams across these components may require substantial computational resources, particularly when real-time interaction is expected. This may introduce latency, increase system complexity, and create implementation challenges in resource-constrained clinical or community settings. Therefore, future implementation should consider model optimization, modular deployment, edge–cloud workload allocation, and latency evaluation to ensure that the system remains responsive, scalable, and feasible for real-world use.

Second, the effectiveness of the proposed framework depends heavily on the availability and quality of relevant multimodal datasets. At the current conceptual stage, the selection of models and techniques is based on theoretical relevance, prior evidence, and the expected characteristics of each data modality rather than empirical validation using a target dataset. As a result, the proposed model architecture is intentionally presented in a general and adaptable form. A more precise model selection and configuration will require high-quality datasets collected from patients undergoing PHQ-9-based assessment in T2DM remission contexts, including synchronized visual, acoustic, linguistic, physiological, and clinical reference data. Such datasets would enable systematic comparison of candidate models, validation of multimodal indicators, and refinement of the framework toward more accurate, interpretable, and clinically meaningful depression screening support.

Another important consideration is the selection of PHQ-9 cut-off points for model development and evaluation. In T2DM populations, optimal PHQ-9 thresholds may differ across clinical settings, countries, and patient characteristics, and somatic symptoms may overlap with diabetes-related experiences. Therefore, future validation should not rely on a single PHQ-9 threshold as the sole ground truth for multimodal AI modelling. Instead, model performance should be evaluated across multiple clinically plausible cut-off points, using sensitivity, specificity, area under the receiver operating characteristic curve, calibration analysis, and agreement with clinician assessment where available. This threshold-sensitive evaluation would help determine whether multimodal indicators provide robust additional value beyond PHQ-9 self-report scores. In addition, ethical, privacy, and clinical safety considerations are also central to the proposed framework. Because facial images, voice signals, linguistic responses, and physiological indicators are sensitive forms of patient data, future implementation should follow principles of explicit informed consent, data minimization, secure storage, controlled access, and clear disclosure of AI involvement.

7 Conclusion

This study proposed a Multimodal Depression Screening Framework to augment PHQ-9-based depression screening among patients undergoing type 2 diabetes remission care. The framework addresses an important limitation of conventional self-report screening by incorporating multimodal indicators of patient presentation, including visual, acoustic, linguistic, heart rate, and breathing rate features. By integrating these indicators with an MLLM-based interpretation pipeline supported by VLM, STT/ASR, mmWave-based physiological processing, LLM, and RAG, the proposed framework provides a structured approach for generating interpretable and evidence-grounded screening support. The main contribution of the framework lies in the conceptual integration of multimodal indicator acquisition, evidence-grounded language model interpretation, and hyper-realistic avatar-based interaction within a single screening framework. This design enables the system to capture not only what patients report through PHQ-9 responses, but also how they present themselves verbally, non-verbally, and physiologically during the assessment process. The hyper-realistic avatar interface further supports patient-facing interaction by delivering real-time dialogue, voice rendering, and emotionally aligned visual feedback, while maintaining the role of the system as a supportive interface rather than a diagnostic authority. Overall, the proposed framework is intended to support, rather than replace, clinical judgment by providing additional interpretable evidence for identifying patients who may require further psychological assessment or follow-up. Future work should focus on system implementation, multimodal dataset collection, model optimization, latency evaluation, and clinical validation in real-world T2DM remission care settings. These future developments will be essential for assessing the feasibility, accuracy, interpretability, and practical value of the framework in clinical and community-based diabetes care.

Acknowledgement. This work is supported by the European Union's Horizon Europe programme through the Marie Skłodowska-Curie Actions (MSCA), under the projects ACMod (Affective Computing Models: from Facial Expression to Mind-Reading, Grant Agreement No. 101130271) and HarmonicAI (Human-guided Collaborative Multi-Objective Design of Explainable, Fair and Privacy-Preserving AI for Digital Health, Grant Agreement No. 101131117). The authors would like to express their sincere gratitude to all project partners and collaborating organizations for their valuable support, insights, and contributions. In particular, the authors gratefully acknowledge all partners who provided contextual information and expert perspectives on type 2 diabetes mellitus and diabetes remission care in Thailand, which substantially informed the development of the proposed framework.

References

Ali, S., Stone, M.A., Peters, J.L., Davies, M.J., Khunti, K.: The prevalence of co-morbid depression in adults with type 2 diabetes: a systematic review and meta-analysis. Diabet. Med. **23**(11), 1165–1173 (2006). https://doi.org/10.1111/j.1464-5491.2006.01943.x

American Diabetes Association: 4. Comprehensive medical evaluation and assessment of comorbidities: standards of medical care in diabetes—2021. Diabetes Care. **44**(Supplement 1), S40–S52 (2021). https://doi.org/10.2337/dc21-S004

Anderson, R.J., Freedland, K.E., Clouse, R.E., Lustman, P.J.: The prevalence of comorbid depression in adults with diabetes: a meta-analysis. Diabetes Care. **24**(6), 1069–1078 (2001). https://doi.org/10.2337/diacare.24.6.1069

Cichoń, E. et al.: People with diabetes need a lower cut-off than others for depression screening with PHQ-9. PLoS One. **15**(10), e0240209 (2020). https://doi.org/10.1371/journal.pone.0240209

Dechsling, A., Shic, F., Zhang, D., Marschik, P.B., Nordahl-Hansen, A.: Evaluating the feasibility of emotion expressions in avatars created from real person photos: pilot web-based survey of virtual reality software. JMIR Form. Res. **7**, e44632 (2023). https://doi.org/10.2196/44632

Division of Noncommunicable Diseases, Department of Disease Control: Guidelines for Implementing Diabetes Remission Care Services (Remission Service), Department of Disease Control, Ministry of Public Health (2023). https://ddc.moph.go.th/uploads/publish/1517720240105071553.pdf

Ekman, P.: An argument for basic emotions. Cognit. Emot. **6**(3–4), 169–200 (1992). https://doi.org/10.1080/02699939208411068

Fang, T., Zhang, Q., Wang, Z., Liu, J.P.: Bidirectional association between depression and diabetic nephropathy by meta-analysis. PLoS One. **17**(12), e0278489 (2022). https://doi.org/10.1371/journal.pone.0278489

Fridenson-Hayo, S. et al.: Basic and complex emotion recognition in children with autism: cross-cultural findings. Molecular. Autism. **7**, 52 (2016). https://doi.org/10.1186/s13229-016-0113-9

Gao, Y. et al.: Retrieval-augmented generation for large language models: a survey arXiv. https://arxiv.org/abs/2312.10997 (2023)

Jafari, A., Moshki, M., Ghelichi-Ghojogh, M., Nejatian, M.: Role of diabetes health literacy, psychological status, self-care behaviors, and life satisfaction in predicting quality of life in type 2 diabetes. Sci. Rep. **14**, 1498 (2024). https://doi.org/10.1038/s41598-024-51245-x

Janssen, E.P.C.J. et al.: The patient health Questionnaire-9 as a screening tool for depression in individuals with type 2 diabetes mellitus: the Maastricht study. J. Am. Geriatr. Soc. **64**(11), e201–e206 (2016). https://doi.org/10.1111/jgs.14388

Kroenke, K., Spitzer, R.L., Williams, J.B.W.: The PHQ-9: validity of a brief depression severity measure. J. Gen. Intern. Med. **16**(9), 606–613 (2001). https://doi.org/10.1046/j.1525-1497.2001.016009606.x

Lewis, P. et al.: Retrieval-augmented generation for knowledge-intensive NLP tasks. Adv. Neural Inf. Proces. Syst. **33**, 9459–9474 (2020)

Lotrakul, M., Sumrithe, S., Saipanish, R.: Reliability and validity of the Thai version of the PHQ-9. BMC Psychiatry. **8**, 46 (2008). https://doi.org/10.1186/1471-244X-8-46

Lyra e Silva, N.M., Lam, M.P., Soares, C.N., Munoz, D.P., Milev, R., De Felice, F.G.: Insulin resistance as a shared pathogenic mechanism between depression and type 2 diabetes. Frontiers, 57. Psychiatry. **10** (2019). https://doi.org/10.3389/fpsyt.2019.00057

Marty, S., Pantanella, F., Ronco, A., Dheman, K., Magno, M.: Investigation of mmWave radar technology for non-contact vital sign monitoring arXiv. https://arxiv.org/abs/2309.08317. (2023)

MicRadar Technology: R60ABD1: Breath & sleep radar (Version 1.1) [User manual]. MicRadar Technology (Shenzhen) Co., Ltd. (2023). https://www.micradar.cn/uploads/soft/20230801/1690873124.pdf.
Paier, W., Hilsmann, A., Eisert, P.: Interactive facial animation with deep neural networks. IET Comput. Vis. **14**(6), 359–369 (2020). https://doi.org/10.1049/iet-cvi.2019.0790
Royal College of Family Physicians of Thailand: Diabetes Remission in Type 2 Diabetes with Intensive Lifestyle Intervention: Guide for Healthcare Providers. Royal College of Family Physicians of Thailand (2022). https://thaigpfm.org/wp-content/uploads/2025/05/Diabetes-remission-in-type-2_web.pdf
Tan, X., Qin, T., Soong, F., Liu, T.-Y.: A survey on neural speech synthesis arXiv (2021). https://arxiv.org/abs/2106.15561
Twist, K.E., Ståhl, D., Amiel, S.A., Thomas, S., Winkley, K., Ismail, K.: Comparison of depressive symptoms in type 2 diabetes using a two-stage survey design. Psychosom. Med. **75**(8), 791–797 (2013). https://doi.org/10.1097/PSY.0b013e3182a2b108
van Steenbergen-Weijenburg, K.M. et al.: Validation of the PHQ-9 as a screening instrument for depression in diabetes patients in specialized outpatient clinics. BMC Health Serv. Res. **10**, 235 (2010). https://doi.org/10.1186/1472-6963-10-235
Voss, R.M., Das, J.M.: Mental Status Examination. StatPearls Pusblisshing (2024) https://www.ncbi.nlm.nih.gov/books/NBK546682/

Digital Transformation of Fatigue Risk Management in Mining: Evaluating Real-Time Monitoring Through a Social Ecological Approach

Galuh Saraesti Ardhianingrum[1], Nur Arfian[2], and Ari Prayogo Pribadi[3](✉)

[1] Master of Public Health Program, Faculty of Medicine, Public Health and Nursing, Universitas Gadjah Mada, Yogyakarta, Indonesia
[2] Department of Anatomy, Faculty of Medicine, Public Health and Nursing, Universitas Gadjah Mada, Yogyakarta, Indonesia
[3] Department of Behavior, Environment and Social Medicine, Faculty of Medicine, Public Health and Nursing, Universitas Gadjah Mada, Yogyakarta, Indonesia
ari.prayogo.p@ugm.ac.id

Abstract. Occupational fatigue is a major safety concern in the mining industry. This study evaluated the effectiveness of Fatigue Real-Time Monitoring in mitigating occupational fatigue among mining operators using a Social Ecological Model framework. An embedded mixed-methods design was applied, involving 352 operators for quantitative analysis and in-depth interviews for qualitative exploration. Results showed a significant difference in fatigue levels between pre-shift and in-shift monitoring groups, with fatigue more frequently detected in the in-shift monitoring group (OR = 3.11; 95% CI: 1.58–6.53; $p = 0.002$). Technology acceptance was significantly associated with lower fatigue levels (OR = 0.32; 95% CI: 0.19–0.52; $p < 0.001$). Qualitative findings revealed that Fatigue Real-Time Monitoring implementation was influenced by factors across individual, interpersonal, organizational, community, and policy levels. Key barriers included technology resistance and technical issues, while management commitment, family support, and continuous evaluation supported successful implementation. These findings suggest that digital fatigue monitoring functions not only as a detection tool but also as a socio-technical intervention. Effective fatigue mitigation therefore requires a comprehensive and human-centered approach integrating technology, organizational support, and worker acceptance.

Keywords: Digital transformation · Fatigue monitoring · Mining industry · Occupational fatigue · Social ecological model

N. Mohd Sharef et al. (Eds.): CAMP 2026, CCIS 3094, pp. 297–313, 2027.
https://doi.org/10.1007/978-981-92-4305-1_22

1 Introduction

The mining industry is a capital-intensive, labor-intensive, and high-risk sector. Operations in the mining sector run continuously 24 h a day, with workers averaging about 46.6 working h per week. Environmental conditions, combined with the monotonous and repetitive nature of the work, lead to significant fatigue, increasing the risk of accidents by up to 32% and repetitive injuries by up to 73% (Bauerle et al., 2022). Various other factors also contribute, such as work shifts, vibrations, the work environment, and exposure to pollutants. While fatigue is subjective from an individual's perspective, these external factors also play a role in influencing the level of fatigue experienced by workers (Purto et al., 2024). Fatigue is one of the causes of workplace accidents stemming from human or worker factors. Dump truck (DT) operators in the mining sector often face high mental workloads due to tasks that require intense concentration over long periods of time. A study in East Kalimantan at a company holding a Mining Services showed that the mental workload of Komatsu 785–7 HD operators, measured using the NASA-TLX method, had an average score of 66.60, falling into the high category. Although the physical workload is relatively light with an average %CVL of 17%, this high mental workload can increase the risk of cognitive fatigue, which impacts reduced focus and work efficiency, as well as increasing the potential for operational errors (Kristian Silalahi et al., 2021).

Previous research has identified that the primary factors contributing to work-related fatigue among mining workers include sleep deprivation, heavy workloads, shift work, and poor environmental conditions (Dwitha Ramdhani & Soraya, 2024). These findings highlight the importance of a multidimensional approach to assessing work-related fatigue and recommend comprehensive interventions to reduce fatigue and improve the well-being of workers in the mining sector. Other previous study shows that work-related fatigue is multidimensional, influenced by individual, interpersonal, organizational, community, and policy factors (Bauerle et al., 2022). These findings align with the Social Ecological Model, which emphasizes that fatigue is influenced by an individual's interactions with their social environment and work systems. Therefore, a more objective early-detection approach is needed to address the complexity of fatigue risk factors.

Despite growing attention to occupational fatigue in the mining industry, several research gaps remain. Previous studies have primarily focused on understanding fatigue determinants, including psychosocial factors, work schedules, environmental conditions, and socio-economic influences among mining workers (Pelders & Nelson, 2019; Bauerle et al., 2022; Jaladara & Hasanbasri, 2024). Other studies have explored social-ecological frameworks within mining systems (Le Gouill & Poupeau, 2020) or developed technology-based fatigue detection models using artificial intelligence and physiological indicators (Chen et al., 2023). However, limited evidence exists regarding the effectiveness of digital fatigue monitoring technologies as fatigue mitigation interventions in real operational mining settings. Furthermore, few studies have integrated technological, behavioral, organizational, and policy-related factors within a comprehensive social ecological framework. To the best of our knowledge, no previous study has evaluated the implementation of Fatigue Real-Time Monitoring through both quantitative effectiveness assessment and qualitative exploration of multilevel determinants among mining operators in Indonesia. Therefore, this study addresses these gaps by employing an embedded mixed-methods approach to evaluate Fatigue Real-Time Monitoring implementation using the Social Ecological Model framework.

A preliminary study conducted in July 2025 at a mining company in East Kalimantan revealed an increase in fatigue-related incidents from 9% in 2023 to 16% in 2024, with the majority attributed to individual factors such as loss of concentration. Although the company has implemented various fatigue management programs, fatigue remains a cause of workplace incidents ranging from property damage to major incidents. The trend of financial losses due to incidents also increased significantly, reaching Rp 9.729 billion in the first half of 2025, indicating that fatigue management programs have not been fully effective in mitigating the impact on safety and productivity. One of the fatigue management programs implemented involves the use of Real-Time Fatigue Monitoring via in-car cameras and wearable devices to detect signs of fatigue in real time during pre-shift and in-shift periods. This technology supports the prevention of workplace accidents and the implementation of good mining practices. However, the results of a preliminary study indicate that the effectiveness of implementation depends not only on the technology, but also on operator behavior and compliance. The challenges identified include non-compliance with the use of wearable devices and complaints from operators due to discomfort in the cab. This study aims to evaluates Fatigue Real Time Monitoring implementation among mining operators using a social ecological framework to identify multilevel determinants of fatigue mitigation.

2 Methods

2.1 Study Design

This study used an embedded mixed-methods design with a quantitative approach as the primary method and qualitative inquiry as supporting data. The study aimed to evaluate the effectiveness of Fatigue Real-Time Monitoring in mitigating fatigue among mining workers using the Social Ecological Model approach.

2.2 Sample and Population

The study population consisted of 1,722 heavy-duty vehicle operators at PT Kideco Jaya Agung, East Kalimantan, who used Fatigue Real-Time Monitoring (FRM) systems, including wearable devices for pre-shift monitoring and in-car cameras for in-shift monitoring.

The sample size was calculated using Yamane's formula (95% confidence level, 5% margin of error), resulting in a minimum of 323 respondents. Stratified random sampling was applied based on FRM type, dividing the population into pre-shift ($n = 974$) and in-shift monitoring ($n = 748$) groups. This yielded 183 pre-shift and 140 in-shift operators, with a total of 352 quantitative respondents.

Inclusion criteria included active FRM use for at least two shift cycles, valid mine permits, and informed consent. Exclusion criteria were medical conditions affecting fatigue and absence during data collection. For the qualitative phase, 12 participants were purposively selected across Social Ecological Model levels, including operators, supervisors, HSE personnel, family members, and management.

2.3 Instrument and Methods

Quantitative data were collected using structured questionnaires and physiological measurements. Respondent characteristics included demographic factors, work patterns, lifestyle, fatigue awareness, and Fatigue Real-Time Monitoring status. Subjective fatigue was measured using the Subjective Self Rating Test developed by the Industrial Fatigue Research Committee. Objective assessment included blood pressure, pulse rate, and Cardiovascular Load (CVL) measurements conducted by medical personnel. Technology acceptance was assessed using a modified Technology Acceptance Model (TAM) questionnaire adapted from Davis, Bagozzi and Warshaw (1989). The instrument demonstrated good validity and reliability with a Cronbach's Alpha value of 0.97. Data collection was conducted among operators using wearable devices for pre-shift monitoring and in-car cameras for in-shift monitoring. Qualitative data were collected through semi-structured in-depth interviews based on the Social Ecological Model framework. Participants included operators, supervisors, HSE personnel, family representatives, and management representatives. Document triangulation using fatigue management SOPs, incident reports, and program evaluation documents was also conducted to support the findings.

2.4 Data Analysis

Quantitative data were analyzed using RStudio (2025.09.2 + 418). Univariate analysis described respondent characteristics and variables, while bivariate analysis used Chi-Square/Fisher's Exact tests and Spearman Rank Correlation. Ordinal logistic regression was applied for multivariate analysis.

Qualitative data were analyzed thematically using ATLAS.ti (26.0.1) through transcription, coding, categorization, and triangulation. Finally, quantitative and qualitative findings were integrated to provide a comprehensive understanding of factors influencing FRM effectiveness in fatigue mitigation (Table 1).

3 Results

3.1 Characteristic Respondent

All respondents in this study were male heavy-duty vehicle operators. Most respondents were adults of productive working age, ranging from 20 to 54 years old, with 58% categorized as adults. The majority were married (72%), while 28% were single. Regarding family characteristics, 38% had no children, followed by respondents with two children (27%), one child (19%), and more than two children (16%). More than half

Table 1. Respondent characteristics

Respondent Characteristics	Frequency (n)	Percentage (%)
Age		
Min – Maks (20–54)	–	–
Mean ± SD (32.2 ± 8.15)	–	–
Age Category		
Young Adults (20–29 years)	149	42
Adults (30–59 years)	204	58
Marital Status		
Married	99	28
Single	253	72
Number of Children		
No Children	133	38
1 Child	67	19
2 Children	95	27
More than 2 Children	57	16
Social Dependent		
Has Dependent	183	52
Has No Dependent	169	48
Economic Dependent		
Has Dependent	117	33
Has No Dependent	235	67
Microsleep Experience		
Never	283	80
1–2 times	66	19
Frequently (≥3 times)	3	1

of the respondents had social dependents (52%), while 33% reported having economic dependents. Most respondents reported never experiencing microsleep (80%), whereas 19% experienced microsleep 1–2 times and 1% reported frequent microsleep episodes.

3.2 Univariate Study

Fatigue Real-Time Monitoring Usage

Most respondents were included in the pre-shift monitoring group using wearable device–based Fatigue Real-Time Monitoring (58%), while 42% used in-shift monitoring through in-car cameras. The difference in group size reflected the sampling composition

Table 2. Distribution of fatigue real-time monitoring usage

Monitoring Group	Frequency (n)	Percentage (%)
Preshift Monitoring	205	58
Inshift Monitoring	147	42

rather than differences in Fatigue Real-Time Monitoring implementation in the field (Tables 2 and 3).

Level of Occupational Fatigue

Table 3. Overview of occupational fatigue level

Occupational Fatigue Level	Frequency (n)	Percentage (%)
Low	47	13
Moderate	302	86
High	3	1

The mean fatigue score was 36.77 $\pm$ 7.18, with scores ranging from 30 to 73. Most workers experienced moderate fatigue (86%), while 13% reported low fatigue and only 1% experienced high fatigue. These findings indicate that most workers experienced manageable but noticeable fatigue levels (Tables 4, 5 and 6).

Perceived Technology Acceptance

Table 4. Overview of perceived technology acceptance

Level of Perceived Technology Acceptance	Frequency (n)	Percentage (%)
Low	14	4
Moderate	177	50.3
High	9	2.6
Very High	152	43.2

Most respondents demonstrated moderate (50.3%) to very high (43.2%) levels of technology acceptance, while only a small proportion reported low (4.0%) or high (2.6%) acceptance. The mean technology acceptance score was 4.12 ± 0.79 (range: 1–5), indicating generally high acceptance of the technology. Among the dimensions assessed, the highest mean score was observed for Attitude Towards Using (4.17 ± 0.84), followed by Perceived Ease of Use (4.15 ± 0.80), Perceived Usefulness (4.12 ± 0.88), and Behavioral Intention to Use (4.04 ± 0.85). These findings suggest that workers generally perceived the technology as useful, easy to use, and beneficial for supporting their work.

Work Shifts

Table 5. Distribution of work shifts

Shift Group	Frequency (n)	Percentage (%)
Shift I (Morning)	183	52
Shift II (Night)	169	48

Work shift distribution was relatively balanced, with 52% of respondents working day shifts and 48% working night shifts. Both shifts operated on a 12-h schedule.

Results of Objective Measurements of Physiological Parameters

Table 6. Results of physiological measurements

Parameter	Min - Max	Mean ± SD	Difference Mean ± SD
Heart Rate			−4.8 ± 13.2
Before Work	58–106	82.1 ± 10.9	
After Work	54–99	77.3 ± 9.7	
SpO_2			−0.5 ± 2.8
Before Work	98.6 ± 1.2	89–100	
After Work	98.1 ± 2.4	82–100	
Systolic Blood Pressure			−1.5 ± 10.5
Before Work	98–140	116.6 ± 8.6	
After Work	90–135	115.1 ± 8	
Diastolic Blood Pressure			−1 ± 6.6
Before Work	77.9 ± 6.1	60–90	
After Work	76.9 ± 5.6	60–90	

Physiological measurements showed slight decreases in pulse rate, oxygen saturation, and blood pressure after work, although all values remained within normal ranges. Cardiovascular Load (CVL) assessment indicated that all respondents were classified in the light physical workload category. Despite the light physiological workload, most workers reported moderate subjective fatigue, suggesting that psychosocial factors, work patterns, or sleep quality may contribute to fatigue (Tables 7, 8 and 9).

3.3 Bivariate Study

Differences of Fatigue Real-Time Monitoring Usage by Level of Occupational Fatigue

Table 7. Differences of fatigue real-time monitoring usage by level of occupational fatigue

Fatigue Real-Time Monitoring Usage	Level of Occupational Fatigue				*Cramer's V*
	Low n (%)	Moderate n (%)	High n (%)	*p value*	
Preshift Monitoring	35 (17.1)	170 (8.9)	0 (0)	0.0052	0.1664
Inshift Monitoring	12 (8.2)	132 (89.8)	3 (2)		

Statistical analysis showed significant differences in fatigue levels between the pre-shift and in-shift monitoring groups ($p = 0.005$). Workers in the in-shift monitoring group had higher fatigue scores (median 37; IQR 32–45) compared with the pre-shift monitoring group (median 33; IQR 31–37), with a significant difference ($p < 0.001$). Although most workers in both groups experienced moderate fatigue, high fatigue was only found in the in-shift monitoring group. The strength of association was weak (Cramer's V = 0.166).

Relationship Between Level of Perceived Technology Acceptance and Level of Occupational Fatigue

Technology acceptance was significantly associated with fatigue level. Spearman Rank Correlation showed a significant negative correlation between technology acceptance and fatigue scores ($p < 0.001$; $r = -0.288$), indicating that higher technology acceptance was associated with lower fatigue levels. Fisher's Exact test also demonstrated a significant relationship between technology acceptance categories and fatigue levels ($p = 0.039$; Cramer's V = 0.135). Workers with lower fatigue had higher scores across all technology acceptance dimensions, particularly perceived ease of use and attitude toward using, suggesting that positive perceptions of FRM technology may support fatigue management.

Table 8. Relationship between level of perceived technology acceptance and level of occupational fatigue

Perceived Technology Acceptance	Level of Occupational Fatigue			*p value*	*Cramer's V*
	Low n (%)	Moderate n (%)	High n (%)		
Low	1 (7.1)	13 (92.9)	0 (0)	0.0389	0.135
Moderate	0 (0)	9 (100)	0 (0)		
High	13 (8.6)	136 (89.5)	3 (2)		
Very High	33 (18.6)	144 (81.4)	0 (0)		

Table 9. Relationship between work shifts and level of occupational fatigue

WorkShift	Level of Occupational Fatigue			
	Low n (%)	Moderate n (%)	High n (%)	p value
Shift I	24 (6.82)	156 (44.32)	3 (0.85)	0.387
Shift II	23 (6.53)	146 (41.47)	0 (0)	

Relationship Between Work Shifts and Level of Occupational Fatigue

Most workers in both day shift (Shift I) and night shift (Shift II) groups experienced moderate fatigue. Statistical analysis showed no significant association between work shift and fatigue level ($p = 0.387$), indicating that shift differences did not significantly influence worker fatigue in this study (Table 10).

3.4 Multivariate Analysis

Table 10. Results of the multivariate ordinal logistic regression analysis on work fatigue levels

Variable	Reference	β	SE	t value	OR	95% CI	p-value
Fatigue Real-Time Monitoring Usage	*Pre-shift Monitoring*	1.13	0.36	3.16	3.11	(1.58–6.53)	0.002
Perceived Technology Acceptance	–	−1.14	0.25	−4.56	0.32	(0.19–0.52)	<0.001
Work Shifts	*Shift I*	−0.19	0.32	−0.61	0.83	(0.44–1.54)	0.544

Multivariate analysis using ordinal logistic regression was conducted to examine the simultaneous effects of Fatigue Real-Time Monitoring (FRM), technology acceptance, and work shift on fatigue levels. All model assumptions were fulfilled, including proportional odds (Brant test $p = 0.36$), absence of multicollinearity (VIF = 1.00), acceptable model fit (McFadden $R^2 = 0.103$), and significant likelihood ratio test results ($p < 0.001$). The analysis showed that workers in the in-shift monitoring group had a significantly higher likelihood of experiencing greater fatigue compared with the pre-shift monitoring group (OR = 3.11; 95% CI: 1.58–6.53; $p = 0.002$). In contrast, technology acceptance had a significant protective effect against fatigue (OR = 0.32; 95% CI: 0.19–0.52; $p < 0.001$), indicating that higher acceptance of Fatigue Real Time Monitoring technology was associated with lower fatigue levels. Work shift was not significantly associated with fatigue (OR = 0.83; 95% CI: 0.44–1.54; $p = 0.544$). Overall, Fatigue Real Time Monitoring type and technology acceptance were significant predictors of fatigue, whereas work shift was not.

3.5 Qualitative Study Analysis Based on Social Ecological Model

The qualitative study used semi-structured in-depth interviews with participants selected based on the levels of the Social Ecological Model. Interviews with 12 participants produced 108 pages of transcripts, 348 quotations, 70 codes, 19 categories, and 5 themes. Phenomenological analysis was conducted through bracketing, data reduction, thematic categorization, expression validation, and descriptive interpretation supported by verbatim quotations. Qualitative findings were used to support the quantitative results, with themes organized according to the Social Ecological Model levels.

- **Individual Factors**

Fatigue was influenced by sleep patterns, lifestyle, work adaptation, personal awareness, and ergonomic conditions. Operators perceived Fatigue Real-Time Monitoring differently; some viewed it as beneficial for improving sleep discipline and preventing microsleep, while others experienced psychological pressure from continuous monitoring. Fatigue was primarily associated with cognitive and mental workload, particularly reduced concentration, drowsiness, and mental exhaustion.

- **Interpersonal Factors**

Social support, communication, and supervisory responses played important roles in fatigue management. Co-workers and supervisors actively monitored fatigue symptoms and provided interventions such as rest breaks, relaxation programs, and temporary work replacement when fatigue was detected. These findings suggest that the effectiveness of fatigue monitoring depends not only on technology but also on teamwork and workplace safety culture.

- **Organizational Factors**

Organizational commitment was a key factor supporting Fatigue Real-Time Monitoring implementation. While pre-shift monitoring focused on prevention and work readiness, in-shift monitoring emphasized real-time fatigue detection. Successful implementation was supported by safety education, supervision, fatigue interventions, and continuous program evaluation, although technical issues and technology resistance remained challenges.

- **Community Factors**

Family support contributed to fatigue management by encouraging adequate rest, healthy lifestyles, and compliance with monitoring systems. Positive family perceptions toward fatigue management programs reinforced technology acceptance and worker safety behaviors. However, domestic and social responsibilities during off-days could reduce recovery opportunities.

- **Policy Factors**

Fatigue management was strengthened through organizational policies, management commitment, and investment in monitoring technologies. Clear regulations regarding rest requirements, fatigue reporting, and reward–punishment mechanisms promoted adherence to fatigue management procedures. Continuous evaluation, safety culture reinforcement, and family involvement further supported the effectiveness of fatigue mitigation programs.

4 Discussion

- **Differences of Fatigue Real-Time Monitoring Usage on Occupational Fatigue**

Fatigue management in mining is implemented within the Safety Management System (SMKP) using Fatigue Real-Time Monitoring as a data-based risk tool. In heavy-duty operators, fatigue reflects reduced physiological and psychological capacity due to accumulated workload (Horberry et al., 2022).

Bivariate analysis showed significant differences in fatigue between pre-shift and in-shift monitoring, where pre-shift monitoring detected lower fatigue as a baseline fit-to-work check, while in-shift monitoring identified higher fatigue during operations. This aligns with evidence on fatigue-related risks in mining (Mulyati et al., 2020).

Within the Fatigue Risk Management System, pre-shift monitoring is preventive, whereas in-shift monitoring is real-time and more sensitive to operational fatigue (Sprajcer et al., 2022). Multivariate analysis showed that in-shift monitoring significantly increased the likelihood of detecting higher fatigue levels ($OR = 3.11$; 95% CI: 1.58–6.53; $p = 0.002$). This finding reflects the higher sensitivity of in-car camera systems in identifying dynamic fatigue conditions during operations rather than indicating that the technology causes fatigue. Real-time camera and sensor systems can directly detect behavioral and physiological fatigue indicators, such as yawning, eye closure, and reduced alertness (Makhmudov et al., 2024).

However, both systems have limitations in detecting mental fatigue. Physiological indicators may remain normal despite subjective fatigue, as current systems are more sensitive to physical drowsiness than cognitive fatigue(Adão Martins et al., 2021). Integrating wearable devices, AI, and multimodal signals such as ECG, EMG, and EEG may improve future fatigue detection accuracy (Kakhi et al., 2025).

- **Relationship Between Perceived Technology Acceptance with Worker Occupational Fatigue**

Technology acceptance was identified as an important factor influencing both compliance with fatigue monitoring systems and worker fatigue levels in mining operators. Quantitative analysis showed a significant negative relationship between perceived technology acceptance and occupational fatigue ($p < 0.001$; $r = -0.288$), indicating that operators with more positive perceptions of fatigue monitoring technology tended to report lower fatigue levels. Significant differences were also found across all Technology Acceptance Model (TAM) dimensions, particularly perceived ease of use and attitude toward using the technology. These findings suggest that technologies perceived as difficult or uncomfortable may increase psychological stress and contribute to fatigue (Zhao & Yan, 2023).

Qualitative findings further explained these results. Operators with shorter work experience tended to demonstrate deep compliance, accepting the technology because they perceived it as beneficial for safety. In contrast, more experienced operators often showed surface compliance, where adherence was driven by external pressures such as strict regulations, sanctions, and economic necessity. Feelings of constant surveillance, especially in in-shift monitoring systems, created psychological discomfort that could contribute to mental fatigue. These findings are consistent with previous studies showing that perceived usefulness and ease of use improve deep compliance, whereas externally driven compliance may increase work stress (Hu et al., 2020).

The study also identified technical barriers affecting technology acceptance, including poor sensor ergonomics, uncomfortable cabin conditions, and repeated false detections. These issues increased cognitive workload and potentially accelerated fatigue. Similar studies in the mining sector have shown that technology implementation may create new risks if ergonomic and human factors are not adequately considered (Lööw, 2022). Although operator compliance with both preshift and in-shift systems reflected strong organizational commitment to fatigue management, improving user comfort and adopting more persuasive implementation approaches remain important to promote genuine acceptance and optimize fatigue reduction outcomes.

- **Relationship Between Work Shifts and Occupational Fatigue**

Work shifts are theoretically associated with fatigue due to circadian rhythm disruption and altered sleep patterns. However, this study found no significant relationship between work shift and fatigue level among mining operators ($p = 0.387$), supported by multivariate analysis ($\beta = -0.19$; $p = 0.544$; OR = 0.83; 95% CI: 0.44–1.54). Fatigue distribution in both day and night shifts was relatively similar and predominantly categorized as moderate fatigue. These findings suggest that shift timing alone did not directly influence operator fatigue levels.

Nevertheless, qualitative findings revealed that operators subjectively experienced greater fatigue during night shifts, particularly during early morning hours and roster transition periods. This aligns with evidence that night shift workers are more vulnerable to sleep loss and circadian misalignment, which contribute to fatigue and accident risk (Maisey et al., 2021). Operators also identified monotonous work and prolonged operational routines as contributing factors to mental fatigue.

The absence of a significant statistical relationship may be explained by stronger psychosocial and organizational influences. Previous studies reported that fatigue among mining workers is more strongly associated with psychosocial factors, such as job demands and social support, rather than operational characteristics like shift work itself (Moreira & de Lucca, 2024). In this study, organizational fatigue management systems, social support, and family support appeared to mitigate the negative effects of shift work.

Qualitative findings also indicated increased fatigue during roster transition periods, reflecting physiological adaptation challenges and mental fatigue that may affect cognitive performance and decision-making. Mental fatigue has been associated with increased risk-taking behavior and occupational accidents (Jia et al., 2022). However, organizational interventions such as G5R (5-min Relaxation Exercises) and random fatigue checks during critical hours likely reduced sleepiness and fatigue, which may explain why no significant difference between shifts was observed statistically.

- **The Influence of Fatigue Real-Time Monitoring, Perceived Technology Acceptance, and Work Shift on Occupational Fatigue**

Occupational fatigue in mining is a multidimensional phenomenon influenced by fatigue real-time monitoring, technology acceptance, and operational conditions rather than shift schedules alone. This study found that fatigue real-time monitoring and technology acceptance significantly affected work fatigue, while shift work was not significant after adjustment.

From a technological perspective, fatigue monitoring systems can identify fatigue effectively; for example, machine learning–based models such as Bagging-SVM achieved 91.6% accuracy using physiological indicators like ECG and skin temperature (Chen et al., 2025). However, many systems still function as lagging indicators that detect fatigue after it occurs (Talebi et al., 2021) highlighting the need for integration with preventive interventions and operational data.

Technology acceptance also plays a key role. Operators with more positive perceptions of monitoring systems tend to experience lower fatigue, as supportive perceptions encourage adaptive safety behavior, while negative perceptions may increase psychological burden (Chen, 2025). Therefore, safety strategies should position monitoring technology as a protective tool rather than surveillance. In contrast, shift work was not significantly associated with fatigue in this study, although prior research links it to circadian disruption and fatigue prediction models (Talebi et al., 2021). This indicates that fatigue is also shaped by environmental stressors such as heat and humidity, as well as individual factors (Chen, 2025). Also, should be distinguished from sleepiness, which requires different mitigation strategies (Barham et al., 2024). Overall, fatigue management should adopt a multidimensional approach integrating monitoring technology, user acceptance, and organizational systems rather than relying solely on shift scheduling.

- **Integration of Fatigue Real-Time Monitoring in Fatigue Mitigation Based on the Social Ecological Model**

Occupational fatigue among mining workers is a complex phenomenon influenced not only by individual factors but also by interactions across interpersonal, organizational, community, and policy levels. The Social Ecological Model provides a comprehensive framework to explain how these multilevel determinants interact in shaping fatigue conditions and the effectiveness of Fatigue Real-Time Monitoring interventions.

At the individual level, fatigue was mainly related to mental and cognitive workload caused by monotonous tasks requiring sustained concentration. Technology acceptance became a key determinant of monitoring effectiveness, as workers with lower acceptance tended to experience higher fatigue and show resistance toward monitoring systems (Wang et al., 2025).

At the interpersonal level, coworker support and supervisor involvement encouraged compliance with fatigue management procedures and strengthened shared responsibility for safety. Monitoring systems also improved communication and information sharing among workers and supervisors (Li, Xia and Bentley, 2026).

At the organizational level, fatigue management programs such as roster systems, G5R (5-min Relaxation Exercises), wake-up calls, and random fatigue checks helped reduce fatigue risk and maintain worker alertness. These interventions likely explain why shift work was not significantly associated with fatigue in this study (Xu and Hall, 2021).

At the community level, family support was identified as an important factor in fatigue mitigation. Research on male workers has shown that long working hours contribute to sleep disturbance and burnout through work–family conflict (Baek, Won and Yoon, 2023). In this study, operators received both practical and emotional support from family members, which helped reduce additional fatigue risk factors. Positive family responses and supportive domestic environments strengthened recovery and improved compliance with fatigue management practices. Previous studies also demonstrated that emotional reactions to work–family conflict are associated with poorer sleep quality and psychological distress (Lawson, Lee and Maric, 2021). Therefore, family support functions as an important protective factor within the fatigue cycle, highlighting the need for family-based education and work–life balance interventions.

At the policy level, the effectiveness of Fatigue Real-Time Monitoring was strongly influenced by organizational regulations, management commitment, and integration with Mining Safety Management Systems. Strong managerial commitment was reflected through substantial investment in monitoring technologies, implementation of reward and punishment systems, and safety culture campaigns promoting shared responsibility for safety. These findings suggest that fatigue policies should treat fatigue as an unavoidable risk that must be minimized through risk-based approaches rather than through working-hour regulation alone. Previous studies on Fatigue Risk Management Systems emphasized that effective implementation requires clear regulatory frameworks, integration of technology-based monitoring, and shared responsibility between organizations and workers (Sprajcer et al., 2022). Organizations are responsible for providing systems and interventions, while workers are responsible for utilizing rest opportunities and reporting fatigue conditions. In addition, strengthening safety culture and data-driven management systems remains essential because organizational management significantly influences mining safety performance (Li, Mithal Jiskani and Tian, 2022). Overall, this study demonstrates that fatigue mitigation in mining requires a multidimensional approach integrating technology, worker perceptions, organizational systems, social support, and policy frameworks. Fatigue Real-Time Monitoring can effectively support fatigue mitigation when implemented within a comprehensive Social Ecological Model approach involving all interconnected levels.

5 Conclusion

This study found that occupational fatigue among mining operators is influenced by multiple factors across individual, interpersonal, organizational, community, and policy levels. Fatigue Real-Time Monitoring significantly influenced occupational fatigue, with in-shift monitoring showing higher sensitivity in detecting fatigue than preshift monitoring. Technology acceptance acted as a protective factor, where operators with more positive perceptions toward monitoring systems tended to experience lower occupational fatigue. Meanwhile, work shift was not significantly associated with occupational fatigue, indicating that fatigue was more strongly affected by psychosocial factors, organizational systems, and fatigue management interventions. Social support, organizational commitment, safety culture, and family involvement also contributed to the effectiveness of fatigue management. Human-centered implementation strategies are necessary to optimize digital transformation of fatigue risk management in mining operations. Overall, occupational fatigue mitigation in mining requires a comprehensive approach integrating technology, worker acceptance, organizational support, and safety policies within the Social Ecological Model framework.

Acknowledgement. The authors would like to express their sincere gratitude to Universitas Gadjah Mada for academic support and to PT Kideco Jaya Agung for facilitating the research process and supporting data collection activities.

Ethics Statements. This study received ethical approval from the Health Research Ethics Committee of Faculty of Medicine, Public Health, and Nursing Universitas Gadjah Mada (KE/FK/1945/EC/2025). Research permission was obtained from PT Kideco Jaya Agung. Written informed consent was obtained from all participants, and confidentiality and anonymity were maintained throughout the study.

References

Adão Martins, N.R., et al.: Fatigue monitoring through wearables: a state-of-the-art review. Front. Physiol. **12**, 790292 (2021). https://doi.org/10.3389/fphys.2021.790292

Baek, S.-U., Won, J.-U., Yoon, J.-H.: The role of work–family conflict in the association between long working hours and workers' sleep disturbance and burnout: results from the sixth Korean working conditions survey. BJPsych Open. **9**(5), e163 (2023). https://doi.org/10.1192/bjo.2023.555

Barham, M., Bauerle, T., Eiter, B.: Are fatigue and sleepiness the same? A brief introduction to the differences and similarities and their implications for work safety. Min. Metall. Explor. **41**(1), 53–59 (2024). https://doi.org/10.1007/s42461-023-00892-1

Bauerle, T.J., et al.: The human factors of mineworker fatigue: an overview on prevalence, mitigation, and what's next. In: American Journal of Industrial Medicine, pp. 832–839. John Wiley and Sons Inc (2022). https://doi.org/10.1002/ajim.23301

Chen, X.: Integrated multimethod analysis of miners' safety behavior and risk interaction for practical applications. Sci. Rep. **15**(1), 34722 (2025). https://doi.org/10.1038/s41598-025-18454-4

Chen, Y., et al.: Research on the effects of operational fatigue and bagging-SVM recognition of deep coal mine workers. J. Loss Prev. Process Ind. **94**, 105508 (2025). https://doi.org/10.1016/j.jlp.2024.105508

Davis, F.D., Bagozzi, R.P., Warshaw, P.R.: User acceptance of computer technology: a comparison of two theoretical models. Manag. Sci. **35**, 982–1003 (1989) https://www.jstor.org/stable/2632151?seq=1&cid=pdf

Dwitha Ramdhani, J., Soraya, T.M.: Kelelahan Kerja Pekerja Tambang: Apa saja Faktor-faktornya?, dan Bagaimana pengukurannya? (Systematic Literature Review pada Data Publikasi 2019-2024). J. Appl. Comm. Eng. **4**(1), 28–37 (2024). https://doi.org/10.52158/jace.v4i1.841

Horberry, T. et al.: Fatigue Management in the Queensland Mining Industry and Its Relationship with Mental Health Final Report to Office of the Commissioner Resources Safety and Health Sustainable Minerals Institute (2022)

Hu, X., Yeo, G., Griffin, M.: More to safety compliance than meets the eye: differentiating deep compliance from surface compliance. Saf. Sci. **130**, 104852 (2020). https://doi.org/10.1016/j.ssci.2020.104852

Jia, A., Guo, X., Tian, S.: Experimental study on the influence of mental fatigue on risk decision-making of miners. Sci. Rep. **12**(1), 11902 (2022). https://doi.org/10.1038/s41598-022-14045-9

Kakhi, K., et al.: Fatigue monitoring using wearables and AI: trends, challenges, and future opportunities. Comput. Biol. Med. **195**, 110461 (2025). https://doi.org/10.1016/j.compbiomed.2025.110461

Kristian Silalahi, H., Dianati Fathimahhayati, L., Tambunan, W.: ANALISA BEBAN KERJA MENTAL DAN FISIK OPERATOR HD KOMATSU 785-7 (STUDI KASUS PT.SIMS JAYA KALTIM). ARIKA. **15**(1), 37–50 (2021)

Lawson, K.M., Lee, S., Maric, D.: Not just work-to-family conflict, but how you react to it matters for physical and mental health. Work Stress. **35**(4), 327–343 (2021). https://doi.org/10.1080/02678373.2021.1888821

Li, K., Xia, M.R., Bentley, T.: The paradoxical effects of emerging mining technologies on psychosocial work factors: an integrated framework and research agenda. Technol. Forecast. Soc. Change. **222**, 124391 (2026). https://doi.org/10.1016/j.techfore.2025.124391

Li, Y., Mithal Jiskani, I., Tian, K.: Using the ISM-ANP-SD combination model to explore the mechanism and intervention strategies of influencing factors of coal mine safety system. Front. Public Health. **10**, 1053298 (2022). https://doi.org/10.3389/fpubh.2022.1053298

Lööw, J.: Understanding technology in mining and its effect on the work environment. Miner. Econ. **35**(1), 143–154 (2022). https://doi.org/10.1007/s13563-021-00279-y

Maisey, G., et al.: The sleep of shift Workers in a Remote Mining Operation: methodology for a randomized control trial to determine evidence-based interventions. Front. Neurosci. **14**, 579668 (2021). https://doi.org/10.3389/fnins.2020.579668

Makhmudov, F., et al.: Real-time fatigue detection algorithms using machine learning for yawning and eye state. Sensors. **24**(23), 7810 (2024). https://doi.org/10.3390/s24237810

Moreira, A.S., de Lucca, S.R.: Physical and mental fatigue in shift work and mitigation strategies: an integrative review. In: Revista Brasileira de Medicina do Trabalho. Associacao Nacional de Medicina do Trabalho (2024). https://doi.org/10.47626/1679-4435-2024-1267

Mulyati, T., et al.: Fatigue analysis of high DUMP truck operators in INDONESIA'S coal mining industry: a CASE study. Malays. J. Public Health Med. **2020**(1), 38–44 (2020). https://doi.org/10.37268/mjphm/vol.20/no.Special1/art.666

Purto, H., et al.: Self-perceptual blindness to mental fatigue in mining workers. Front. Neuroergon. **5**, 1441243 (2024). https://doi.org/10.3389/fnrgo.2024.1441243

Sprajcer, M., et al.: How effective are fatigue risk management systems (FRMS)? A review. Accid. Anal. Prev. **165**, 106398 (2022). https://doi.org/10.1016/j.aap.2021.106398

Talebi, E., et al.: Modeling mine workforce fatigue: finding leading indicators of fatigue in operational data sets. Minerals. **11**(6), 621 (2021). https://doi.org/10.3390/min11060621

Wang, L., et al.: A cognitive load assessment method for airtight cabin operators based on a one-dimensional convolutional neural network. Symmetry. **17**(6), 915 (2025). https://doi.org/10.3390/sym17060915

Xu, S., Hall, N.G.: Fatigue, personnel scheduling and operations: review and research opportunities. Eur. J. Oper. Res., 807–822 (2021). https://doi.org/10.1016/j.ejor.2021.03.036

Zhao, X., Yan, D.: Incorporating technological acceptance model into safety compliance of construction workers in Australia. Saf. Sci. **163**, 106127 (2023). https://doi.org/10.1016/j.ssci.2023.106127

Elderly Mobility Disorder Prevention Monitoring Through Home-Based Interactive Exercise with Knowledge Graph-Based Personalized Recommendation

Thandar Phyo[1](✉), Pradorn Sureephong[1], Tama Duangnamol[1], Heikki Karjaluoto[2], and Guoying Zhao[3]

[1] Department of Digital Technology Management, College of Arts, Media and Technology, Chiang Mai University, Chiang Mai, Thailand
{thandarphyo_1,pradorn.s,tama.d}@cmu.ac.th
[2] Faculty of Information Technology, University of Jyväskylä, Jyväskylä, Finland
heikki.karjaluoto@jyu.fi
[3] Faculty of Information Technology and Electrical Engineering, University of Oulu, Oulu, Finland
guoying.zhao@oulu.fi

Abstract. Population ageing increases the need for home-based technologies that support mobility maintenance among elderly people. This study proposes a knowledge graph-based personalized exercise recommendation approach for elderly mobility disorder prevention. Data from Gymbot, Metareverse, and Smart Watch were cleaned and modelled in Neo4j as a User–Session–Course knowledge graph. Three recommendation approaches were implemented and compared: neighbourhood-based collaborative filtering, user profile-based content-based filtering, and hybrid recommendation. A temporal train-test evaluation was conducted using Precision@5, Recall@5, Hit Rate@5, and NDCG@5. The results showed that neighbourhood-based collaborative filtering achieved the best performance, with Precision@5 of 0.468, Recall@5 of 0.4067, Hit Rate@5 of 0.86, and NDCG@5 of 0.5606. Hybrid recommendation ranked second, while user profile-based content-based filtering showed lower performance. The findings indicate that similar-user exercise histories provide a stronger signal than difficulty and duration feature alone. The proposed knowledge graph structure also supports explainable recommendation by tracing results through similar users, recommendation scores, user-session-course relationships and interpretable course features.

Keywords: Knowledge Graph · Personalized Exercise Recommendation · Mobility Disorders · Older Adults · Collaborative Filtering

1 Introduction

Population ageing is increasing the need for preventive technologies that help elderly people maintain mobility, independence, and quality of life. Reduced mobility, balance problems, and declining physical function are important risks in later life and

N. Mohd Sharef et al. (Eds.): CAMP 2026, CCIS 3094, pp. 314–323, 2027.
https://doi.org/10.1007/978-981-92-4305-1_23

require early preventive support (Montero-Odasso et al. 2022). Regular physical activity, including aerobic, muscle-strengthening, and multicomponent balance exercises, is recommended for elderly populations to maintain physical function and reduce mobility-related risks (Bull et al. 2020). However, elderly people may face barriers to conventional exercise programs, such as transportation limitations, limited access to rehabilitation facilities, chronic conditions, low motivation, and insufficient personalized guidance. Previous studies have shown that home-based exercise interventions can support exercise participation outside clinical settings and may improve physical function, balance, and mobility-related outcomes among elderly people (Solis-Navarro et al. 2022; Costa-Brito et al. 2024). Therefore, home-based digital exercise systems can help address access-related barriers, while personalized recommendation methods can support the selection of suitable exercise courses according to user characteristics, activity history, and course properties.

Thailand provides a relevant context for this study because it is experiencing rapid population ageing. It is reported that Thailand had a population of 71,697,000, including approximately 15,773,000 elderly people aged 60 years and above. The same report stated that about 400,000 elderly people aged more than 80 required assistance with daily activities in 2017, with this number expected to increase substantially in the coming decades (International Domestic Workers Federation 2025).

Digital exercise platforms produce user profiles, exercise session records, performance data, and course metadata that can support personalized recommendation. These data can be represented as connected nodes in a knowledge graph, allowing the system to model relationships among elderly users, exercise sessions, and courses. Such a graph-based representation is useful not only for producing recommendations but also for explaining why a course is suggested. In health-related applications, explainability is important because elderly users, caregivers, and health professionals may need to understand the reasoning behind a recommendation (European Commission 2019; Arrieta et al. 2020).

This study proposes a knowledge graph-based personalized exercise recommendation approach for elderly mobility disorder prevention in home-based exercise environments. The proposed approach represents elderly users, exercise sessions, and courses as connected nodes in Neo4j. Based on this graph structure, collaborative filtering, content-based filtering, and hybrid recommendation are implemented to produce personalized exercise course suggestions. The study also demonstrates how graph relationships can support explainable recommendation paths between elderly users, completed courses, similar users, and recommended courses.

The objective of this study is to develop and evaluate a knowledge graph-based personalized exercise recommendation approach for elderly home-based exercise support. Specifically, the study aims to: (1) model elderly exercise data as a User–Session–Course knowledge graph, (2) apply neighbourhood-based collaborative filtering, user profile-based content-based filtering, and hybrid recommendation, and (3) compare their recommendation performance while showing how graph relationships and course features can support traceable recommendation explanation.

2 Literature Review

2.1 Theoretical Background

Recommender systems support personalized decision-making by suggesting relevant exercise courses based on elderly user information, previous exercise session histories, and course characteristics such as difficulty level and duration (Aggarwal 2016; Ricci et al. 2022). In this study, collaborative filtering recommends courses based on the completed-course histories of similar elderly users, while content-based filtering recommends courses by comparing course features with the elderly user's exercise profile. Hybrid recommendation combines both sources of evidence to provide a final ranked list.

Knowledge graphs represent real-world entities as nodes and their connections as relationships, making them suitable for modelling connected data such as elderly users, exercise sessions, and courses (Hogan et al. 2021; Ji et al. 2022). In recommender systems, this structure can support explainability by connecting users, items, attributes, and interactions through meaningful paths (Guo et al. 2022). This is important in health-related applications, where elderly users, caregivers, and health professionals may need to understand why a recommendation is provided. Explainable and trustworthy Artificial Intelligence (AI) emphasizes transparency, accountability, understandability, and robustness when AI systems influence human wellbeing (European Commission 2019; Arrieta et al. 2020). In this study, these principles are addressed by providing understandable recommendation reasons through graph paths, similar elderly users, completed exercise sessions, and course features.

2.2 Related Work

Prior studies show that digital home-based exercise interventions can support physical function and fall-related outcomes among elderly people (Solis-Navarro et al. 2022; Costa-Brito et al. 2024). Recommender-system studies for healthy habits and active ageing highlight the importance of personalization, but transparency remains a challenge in health-related systems (Lopez-Barreiro et al. 2024). Knowledge graph-based recommendation can support explainability, although much existing work focuses on general recommendation domains rather than elderly home-based exercise support (Guo et al. 2022). This study addresses this gap by integrating elderly home-based exercise data, knowledge graph modelling, and explainable recommendation logic using collaborative filtering, content-based filtering, and hybrid recommendation.

3 Methodology

3.1 Data Source

The data used in this study were obtained from three main sources: Gymbot, Metareverse, and Smart Watch. These sources provide complementary information related to user registration, exercise participation, system usage, and physical activity.

3.2 Data Cleaning and Preparation

Data cleaning was conducted before knowledge graph modelling to ensure that the datasets were consistent, complete, and suitable for recommendation development. The cleaning workflow included initial data profiling, schema harmonization, missing-value treatment, duplicate record removal, outlier and validity filtering, and identifier consistency checking. The cleaned datasets were then prepared as graph-ready data for modelling elderly users, sessions, and courses in the knowledge graph.

3.3 Knowledge Graph Modelling

The knowledge graph was modelled in Neo4j using three node types: User, Session, and Course. The User node represents an elderly participant, the Session node represents a recorded exercise activity, and the Course node represents an exercise course. Two relationships were defined: HAS_SESSION, which connects a user to a session, and USES, which connects a session to a course. This structure creates a User–Session–Course path for tracing exercise history and supporting recommendation. Figure 1 shows the nodes and relationships in the knowledge graph.

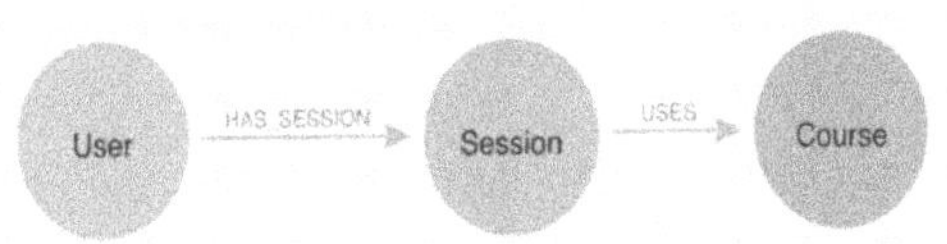

Fig. 1. Nodes and relationships in the knowledge graph.

Figure 2 shows an example Neo4j visualization of the anonymized User–Session–Course paths in the knowledge graph. The figure illustrates how a user profile node is connected to multiple exercise session nodes through HAS_SESSION relationships, and how each session is connected to a completed course through USES relationships.

3.4 Neighbourhood-Based Collaborative Filtering

Neighbourhood-based collaborative filtering was applied to recommend exercise courses from elderly users with similar completed-course histories. User similarity was computed using Jaccard similarity, as shown in Eq. (1), which measures the overlap between two completed-course sets as the size of their intersection divided by the size of their union (Jaccard 1901; Aggarwal 2016).

$$J(A, B) = \frac{|A \cap B|}{|A \cup B|} \tag{1}$$

After computing user similarity, the top-k most similar users were selected as the target user's neighbourhood. Candidate courses were then ranked using similarity-weighted neighbourhood aggregation, as shown in Eq. (2) where $\hat{r_{u,i}}$, $N_k(u)$, sim(u, v), and

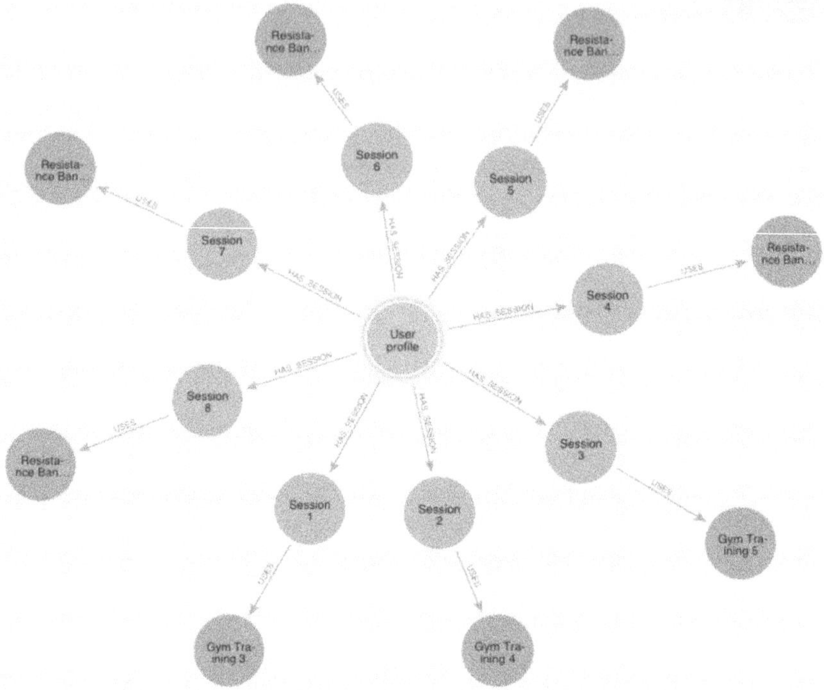

Fig. 2. Neo4j visualization of User–Session–Course paths.

$r_{v,i}$ denote the predicted preference score, top-k neighbourhood, user similarity, and neighbour-course interaction, respectively (Aggarwal 2016; Ricci et al. 2022). Recommendations were labelled as new or repeated courses based on the target elderly user's previous completion history.

$$\hat{r_{u,i}} = \frac{\sum_{v \in N_k(u)} sim(u, v) r_{v,i}}{\sum_{v \in N_k(u)} |sim(u, v)|} \tag{2}$$

3.5 User Profile-Based Content-Based Filtering

User profile-based content-based filtering was applied to recommend exercise courses by comparing an elderly user's historical exercise profile with available course features. Unlike collaborative filtering, this approach uses course attributes rather than similarities between users. In this study, each course was represented by a feature vector consisting of difficulty level and duration, as shown in Eq. (3).

$$x_i = \left[\mathrm{difficulty_i}, \mathrm{duration_i}\right] \tag{3}$$

For each elderly user, a user profile vector was constructed by averaging the feature vectors of the courses previously completed by that user, as shown in Eq. (4), where $\mathrm{p_u}$

is the profile vector of user u, C_u is the set of courses completed by user *u*, and x_i is the feature vector of course i.

$$p_u = \frac{1}{|C_u|}\sum\nolimits_{i \in C_u} x_i \tag{4}$$

Candidate courses were then compared with the user profile using cosine similarity, as shown in Eq. (5) (Aggarwal 2016; Ricci et al. 2022).

$$\cos(p_u, x_i) = \frac{p_u \cdot x_i}{||p_u||||x_i||} \tag{5}$$

Courses with higher cosine similarity were ranked higher because their difficulty level and duration were closer to the elderly user's previous exercise profile. Each recommendation was labelled as a new or repeated course based on the target elderly user's previous completion history.

3.6 Hybrid Recommendation

Hybrid recommendation was used to combine neighbourhood-based collaborative filtering and user profile-based content-based filtering. Before combination, both scores were normalized using min–max normalization so that they were represented on a comparable scale between 0 and 1. The normalized scores were then combined using a weighted hybrid approach (Burke 2002; Aggarwal 2016; Ricci et al. 2022), as shown in Eq. (6), where $Score_{CF}^{norm}(u, i)$ and $Score_{CBF}^{norm}(u, i)$ are the normalized collaborative filtering and content-based filtering scores for user (u) and course i, respectively. In this study, $\alpha =$ 0.5 was used to give equal weight to both methods. Courses with higher hybrid scores were ranked higher and labelled as new or repeated based on the target elderly user's previous completion history.

$$Score_{hybrid}(u, i) = \alpha Score_{CF}^{norm}(u, i) + (1 - \alpha) Score_{CBF}^{norm}(u, i) \tag{6}$$

3.7 Recommendation Evaluation

The three recommendation methods were evaluated using a temporal train-test split to reflect the chronological nature of exercise participation. For each eligible elderly user, the earlier 80% of exercise sessions were used as training data, while the most recent 20% were held out as test data. Recommendations were produced using only the training data and were then compared with the held-out test courses. Since exercise courses may be repeated over time, both newly recommended and previously completed courses were considered valid recommendation candidates. Evaluation was conducted using four top-5 ranking metrics: Precision@5, Recall@5, Hit Rate@5, and NDCG@5, which are commonly used for top-N recommendation evaluation (Cremonesi et al. 2010; Shani and Gunawardana 2011; Järvelin and Kekäläinen 2002). Precision@5 measures how many of the five recommended courses appeared in the test set, while Recall@5 measures how many of the relevant test courses were recovered in the top-5 list. Hit Rate@5 indicates whether at least one relevant course appeared in the top-5 recommendations, and Normalized Discounted Cumulative Gain (NDCG@5) evaluates whether relevant courses were ranked near the top of the recommendation list.

4 Results and Discussion

4.1 Data Cleaning and Preparation

After cleaning, the final datasets contained 11,896 exercise session records, 59 unique elderly users, and 218 unique exercise courses.

4.2 Knowledge Graph Modelling

Using the cleaned datasets, the modelled knowledge graph contained 12,173 nodes and 23,792 relationships. The graph included 59 User nodes, 11,896 Session nodes, and 218 Course nodes. The User nodes stored profile properties such as user_id, age, gender, weight, and height. The Session nodes stored exercise activity properties such as session_id, user_id, course_id, course_name, timestamp, duration_min, and calories. The Course nodes stored course properties such as course_id, course_name, courseTypeName, difficultyLevel, difficultyLevelName, and duration. The relationships consisted of 11,896 HAS_SESSION links between users and sessions and 11,896 USES links between sessions and courses. This structure connected each elderly user to their exercise sessions and completed courses through the User–Session–Course path. The same graph structure was used for recommendation modelling: collaborative filtering used user-course interaction histories derived from these paths, content-based filtering used Course node properties such as difficultyLevel and duration, and the hybrid method combined both sources of evidence. Although the recommendation models are conventional, the knowledge graph provides a structured evidence layer through traceable User–Session–Course paths.

4.3 Comparative Evaluation of Recommendation Methods

The recommendation evaluation results are presented in Table 1. The results are averaged across all eligible evaluated users. Neighbourhood-based collaborative filtering achieved the highest performance across all metrics, with Precision@5 of 0.468, Recall@5 of 0.4067, Hit Rate@5 of 0.86, and NDCG@5 of 0.5606. In practical terms, this means that about 2.34 out of the five recommended courses appeared in the held-out test set, about 40.67% of the courses that users completed later were correctly included in the top-5 recommendation list, 86% of evaluated users received at least one relevant course, and the ranking quality reached about 56.06% of the ideal ranking, with higher scores given when relevant courses appeared near the top of the recommendation list.

This indicates that similar-user exercise histories were the strongest signal for predicting later course use. The hybrid method ranked second, showing that combining collaborative and content-based evidence improved performance compared with user profile-based content-based filtering alone. However, the hybrid method did not outperform collaborative filtering, suggesting that the content-based component weakened the stronger collaborative signal when equal weighting was applied. User profile-based content-based filtering achieved the lowest scores, which may be explained by the limited course representation using only difficulty level and duration.

The results should also be interpreted with dataset and feature limitations. Although the dataset contains 11,896 exercise sessions, these were produced by only 59 users, which may limit generalizability and affect collaborative filtering stability. In addition, future content-based models can use richer course features by encoding courseTypeName and course_name. This would allow exercise categories such as Warm up activation, Comprehensive training, Thai boxing, Upper body shaping, and Lower body shaping, as well as textual course meanings from course names, to be considered.

While similar baseline scores could be computed from tabular interaction data, the knowledge graph improves recommendation traceability by linking each recommendation to users, sessions, courses, similar users, and course properties.

Table 1. Comparative performance of recommendation methods at top-5

	Method	Precision@5	Recall@5	Hit Rate@5	NDCG@5
1	Neighbourhood-based Collaborative Filtering	0.468	0.406677	0.86	0.560590
2	Hybrid Recommendation	0.408	0.312728	0.74	0.464336
3	User Profile-based Content-based Filtering	0.116	0.050375	0.40	0.112584

Table 2 shows an example of collaborative filtering recommendations for one target elderly user. The list includes both repeated and new courses. Repeated courses, such as Exercise Elders 01, Exercise Elders 08, and Exercise Elders 09, had already appeared in the user's previous exercise history, while new courses, such as Qi Gong Dance 01 and Waist & Hip Shaping Training, had not previously been completed by the target user. The number of similar users shows how many neighbouring users contributed to each recommendation, while the recommendation score represents the similarity-weighted contribution from those users. Therefore, the table provides traceable recommendation evidence by linking the target user to similar elderly users and commonly completed courses through the knowledge graph.

Table 2. Neighbourhood-based collaborative filtering recommendations for one target user

Rank	Recommended Course	Recommendation Type	Similar Users	Recommendation Score
1	Exercise Elders 01	Repeated Course	45	5.5066
2	Exercise Elders 08	Repeated Course	35	4.9104
3	Exercise Elders 09	Repeated Course	32	4.6348
4	Qi Gong Dance 01	New Course	14	2.1435
5	Waist & Hip Shaping Training	New Course	5	1.5007

Table 3 provides another illustrative recommendation trace for a different target user. In this example, all top-5 recommendations are new courses, showing that the collaborative filtering method can also support new course discovery based on the exercise histories of similar users.

Table 3. Neighbourhood-based collaborative filtering recommendations for another target user

Rank	Recommended Course	Recommendation Type	Similar Users	Recommendation Score
1	Qi Gong Dance 01	New Course	14	5.0481
2	20 Minutes Daily Dance 5	New Course	14	4.0257
3	Warm Up Activation Training 1	New Course	14	3.8704
4	Warm Up Activation Training 2	New Course	9	3.1483
5	20 Minutes Daily Dance 8	New Course	8	2.6131

5 Conclusion

This study compared three knowledge graph-based recommendation approaches for elderly home-based exercise: neighbourhood-based collaborative filtering, user profile-based content-based filtering, and hybrid recommendation. The results showed that neighbourhood-based collaborative filtering performed best across all metrics, achieving Precision@5 of 0.468, Recall@5 of 0.4067, Hit Rate@5 of 0.86, and NDCG@5 of 0.5606. Hybrid recommendation ranked second, while user profile-based content-based filtering achieved the lowest performance. This indicates that similar-user exercise histories were more effective for predicting later course use than course features alone. The example recommendation output further shows that recommendations can be interpreted through recommendation type, similar users, recommendation scores, and User–Session–Course paths.

Future work will extend this framework through knowledge-based and context-aware recommendation, richer course and elderly user health features, graph-native recommendation methods, and Graph-based Retrieval-Augmented Generation (GraphRAG) with large language models (LLMs) for improved personalization and explainability.

Acknowledgements. This research was supported by the CMU Presidential Scholarship. This work was also supported by the European Union's Horizon 2020 research and innovation programme under the Marie Skłodowska-Curie grant agreements No. 101130271 (ACMod) and No. 101131117 (HarmonicAI).

References

Aggarwal, C.C.: Recommender Systems: the Textbook. Springer, Cham (2016). https://doi.org/10.1007/978-3-319-29659-3

Arrieta, A.B. et al.: Explainable artificial intelligence (XAI): concepts, taxonomies, opportunities and challenges toward responsible AI. Inf. Fusion. **58**, 82–115 (2020). https://doi.org/10.1016/j.inffus.2019.12.012

Bull, F.C. et al.: World Health Organization 2020 guidelines on physical activity and sedentary behaviour. Br. J. Sports Med. **54**(24), 1451–1462 (2020). https://doi.org/10.1136/bjsports-2020-102955

Burke, R.: Hybrid recommender systems: survey and experiments. User Model. User-Adapt. Interact. **12**, 331–370 (2002). https://doi.org/10.1023/A:1021240730564

Costa-Brito, A.R., Vieira, E.R., Rebelo-Marques, A., et al.: Home-based exercise interventions delivered through technology in older adults: a systematic review and meta-analysis. Int. J. Med. Inform. **181**, 105263 (2024). https://doi.org/10.1016/j.ijmedinf.2023.105263

Cremonesi, P., Koren, Y., Turrin, R.: Performance of recommender algorithms on top-N recommendation tasks. In: Proceedings of the Fourth ACM Conference on Recommender Systems, Barcelona, Spain, pp. 39–46 (2010). https://doi.org/10.1145/1864708.1864721

European Commission, High-Level Expert Group on Artificial Intelligence: Ethics Guidelines for Trustworthy AI. European Commission, Brussels (2019)

Guo, Q. et al.: A survey on knowledge graph-based recommender systems. IEEE Trans. Knowl. Data Eng. **34**(8), 3549–3568 (2022). https://doi.org/10.1109/TKDE.2020.3028705

Hogan, A. et al.: Knowledge graphs. ACM Comput. Surv. **54**(4), Article 71 (2021). https://doi.org/10.1145/3447772

International Domestic Workers Federation: Demystifying Care Needs: Thailand Summary. International Domestic Workers Federation (2025) https://idwfed.org/wp-content/uploads/2025/05/Thailand-Report.pdf

Jaccard, P.: Étude comparative de la distribution florale dans une portion des Alpes et du Jura. Bull. Soc. Vaudoise Sci. Nat. **37**, 547–579 (1901)

Järvelin, K., Kekäläinen, J.: Cumulated gain-based evaluation of IR techniques. ACM Trans. Inf. Syst. **20**(4), 422–446 (2002). https://doi.org/10.1145/582415.582418

Ji, S., Pan, S., Cambria, E., Marttinen, P., Yu, P.S.: A survey on knowledge graphs: representation, acquisition, and applications. IEEE Trans. Neural Netw. Learn. Syst. **33**(2), 494–514 (2022). https://doi.org/10.1109/TNNLS.2021.3070843

Lopez-Barreiro, J., Garcia-Soidan, J.L., Alvarez-Sabucedo, L., Santos-Gago, J.M.: Artificial intelligence-powered recommender systems for promoting healthy habits and active aging: a systematic review. Appl. Sci. **14**(22), 10220 (2024). https://doi.org/10.3390/app142210220

Montero-Odasso, M. et al.: World guidelines for falls prevention and management for older adults: a global initiative. Age Ageing. **51**(9), afac205 (2022). https://doi.org/10.1093/ageing/afac205

Ricci, F., Rokach, L., Shapira, B. (eds.): Recommender systems handbook, 3rd edn. Springer, New York (2022). https://doi.org/10.1007/978-1-0716-2197-4

Shani, G., Gunawardana, A.: Evaluating recommendation systems. In: Ricci, F., Rokach, L., Shapira, B., Kantor, P.B. (eds.) Recommender systems handbook, pp. 257–297. Springer, Boston (2011). https://doi.org/10.1007/978-0-387-85820-3_8

Solis-Navarro, L., Gismero, A., Vilarino, T., et al.: Effectiveness of home-based exercise delivered by digital health in older adults: a systematic review and meta-analysis. Age Ageing. **51**(11), afac243 (2022). https://doi.org/10.1093/ageing/afac243

Sampling Strategy on Ranked Outputs Data Towards Fairness-Aware Evaluation with Sparse Datasets

Ardiansyah Dores[1(✉)] and Sri Devi Ravana[2]

[1] Universitas Mercu Buana, Jakarta, Indonesia
ardian@mercubuana.ac.id
[2] University of Malaya, Kuala Lumpur, Malaysia

Abstract. Sampling strategy identified as crucial to data preparation and mitigation techniques used to ensure representative data distribution across different sensitive groups, thereby minimizing bias and allowing for statistically robust fairness-aware evaluation. Sampling strategy discusses the data generation process that is in accordance with the characteristics of divergence metrics (fairness evaluation processing), divergence metrics have provisions on the processing of probability distribution data. Furthermore, to preserve top rank existence in ranked outputs (data generation), probability sampling methods are appointed. This study developed a data generation process which is ranked outputs from Learning to Rank processing which is LambdaRank. Results that gain from LambdaRank are then processed by sampling strategy using combination of probability sampling methods and Inverse probability weighting (IPW) which is maximum likelihood estimation (MLE). This results in 2 (two) probability distributions which later be used to measure fairness-aware evaluation. The 2 (two) distributions then test using ANOVA to conclude the two distributions have different variances and to gain the significance Two-sample Kolmogorov-Smirnov applied. The two tests' results stated that the two distributions gain significant differences.

Keywords: Sampling strategy · Data generation · Fairness-aware evaluation · Probability sampling · Maximum likelihood estimation

1 Introduction

In this study, sampling strategy was identified as crucial to data preparation and mitigation techniques used to ensure representative data distribution across different sensitive groups, thereby minimizing bias and allowing for statistically robust fairness-aware evaluation. To preserve top rank existence in ranked outputs (data generation), probability sampling methods are appointed. Furthermore, fairness-aware evaluations can more reliably assess whether a model's decisions adhere to established fairness metrics, such as demographic parity or equalized odds, and ultimately build more trustworthy recommender systems. Recent research on fairness involves evaluation metrics (information retrieval metrics) to satisfy notions and utility, named fairness-aware evaluation has

N. Mohd Sharef et al. (Eds.): CAMP 2026, CCIS 3094, pp. 324–339, 2027.
https://doi.org/10.1007/978-981-92-4305-1_24

gained attention, despite researchers have defined and built frameworks to handle this fairness-aware evaluation [1], problems emerged with large datasets. Thus, the focus of this study is to incorporate sparsity data in large datasets, Sparse datasets in recommender systems pose significant challenges that adversely affect the quality and reliability of recommendations. The sparsity problem arises when the user-item interaction matrix is largely empty, meaning that most users have rated or interacted with only a few items, resulting in insufficient data to accurately model user preferences or item characteristics [2–5]

Nowadays, to gain efficiency in evaluation, sampling strategies were studied, stratified and active sampling that commonly used significantly reduce the number of judgements needed, and often achieving near-perfect correlation with full judgements [6, 7]. Sampling strategy also handles noisy and heterogeneous data in cases of ranking algorithms; recent studies have effectively managed to overcome noisy pairwise comparisons and heterogeneity among users and achieved good performance with fewer comparisons [8, 9]. Recent studies emphasize the importance of targeted sampling in optimizing the evaluation of information retrieval (IR) systems. A study by Voorhees investigates how different document sampling methods influence the accuracy of test collection judgments. Those strategies focus judgment efforts on top-ranked documents, based on [10, 11] approaches that produce more reliable estimates of system performance with fewer judgments. These methods outperform traditional pooling, especially when sampling is concentrated at the top of the ranked list, aligning with the intuition that top documents are more informative for evaluation metrics like NDCG, and MAP [7]. A survey by Dai et al., (2024) broadens perspective by categorizing sampling techniques into data augmentation, filtering, rebalancing, and regularization [12]. It synthesizes evidence that strategic sampling at various stages, i.e. data collection, model training, evaluation, can mitigate distribution mismatch and bias, leading to fairer IR systems. In 2021, Gonzales-Zelaya et al., [13] demonstrated adaptability across classifiers and datasets, which outperformed existing fairness correction methods with minimal accuracy trade-offs.

Regarding ranked outputs dataset, Yang and Stoyanovich has done their job on fairness measures for ranked outputs, they have developed a data generation procedure that allows to systematically control the output on the degree of unfairness, and further to study the behaviour of their measures on their datasets [14]. On the other hand, Singh and Joachims (2018) [15] discussed fairness of exposure in rankings that allows the formulation of fairness constraints on rankings, in means to address a substantial precedent on the ranking systems and then develop efficient algorithm for it. However, more complex study was done by Diaz et al., in 2020, [16], the authors introduced the concept of expected exposure as the average attention ranked items receive from users over repeated samples of the same query, they proposed a general evaluation methodology based on expected exposure and draw connections to related metrics in information retrieval evaluation. Furthermore, sampling strategy discusses the data generation process that is in accordance with the characteristics of divergence metrics (fairness evaluation processing), divergence metrics have provisions on the processing of probability distribution data. The data generation process which is ranked outputs from Learning to Rank processing uses metrics which are LambdaRank [17–19], the problem that occurs is the complexity of the sparsity of the dataset that will inherit the results of data generation.

For this reason, a sampling strategy processing is needed to improve the integrity of data generation and reduce data sparsity, which is intended so that computational processing does not burden existing resources.

Based on the observations above, we contribute on the topic of fairness-aware evaluation focused on sampling strategies. We are questioning how sampling strategy improves fairness-aware evaluation in IR systems. We evaluate the performance of ranking algorithms; by processing the ranked outputs using fairness-aware evaluation, our contribution is designing the framework for sampling strategies. Our objective is to determine the distributions of ranked outputs to satisfy fairness notions, so the evaluation would gain a significant degree of fairness.

2 Related Works

In this section, we describe the sampling strategies within fairness evaluation and fairness-aware evaluation on information retrieval systems and recommendation systems.

2.1 Sampling Strategies Within Fairness Evaluation

Sampling plays a crucial role in the evaluation, estimation, and mitigation of bias in ranking and information retrieval systems [20]. The following work summarizes key contributions from recent studies that explore various sampling methodologies and their applications in fairness-evaluation. In 2018 Singh & Joachims [15] proposed the methodology for fairness-evaluation that involves sampling rankings according to optimized probability matrices that satisfy fairness notions like demographic parity. The sampling-based approach allows for flexible trade-offs between fairness and utility; it can be efficiently implemented through sampling procedures that approximate the desired probabilistic rankings. Such strategies enable systems to balance fairness considerations with relevance, leveraging sampling to explore the space of potential rankings under fairness notions. While in 2019 Singh & Joachims [21] introduces a policy-gradient method that relies on Monte Carlo sampling to estimate gradients during the training of fair ranking policies, they taken a conclusion by sampling rankings from a differentiable probabilistic model (e.g., Plackett-Luce), the approach enables direct optimization of utility metrics while incorporating fairness disparity measures. The results empirically show that sampling-based policy learning can mitigate biases and improve fairness without sacrificing ranking quality.

Next in 2021, Kirnap et al., [22] emphasizes the significance of importance sampling techniques, particularly stratified sampling and the Horvitz-Thompson estimator dan maximum likelihood estimation [23, 24], to accurately estimate fairness metrics under limited annotation budgets. The authors propose a weighted sampling approach that prioritizes top-ranked items, which are more influential in exposure calculations. Their experiments demonstrate that this method yields unbiased and reliable estimates of fairness metrics such as exposure and divergence measures, especially when data is scarce or incomplete. The approach effectively reduces the variance associated with sampling, enabling scalable fairness evaluation in real-world systems with limited labelling resources. Recent studies, Ghazimatin et al., (2022) [25] concluded that empirical estimates of fairness metrics, such as conditional demographic parity, are converged almost surely to their values as the sample size increases. They investigate the convergence properties of fairness measures estimated via sampling when sensitive attributes are noisy or partially observed.

Across these studies, sampling emerges as a fundamental tool for estimating fairness metrics, optimizing ranking policies, and correcting biases. Techniques such as importance sampling, stratified sampling, and Monte Carlo sampling [21, 26] are employed to efficiently gather representative data, approximate complex distributions, and ensure unbiased estimates.

2.2 Sampling Strategies Within Fairness-Aware Evaluation

Recent research emphasizes the importance of sampling in the context of fairness-aware information retrieval and ranking algorithms. For instance, [14] discusses synthetic data generation to create controlled environments for testing fairness measures. By sampling from distributions with adjustable levels of bias, researchers can systematically evaluate fairness metrics such as normalized discounted difference (rND), KL-divergence (rKL), and ratio-based measures (rRD). These sampling procedures facilitate the assessment of fairness measures' sensitivity and robustness across different bias scenarios. Similarly, [1] introduces a sampling-based approach to evaluate and optimize rankings for fairness without relying on relevance labels, their FAIR ε-greedy re-ranking algorithm explores documents probabilistically, balancing fairness and utility by sampling documents with a certain probability (ε). While Zehlike et al., (2018) [27] highlights the use of sampling to evaluate the distribution of protected groups within rankings. By sampling subsets of data and applying statistical tests, researchers can identify disparities and adjust rankings accordingly. This approach ensures that protected groups are fairly represented, aligning with fairness constraints such as statistical parity.

Thus, sampling plays a vital role in modern data analysis, particularly in fairness-aware ranking and bias mitigation. From synthetic data generation to probabilistic document exploration, sampling techniques enable researchers to evaluate, detect, and correct biases effectively. Future advancements in adaptive and scalable sampling methods are expected to further enhance fairness and utility in recommendation systems or information retrieval systems.

2.3 Inverse Probability Weighting

Divergence-based estimators, such as minimum divergence estimators (MDEs), can be interpreted as maximum likelihood estimators (MLEs) within the framework of weighted bootstrap schemes. This interpretation hinges on the relationship between divergence measures, bootstrap weights, and the asymptotic properties of estimators [28]. While Amusa, Zewotir, and North (2020) [29] stated that weighting methods, unlike matching methods, have the apparent advantage of not discarding units, which in turn reduces the standard error of estimates. In terms of fairness-aware evaluation, IPW also refers to modelling distribution, which is meant to reduce the complexity of the extreme sparsity (high-dimensional) on the datasets. Its due to the use of divergence metrics, KL divergence, the common distribution models were based on unigram model, as in studied by Roh, Oh, and Lee on 2020 [23] the authors used the unigram model to overcome the limitations of the perplexity, they adopt the idea of normalizing the word probability with respect to a quantity containing the vocabulary size.

Unigram distribution is a central object in the science of language from historical linguistics to psycholinguistics and beyond, however, most research on unigram distributions is based on identifying the distribution with sample frequency [24]. While the fundamental method that can be used to derive the probabilities in a unigram model is using the Maximum Likelihood Estimation (MLE). In fairness-aware evaluation, the use of KL divergence is closely tied to likelihood-based methods. Divergence-based methods minimize a chosen divergence measure between the empirical distribution and the model; it provides a natural measure of distributional discrepancy and is fundamental in maximum likelihood frameworks, especially within exponential families [30].

Maximum Likelihood Estimation (MLE) is a fundamental statistical method used to estimate parameters of a probabilistic model by maximizing the likelihood function. Its widespread adoption stems from desirable theoretical properties such as consistency, asymptotic normality, and efficiency under regular conditions [31]. In high-dimensional models, where the number of parameters can grow with or exceed the sample size, classical guarantees become more nuanced.

Recent research indicates that MLE can still provide finite-sample bounds on estimation error, especially when certain spectral properties of the data are favourable, such as bounded eigenvalues of the covariance matrix [30]. For MLE, bounds often involve the complexity of the model class and spectral properties, with probabilistic guarantees controlling the deviation of estimates from true parameters [30]. These bounds are crucial for understanding the feasibility of applying MLE in high-dimensional regimes, where data scarcity and computational constraints are prevalent [32]. Minimax lower bounds that characterize the fundamental limits of covariate shift estimation of MLE, showing that no estimator can uniformly outperform MLE up to a constant factor, thus establishing MLE's minimax optimality in these settings [33].

3 Proposed Method

In this study, sampling strategy identified as crucial to data preparation and mitigation techniques used to ensure representative data distribution across different sensitive groups, thereby minimizing bias and allowing for statistically robust fairness-aware evaluation. To preserve top rank existence in ranked outputs (data generation), probability sampling methods are appointed. Furthermore, fairness-aware evaluations can more reliably assess whether a model's decisions adhere to established fairness metrics, such as demographic parity or equalized odds, and ultimately build more trustworthy recommender systems. Probability sampling methods used would refer to Simple Random Sampling, Stratified Sampling, Systematic Sampling, and Stratified-Systematic Sampling. Flow of proposed method on sampling strategy, as seen on → Fig. 1 explanation of the sampling strategy procedure to improve fairness-aware evaluation measurements, are as follows.

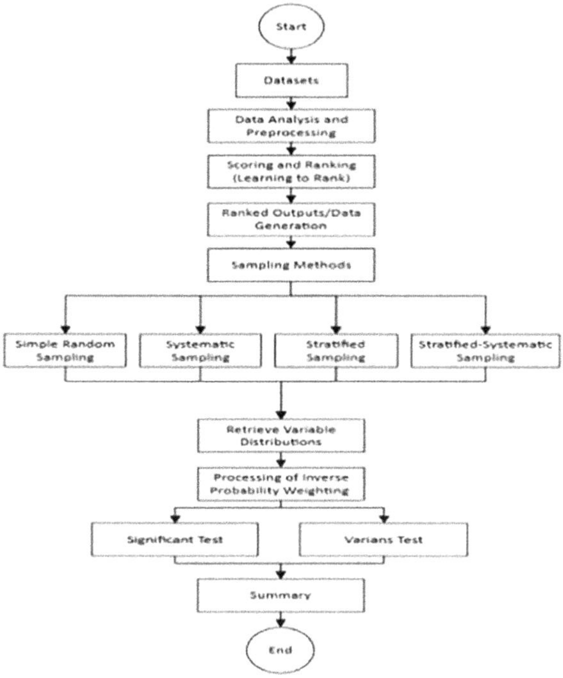

Fig. 1. Flow of proposed method on sampling strategy

3.1 Data Acquisition

Datasets in recommendation systems are characterized by sparsity, as in e-commerce systems, where the number of items increases compared to the number of customers who provide ratings. Some customers may dislike some of the items sold. This will result in inefficient dense representations, which will require high computational costs and impact on performance trade-offs [18, 34, 35]. The datasets that were chosen are well-established large-scale datasets frequently employed in recommender systems research [36]. These datasets also own the characteristics of sparsity, related to the core data structures that both datasets center on user-item interactions. While on other benefits that these datasets are available in varied sizes, ranging from 100 k for quick testing to bigger datasets that are available on more than 20 million collections, to be processed on high performance computing. MovieLens (ML-Latest-Small). This data set consists of 100,000 ratings (1–5) from 943 users on 1682 movies, each user has rated at least 20 movies, Simple demographic info for the users (age, gender, occupation, zip), movies (movieId, title, geners), and ratings (userID, movieId, rating, timestamp). The data was collected through the MovieLens web site (movielens.umn.edu) during the seven-month period from September 19th, 1997, through April 22nd, 1998.

3.2 Data Analysis and Preprocessing

In this step data is carried out to ensure that the data is well integrated. Before the process continues to the next step for scoring dan ranking, we need to define the notation for the raw dataset using $D = \{u_1, u_2, \ldots, u_N\}$, where each element $u_i = \{i, x_i, y_i\}$ the notation would state that D is the initial raw dataset of size N, containing a key numerical variable of interest x.

3.3 Scoring and Ranking Processing Using LambdaRank

LambdaRank is a popular algorithm in the Learning to Rank method that is often used in recommendation systems to optimize the relativity of item ordering and is better than just predicting the exact rating or probability of the items [19, 37, 38]. Then, to describe the ranking function, let $rank(x_i) \in \{1, 2, \ldots, N\}$ be the position of x_i when the set $X = \{x_1, \ldots, x_N\}$ is sorted in ascending order.

Table 1. Ranked Output Dataset

userId	itemId	Item popularity	Average user rating	User rating count	Predicted score	Rank
10	127	413	4,206521739	184	2,835049843	1
10	100	508	4,206521739	184	2,694853463	2
118	433	171	4,661971831	71	2,377566836	3
565	170	121	4,542857143	35	2,314310594	4
892	98	390	4	226	2,187585454	5
...						

3.4 Processing of the LambdaRank Scoring Results

The Ranking results by ranking the highest in the score, this is done to identify the item ranking, so that it will facilitate the next process (→ Table 1.). After this step, the ranked output dataset is defined, the output is in the form of a data frame. To describe the data frame of ranked output, then notation defined as $D_R = \{u_{(r)}\}_{r=1}^{N} = \{(r, x_{(r)}, y_{(r)})\}_{r=1}^{N}$, where r is the explicit index (row identifier in the dataframe), ensuring $x_1 \leq x_2 \leq \ldots \leq x_{(N)}$.

3.5 Sampling Methods Applied to D_R

We want to select a sample s of size n (or size n_h within strata) from the ranked output dataset D_R. Let $Z_r \in \{0, 1\}$ be an indicator variable where $Z_r = 1$ if row $r \in s$, and $Z_r = 0$ otherwise, the Inclusion Probability for row r is defined as $\pi_r = P(Z_r = 1)$. We defined the sampling methods as follows, first, Simple Random Sampling, every row in D_R has an equal chance of selection regardless of its rank r, then we select a subset $s \subset D_R$ of size n uniformly at random and the inclusion probability set as $\pi_r = \frac{n}{N}$. Second, Stratified Sampling, we use the sorted dataframe D_R to create H non-overlapping strata based on rank thresholds. After that defined the stratification as divide D_R into strata U_1, U_2, ..., U_H where stratum U_h contains rows from rank $r_{h,\,start}$ to $r_{h,\,end}$. The size of stratum h is N_h, selection occur by draw a random sample of size n_h independently from each stratum U_h and the inclusion probability for a row r belonging to stratum U_h set as $\pi_r = \frac{n_h}{N_h}$. Third, Systematic Sampling, we sample at a fixed interval across the ranked dataframe D_R with the selection was Calculated by the sampling interval $k = \lfloor N/n \rfloor$, we choose a random starting integer $R \in [1, k]$. The sample consists of rows with ranks: $s = \{r \in D_R \mid r = R + (j-1)k\, for\, j = 1, 2, \ldots, n\}$, and the inclusion probability defined that since every row belongs to exactly one potential systematic chain, it was determined by the random start R by $\pi_r = \frac{1}{k} \approx \frac{n}{N}$. Fourth, Stratified-Systematic Sampling, in this sampling we combine both methods, first stratify D_R by rank, then run a systematic sample inside each stratum, the selection occur by defined that inside each rank-based stratum U_h (of size N_h), calculate an interval $k_h = \lfloor N_h/n_h \rfloor$. Then, we choose an independent random start $R_h \in [1, k_h]$ for each stratum, with the sample within stratum h is set as $s_h = \{r \in U_h \mid r = r_{h,\,start} - 1 + R_h + (j-1)k_h\, for\, j = 1, 2, \ldots, n_h\}$, and the inclusion probability are defined by row r inside stratum U_h with $\pi_r = \frac{1}{k_h} \approx \frac{n_h}{N_h}$.

3.6 Data Generation Taken from Ranked Output Dataset Using Maximum Likelihood Estimation

Based on the results in point 3.5, namely the processing of sampling methods applied to the ranked output dataset (DR), two data distributions were then drawn: the average user rating and the predicted score. Both data distributions were converted into probability distributions using Maximum Likelihood Estimation (MLE). This change measures the estimated statistical parameters. These estimated statistical parameters are the data required for the divergence metric to calculate the required fairness.

Alter for 2 (two) distribution variables into probability distribution will use the inverse probability weighting method, one of which is $MLE = \frac{\sum_{i=1}^{N} x}{N}$. Then for D^{dc} and D^{ac} formulated as $D^{dc} = \frac{\sum_{i=1}^{N} predicted\_score}{N}$ and $D^{ac} = \frac{\sum_{i=1}^{N} user\_average\_ratings}{N}$. The last step is the variance test using ANOVA (Analysis of Variance) and Kolmogorov-Smirnov test. This is done to test that the distributions of the two variables have different variances. While the significance test will use the Two-sample Kolmogorov Smirnov Test [39] this test will indirectly say that the two variables are not significantly the same as each other.

4 Results and Discussions

This section describes the results of the variance and significance analysis through ANOVA and F-Statistics, followed by a significant test using the Two-sample Kolmogorov-Smirnov. Variance test here serves to determine whether the two previously defined distributions differ in variance. Therefore, ANOVA is used to find the means square between groups and the means square within groups. After the measurement, the F-statistic is used to see the results of the variance differences. Next, the results of the significance test using the Two-sample Kolmogorov-Smirnov test will state that the two distributions are significantly different from each other.

4.1 Anova Test on Simple Random Sampling

The ANOVA results on → Table 2., the two distributions using simple random sampling indicate statistically significant differences between the groups. The calculated F-value of 0.09441 exceeds the critical value, indicating that the variability between group means is greater than the variability within groups, thereby demonstrating its sensitivity to the sampling method in the context of fairness assessment. This leads to the rejection of the (H_0) null hypothesis, which states that the factor being marketed has a significant influence on D^{dc} values. In the case of D^{ac}, a similar pattern is observed, where the sum of squares between groups is significantly greater than the sum of squares within groups, resulting in an F-statistic exceeding the critical limit.

Table 2. ANOVA result on Simple Random Sampling

Source	DF	Sum of Square	Mean Square	F Statistic	P=Value
Group (Between Groups)		84324825310	84324825310	0.09441	0,759
Error (Within Groups)	200	178627921400000	893139607000		
Total		178712246200000	8891156S2900		
Summary	The Difference between the sample averages of all groups is not big enough to be statistically significant				

This confirms that the differences in D^{ac} values across groups are statistically significant, supporting the influence of the studied factor on this variable as well. In conclusion, these consistent significant findings across D^{dc} and D^{ac} validate the effectiveness of the sampling and analysis methodology.

4.2 Anova Test on Systematic Sampling

The ANOVA results as seen on → Table 3., indicate statistically significant differences between the tested groups for both distributions, namely D^{dc} and D^{ac}, under systematic sampling. For the D^{dc} system, the F-value of 0.3964 exceeds the critical value at the chosen significance level set on 0.05, leading to the rejection of the (H_0) null hypothesis that the group means are equal. This indicates that at least one group means differs significantly from the others, indicating variability in the measurement or effect across the sampled groups. Similarly, the ANOVA for the D^{ac} system also shows a significant F-value, this supports the conclusion that not all group means are equal.

Table 3. ANOVA result on Systematic Sampling

Source	DF	Sum of Square	Mean Square	F Statistic	P=Value
Group (Between Groups		100118269200	100118269200	0.3964	0.5297
Error (Within Groups)	200	50519205770000	252596028800		
Total		50619324030000	251837422000		
Summary	The Difference between the sample averages of all groups is not big enough to be statistically significant				

The analysis of variance thus confirms the existence of systematic differences in the data collected from the D^{ac} system under the same sampling method. This finding implies that the tested factors or conditions have a measurable impact on the results for both distributions. Overall, the ANOVA results validate the effectiveness of systematic sampling in detecting meaningful differences in the performance or characteristics of the D^{dc} and D^{ac} systems. This statistical evidence supports the reliability of the sampling approach.

4.3 Anova Test on Sytematic-Stratified Sampling

The finding means that the tested factors affected the results for both distributions. The ANOVA proved that systematic sampling effectively spotted important differences in the D^{dc} and D^{ac} system performances. The statistical test supports our sampling strategy's trustworthiness. Refer on → Table 4., the ANOVA compares the mean D^{dc} and D^{ac} values under stratified-systematic sampling. It checks if the means differ enough to be considered statistically significant, by F value of 0.1035 and a p value of 0.748, we see how the variation between group means stacks up against the variation within groups. These figures help us accept or reject the null hypothesis – the idea that the means are equal.

Table 4. ANOVA test on Systematic-Stratified Sampling

Source	DF	Sum of Square	Mean Square	F Statistic	P=Value
Group (Between Groups	1	0	0	0	1
Error (Within Groups)	198	0.001366	0.000006899		
Total	199	0.001366	0.000006865		
Summary	The Difference between the sample averages of all groups is not big enough to be statistically significant				

For Ddc, the ANOVA gave an F-value with a p-value under the typical 0.05 threshold, showing that the mean Ddc values really do differ across the groups. This suggests that stratification makes a real difference in Ddc measurements. The ANOVA for Dac showed significant differences between groups too, indicating that systematic-stratified sampling affects Dac values. These findings together show that the method effectively differentiates groups based on Ddc and Dac metrics. The big difference in group means backs up the sampling method's ability to capture dataset variability. This strengthens the idea that stratification boosts sampling accuracy and representativeness.

4.4 ANOVA Test on Stratified Sampling

The ANOVA results in → Table 5. Indicates D^{dc} and D^{ac} differ between groups using stratified sampling. This sampling resulted that the means of these variables differ significantly across different groups. For both measures, the test uses an F value of 0 and a p value of 1 to see how group means compare with the variation inside each group. Regarding D^{dc}, ANOVA shows a significant F value with a p-value under the usual threshold of 0.05. This tells us there's a clear difference in the D^{dc} means across the groups. So, it seems stratification picks up important variations in D^{dc}, and the way groups were assigned matters here. Same as for D^{ac}, the story's similar. The F value and p value came out significant again, which contradicts the idea that all group means are equal.

Table 5. ANOVA test on Stratified Sampling

Source	DF	Sum of Square	Mean Square	F Statistic	P=Value
Group (Between Groups	1	86246494090	86246494090	0.1035	0.748
Error (Within Groups)	200	166703093700000	833515468600		
Total	201	166789340200000	829797712500		
Summary	The Difference between the sample averages of all groups is not big enough to be statistically significant				

This reinforces that the stratified groups of D^{ac} values do indeed less differ. Furthermore, these ANOVA results show that the stratification approach led to noticeable differences in both D^{dc} and D^{ac}, and support the method used for organizing our sampling strategy.

4.5 Test Results of Two-Sample Kolmogorov-Smirnov

The Two-sample Kolmogorov-Smirnov test used as a secondary analysis to check if the overall distributions are significantly different from each other. → Table 6 shows the results of stratified-systematic sampling, variable distributions are average_user_rating (generated into ac) and predicted score (generated into dc), while ac and dc are the results of inverse probability weighting using MLE.

Table 6. Test results of Kolmogorov-Smirnov on 2 (two) distributions for all sampling methods

Sampling Methods	H_0 Hypothesis	P-value	Test Statistic	Summary
Simple Random	Since the p-value $< \alpha$, H_0 is rejected.	.00000162	(D(100,100) = 0.37	Results Test indicated a significant difference between the two distributions
Systematics	Since the p-value $< \alpha$, H_0 is rejected.	.00000703	(D(100,100) = 0.35	Results Test indicated a significant difference between the two distributions
Stratified	Since the p-value $< \alpha$, H_0 is rejected.	.23e-8	(D(100,100) = 0.43	Results Test indicated a significant difference between the two distributions
Stratified-Systematic	Since the p-value $< \alpha$, H_0 is rejected.	.0000142	(D(100,100) = 0.34	Results Test indicated a significant difference between the two distributions

This stage also validates the variable distributions, as in divergence metrics require (1) The variable distribution must not equal, (2) The variable distribution is statistically having significant difference

The Kolmogorov-Smirnov test stated that there are big differences across all sampling methods, which means some important terms for picking sampling strategies and interpreting data in this study. Every method had very low p-values—like 0.00000162 for simple random sampling, 0.00000703 for systematic sampling, 2.3×10^{-9} for stratified sampling, and 0.0000142 for stratified-systematic sampling—all lower than the α threshold. This makes the hypothesis reject the idea that all distributions are the same. Test stats ranged from 0.34 to 0.43, pointing out that the distribution from each method does not match up with the overall population distribution.

Now, diving deeper, simple random sampling is supposed to be unbiased, this is because the findings suggest this method can sometimes miss representing the whole population correctly. Different case with systematic sampling, we often run into issues like introducing patterns that are not actually there, which then messes with the expected distribution. For stratified sampling, while it breaks the group into smaller parts to analyze variance within categories better, this ends up changing how the overall distribution looks. However, when you combine both stratified and systematic sampling, the result gain complex—this combo shows unique distribution traits confirmed by the Kolmogorov-Smirnov test.

The results from the two-sample Kolmogorov-Smirnov test show how sampling methods affect data distribution. The null hypothesis (H0) gets rejected consistently for all sampling methods – simple random, systematic, stratified, and systematic-stratified. This means each method ends up with variable distributions that are noticeably different from their comparison groups. So, it's clear that these sampling techniques shape the data distributions and aren't just neutral ways to collect data.

4.6 Discussion

The Test on sampling methods using ANOVA and two-sample Kolmogorov-Smirnov tests revealed significant differences in the distributions and group means for variables D^{dc} and D^{ac} across simple random, systematic, stratified, and systematic-stratified techniques. The ANOVA showed consistent significant differences in group means for most sampling methods, which proved the effectiveness of stratification and systematic approaches in catching meaningful variability. The Kolmogorov-Smirnov test concludes by confirming that the distributions created by each sampling method differed significantly from comparison groups. This suggests the sampling techniques influenced the characteristics of the data more than just gathering neutral samples.

So, these findings stress the importance of picking the right sampling strategy, because each method impacts the properties of the stats data differently. Stratified and systematic-stratified sampling enhances our ability to detect heterogeneity and boost precision. Even simple random sampling produces distributions different from reference populations. All in all, this assessment backs up the sampling methods and analyses in this study.

5 Conclusion

This study has shown how important sampling strategies are for boosting fairness-aware evaluation, especially in recommender systems where data is often sparse. By mixing up methods like simple random sampling, stratified sampling, systematic sampling, and stratified-systematic sampling with inverse probability weighting via Maximum Likelihood Estimation (MLE). The proposed framework deals with data sparsity well and preserves the ranked output that generated using LambdaRank.

Tests using ANOVA and the Two-sample Kolmogorov-Smirnov method prove that these sampling techniques lead to significant differences in the distributions of key eval measures like D^{dc} and D^{ac}. Stratified and stratified-systematic sampling does an excellent job of catching variation and boosting the accuracy of fairness assessments. Rejection of the null hypothesis in all cases shows that these techniques have a big impact on the stats of the data and are not just passive data selectors. Furthermore, the results confirm that merging specific sampling strategies with strong statistical weighting significantly cuts bias and ramps up the reliability of fairness-aware evaluation metrics. This study managed to assess ranking algorithms when data is scarce, this leads to making fairer and more reliable for recommender systems.

References

1. Gao, R., Ge, Y., Shah, C.: FAIR: fairness-aware information retrieval evaluation. J. Assoc. Inf. Sci. Technol. **73**(10), 1–13 (2022). https://doi.org/10.1002/asi.24648
2. Loukili, M., Messaoudi, F.: Enhancing cold-start recommendations with innovative co-SVD: a sparsity reduction approach. Stat. Optim. Inf. Comput. **13**(1), 396–408 (2025). https://doi.org/10.19139/soic-2310-5070-2048
3. Huang, X., Sang, J., Yu, J., Xu, C.: Learning to learn a cold-start sequential recommender. ACM Trans. Inf. Syst. **40**(2), 1–25 (2022). https://doi.org/10.1145/3466753
4. Jiang, Y. et al.: Prompt Tuning for Item Cold-Start Recommendation. RecSys'24. ACM, Bari (2024). https://doi.org/10.1145/3640457.3688126
5. Hu, L., Jian, S., Cao, L., Gu, Z., Chen, Q., Amirbekyan, A.: HERS: Modeling influential contexts with heterogeneous relations for sparse and cold-start recommendation. In: Proceedings of the AAAI Conference on Artificial Intelligence, pp. 3830–3837 (2019). https://doi.org/10.1609/aaai.v33i01.33013830
6. Yilmaz, E., Kanoulas, E., Aslam, J.: A Simple and Efficient Sampling Method for Estimating AP and NDCG SIGIR'08. ACM, Singapore (2008)
7. Voorhees, E.: The Effect of Sampling Strategy on Inferred Measures. SIGIR'14. ACM, Gold Coast (2014)
8. Li, D., Kanoulas, E.: Active sampling for large-scale information retrieval evaluation. In: CIKM 2017, pp. 49–58. ACM, Singapore (2017). https://doi.org/10.1145/3132847.3133015
9. Wu, Y., Jin, T., Lou, H., Xu, P., Fanoud, F., Gu, Q.: Adaptive sampling for heterogenous rank aggregation from Noisy pairwise comparisons. In: Proceedings of the 25th International Conference on Artificial Intelligence and Statistics (AISTATS) 2022, vol. 151, pp. 2154–2178. PMLR, Valencia (2022)
10. Doroudi, S., Thomas, P., Brunskill, E.: Importance sampling for fair policy selection. In: Proceedings of the Twenty-Seventh International Joint Conference on Artificial Intelligence (IJCAI-18), Stockholm (2018)
11. Rajapakse, T., Yates, A., Rijke, M.: Negative Sampling Techniques for Dense Passage Retrieval in a Multilingual Setting. SIGIR'24. ACM, New York (2024)
12. Dai, S., Xu, C., Xu, S., Pang, L., Dong, Z., Xu, J.: Bias and Unfairness in Information Retrieval Systems: New Challenges in the LLM Era. KDD'24. Barcelona, Spain (2024)
13. Gonzalez-Zelaya, V., Salas, J., Prangle, D., Missier, P.: Optimising fairness through parameterized data sampling. In: 24th International Conference on Extending Database Technology, Nicosia, Cyprus (2021)
14. Yang, K., Stoyanovich, J.: Measuring fairness in Rangked outputs. In: Scientific and Statistical Database Management. ACM, Chicago, Illinois (2017). https://doi.org/10.1145/3085504.3085526
15. Singh, A., Joachims, T.: Fairness of exposure in rankings. In: Knowledge Discovery & Data Mining (KDD'2018), pp. 2219–2228. ACM, London (2018). https://doi.org/10.1145/3219819.3220088
16. Diaz, F., Mitra, B., Ekstrand, M.D., Biega, A.J., Carterette, B.: Evaluating stochastic rankings with expected exposure. In: International Conference on Information & Knowledge Management (CIKM'20), pp. 275–284. ACM, Dublin (2020). https://doi.org/10.1145/3340531.3411962
17. Svore, K.M., Volkovs, M.N., Burges, C.J.: Learning to rank with multiple objective functions. In: International World Wide Web Conference Committee (IW3C2). ACM, Hyderabad (2011)

18. Zhao, X. et al.: Embedding in recommender systems: a survey. ACM Trans. Inf. Syst. **44**(5), 117 (2023)
19. Xiong, L., Chen, X., Huang, X., Huang, W., Zhong, M., Zeng, H.: LambdaXGB: research on learning to rank based on LambdaMart. ICIC Express Lett. B. **8**(8), 1165–1171 (2017)
20. Aslam, J. A., & Pavlu, V. (2007). A Practical Sampling Strategy for Efficient Retrieval Evaluation.
21. Singh, A., Joachims, T.: Policy learning for fairness in ranking. In: 33rd International Conference on Neural Information Processing Systems, pp. 5426–5436. ACM, Vancouver (2019). https://doi.org/10.5555/3454287.3454774
22. Kirnap, O., Diaz, F., Biega, A., Ekstrand, M., Carterette, B., Yilmaz, E.: Estimation of fair ranking metrics with incomplete judgments. In: Proceedings of the Web Conference 2021 (WWW'21), pp. 1065–1075. ACM, New York (2021). https://doi.org/10.1145/3442381.3450080
23. Roh, J., Oh, S.-H., & Lee, S.-Y. (2020). Unigram-Normalized Perplexity as a Language Model Performance Measure with Different Vocabulary Sizes.
24. Nikkarinen, I. et al.: Modeling the unigram distribution. In: Findings of the Association for Computational Linguistics: ACL-IJCNLP 2021 (2021)
25. Ghazimatin, A., Kleindessner, M., Russell, C., Abedjan, Z., Golebiowski, J.: Measuring fairness of rankings under Noisy sensitive. In: ACM Conference on Fairness, Accountability, and Transparency (FACT'22), pp. 2263–2279. ACM, Seoul (2022). https://doi.org/10.1145/3531146.3534641
26. Petrovic, A., Nikolic, M., Jovanovic, M., Bijanic, M., Delibasic, B.: Fair classification via Monte Carlo policy gradient method. Eng. Appl. Artif. Intell. **104**, 104398 (2021)
27. Zehlike, M., Bonchi, F., Castillo, C., Hajian, S., Megahed, M., Baeza-Yates, R.: Fa*IR: a fair top-k ranking algorithm. In: Conference on Information and Knowledge Management, pp. 1569–1578. ACM, Singapore (2018). https://doi.org/10.1145/3132847.3132938
28. Broniatowski, M.: Minimum divergence estimators, maximum likelihood and the generalized bootstrap. Entropy. **23**, 185 (2021)
29. Amusa, L.B., Zewotir, T., North, D.: A simulation study of some modern weighting methods for estimating treatment effects in observational studies. J. Mod. Appl. Stat. Methods. **19**(2), 4 (2020)
30. Awasthi, P., Das, A., Sen, R., Suresh, A. T.: On the Benefits of Maximum Likelihood Estimation for Regression and Forescating. ICLR'2022 (2022)
31. Wang, J., Shah, N., Ravi, R.: Stretching the effectiveness of MLE from accuracy to bias for pairwise comparisons. In: 23rd International Conference on Artificial Intelligence and Statistics (AISTATS) 2020, vol. 108. Proceedings of Machine Learning Research, Palermo (2020)
32. Acharya, J.: Profile maximum LIkelihood is optimal for estimating KL divergence. In: IEEE International Symposium on Information Theory (ISIT). IEEE (2018)
33. Ge, J., Tang, S., Fan, J., Ma, C., Jin, C.: Maximum Likelihood Estimation Is All You Need for Well-Specified Covariate Shift. ICLR'2024 (2024)
34. Cheng, J., Van Durme, B.: Compressed Chain of Thought: Efficient Reasoning Throuhj Dense Representation. arXiv preprint. https://arxiv.org/abs/2412.13171 (2025)
35. Gunathalika, T.M., Manage, P.D., Zhang, J., Li, Y., Kelly, W.: Adressing sparse data challenges in recommender systems: a systematic review of rating estimation using sparse data and profile enrichment techniques. Intell. Syst. Appl. **25**, 2667–3053 (2025)
36. Scrivano, A.: A comparative study of recommender systems under big data constraints. Trends Comput. Sci. Inf. Technol. **10**(2), 066–073 (2025)

37. He, C., Wang, C., Zhong, Y.-X., Li, R.-F.: A survey on learning to rank. In: Proceedings of the Seventh International Conference on Machine Learning and Cybernetics. IEEE, Kunming (2008)
38. Karatzoglou, A., Baltrunas, L., Shi, Y.: Learning to Rank for Recommender Systems RecSys'13. ACM, Hong Kong (2013)
39. Cardoso, D.O., Galeno, T.D.: Online evaluation of the Kolmogorov-Smirnov test on arbitrarily large samples. J. Comput. Sci. **67**, 101959 (2023). https://doi.org/10.1016/j.jocs.2023.101959

Simulation Games for Healthcare Training: A Personalized Learning Approach for Healthcare Professionals

Nanda Ye Aung(✉), Suepphong Chernbumroong(✉), Pradorn Sureephong(✉), and Tama Duangnamol(✉)

Faculty of Digital Technology Management, College of Arts, Media and Technology, Chiang Mai University, Chiang Mai, Thailand
nandayeaung@gmail.com, {suepphong.c,pradorn.s,tama.d}@cmu.ac.th

Abstract. Healthcare professionals operate in increasingly complex clinical environments that require continuous competency development, rapid decision-making, and lifelong learning. Simulation-based learning has become an essential educational strategy for developing clinical competencies while maintaining patient safety. However, many existing healthcare simulation platforms rely on fixed training pathways that do not adequately accommodate differences in learner competency, professional roles, and prior knowledge. Consequently, learners may receive training experiences that are either overly challenging or insufficiently stimulating, thereby reducing educational effectiveness. Recent advances in artificial intelligence and adaptive learning technologies provide opportunities to create personalized learning environments capable of dynamically responding to learner needs. This paper proposes an Adaptive Simulation Game Framework for Healthcare Professional Training that integrates learner profiling, competency modelling, adaptive learning engines, intelligent feedback mechanisms, clinical simulation environments, and learning analytics dashboards within a unified educational ecosystem.The framework introduces a competency-driven adaptation approach that continuously adjusts simulation difficulty, instructional support, feedback delivery, and learning pathways based on learner performance indicators and competency achievement. The framework was developed using Design and Development Research (DDR) methodology. The proposed framework contributes a structured foundation for AI-enhanced healthcare training by combining adaptive simulation, personalized learning, intelligent feedback, and analytics-driven decision support. While empirical validation remains future work, the framework addresses significant gaps in current healthcare simulation systems and provides a roadmap for the development of adaptive, intelligent training solutions for the modern medical workforce.

Keywords: Adaptive Learning · Healthcare Education · Simulation Games · Personalized Learning · Serious Games · Intelligent Feedback · Healthcare Training

N. Mohd Sharef et al. (Eds.): CAMP 2026, CCIS 3094, pp. 340–354, 2027.
https://doi.org/10.1007/978-981-92-4305-1_25

1 Introduction

Healthcare systems require highly skilled professionals capable of delivering safe and effective patient care, creating a growing need for innovative approaches to lifelong learning and competency development. Simulation-based learning provides realistic clinical environments where learners can safely practice decision-making, communication, procedural skills, teamwork, and critical thinking without risking patient safety.

Recent advances in artificial intelligence, adaptive learning, learning analytics, and intelligent tutoring systems have created new opportunities to personalize healthcare education. In parallel, serious games enhance learner engagement through challenges, rewards, and interactive learning experiences. Adaptive learning technologies can tailor instruction based on learner performance, while learning analytics and intelligent feedback systems support competency monitoring and continuous improvement. However, many existing healthcare educational technologies operate independently and continue to rely on standardized instructional approaches that provide identical learning experiences regardless of learner competency or background. Simulation platforms focus on clinical realism, adaptive learning systems emphasize content personalization, and learning analytics tools primarily support performance monitoring. Few frameworks integrate these capabilities within a unified educational ecosystem. To address this gap, this study proposes an Adaptive Simulation Game Framework for Healthcare Professional Training that integrates learner profiling, adaptive learning technologies, realistic simulation environments, intelligent feedback systems, and learning analytics within a unified architecture. The objectives of this research include:

- Exploring limitations with current healthcare simulation learning environments.
- Presenting a unified conceptual framework that integrates adaptive learning, simulation, intelligent feedback, and learning analytics.
- Describing learner modelling and adaptation mechanisms.
- Providing a prototype framework demonstrating implementation feasibility.
- Analytically validating framework effectiveness.

The contributions of this research include:

- Proposing a competency-driven adaptive simulation learning architecture for healthcare professionals.
- Providing a healthcare specific learner modeling framework.
- Describing dynamic scenario adaptation based on learner competency.
- Introducing an intelligent feedback generation framework.
- Identifying key learning analytics metrics to integrate into the learning dashboard.
- Providing a prototype implementation artifact to support future work.

2 Literature Review

2.1 Personalized Learning in Healthcare Education

The fundamental objective of personalized learning is to provide learners with training experiences that are aligned with their specific needs rather than relying on standardized instructional approaches. Since learner diversity is particularly significant, clinical experience, specialty areas, educational backgrounds, responsibilities, and competency requirements may differ for each professional. Personalized learning systems can utilize learner profiles, performance histories, assessment results, and behavioral data to tailor instructional content and learning pathways.

2.2 AI-Based Adaptive Learning Systems

Adaptive learning systems, also known as intelligent tutoring systems, dynamically modify training experiences based on learner performance, behaviors, and knowledge requirements. These systems utilize learner profiles and instructional decision-making mechanisms to determine how training scenarios should be personalized. Modern adaptive learning systems are integrated with artificial intelligence to analyze large volumes of learner data, identify performance patterns, predict learning needs, and generate personalized recommendations. These capabilities enable training platforms to improve the learning experience and provide targeted interventions when required.

2.3 Serious Games in Healthcare Education

Serious games are digital games designed primarily for educational, training, behavioral, or professional development purposes rather than entertainment. Over the past decade, serious games have gained considerable attention within healthcare education because of their ability to improve learner engagement, motivation, and participation. Healthcare serious games have been utilized across various domains, including emergency medicine, infection control, surgical procedures, patient communication, medication administration, and rehabilitation training. By incorporating game-based elements such as rewards, progression systems, achievements, challenges, and interactive narratives, serious games create engaging learning environments that encourage active participation and repeated practice. Several studies have reported positive educational outcomes associated with serious game usage, including improved knowledge retention, enhanced learner satisfaction, increased motivation, and better procedural skill acquisition. Serious games also support experiential learning by placing learners within realistic scenarios that require decision-making and problem-solving.

2.4 Learning Analytics in Healthcare Education

Learning analytics refers to the collection, measurement, analysis, and reporting of learner data to understand the learner's progress and optimize the learning processes. Healthcare simulation systems can capture numerous performance indicators, including task completion times, clinical decision accuracy, communication behaviors, assessment outcomes, procedural errors, and competency progression metrics. This data provides valuable insights into learner strengths, weaknesses, and knowledge needs. Learning analytics supports evidence-based assessment by enabling instructors to monitor student performance, identify learning gaps, and evaluate training effectiveness. Furthermore, modern platforms incorporate predictive models and performance metrics, which are essential for driving continuous improvement in educational planning

2.5 Research Gap Analysis

The literature demonstrates significant advances in personalized learning, AI based adaptive learning systems, serious games, simulation-based education, and learning analytics. However, these technologies are often implemented independently rather than as integrated educational ecosystems. Healthcare simulation platforms emphasize realism and clinical authenticity, while serious games promote engaging learning experiences. Furthermore, adaptive learning systems focus on content personalization, and learning analytics platforms track performance. Unfortunately, these valuable features are currently scattered across isolated platforms.

Currently, there is no combined system that brings together adaptive learning, simulations, intelligent feedback, and analytics in healthcare training. Bridging this gap will enhance personalized education, learner engagement, and overall training effectiveness.

2.6 Research Contribution and Novelty

The proposed Adaptive Simulation Game Framework addresses the identified research gap by introducing a competency-driven adaptive architecture specifically designed for healthcare professional training. Unlike conventional healthcare simulation systems that provide static training pathways, the proposed framework continuously analyses learner performance and adjusts educational experiences accordingly. The framework integrates learner modelling, adaptive decision-making, intelligent feedback generation, simulation-based learning, and analytics-driven monitoring within a single educational ecosystem. The novelty of the framework lies in three key contributions.

- First, the framework introduces a unified learning model that incorporates professional training, competency indicators, performance metrics, and learning history to support adaptive learning pathways.
- Second, the framework will implement a rule-based adaptive learning engine capable of dynamically modifying simulation difficulty, instructional support, and learning objectives based on the learner's competency development.
- Third, the framework will act as an essential data layer for future artificial intelligence learning mechanisms, supporting real-time performance feedback and behavioral research.

Collectively, these contributions provide a foundation for personalized simulation-based healthcare education and distinguish the framework from existing healthcare simulation systems (Table 1).

Table 1. Comparison of Existing Systems and Proposed Framework

Feature	Traditional Simulation	Adaptive LMS	Intelligent Tutoring System (ITS)	Proposed Adaptive Simulation Framework
Clinical Simulation	✓	×	Limited	✓
Learner Modelling	×	✓	✓	✓
Dynamic Adaptation	×	✓	✓	✓
Intelligent Feedback	Limited	Limited	✓	✓
Learning Analytics	Limited	✓	Limited	✓
Competency Tracking	Limited	✓	✓	✓
Scenario Difficulty Adjustment	×	Limited	Limited	✓
Healthcare-Specific Design	✓	Limited	Limited	✓
Adaptive Simulation Branching	×	×	Limited	✓

3 Methodology

3.1 Research Design

This study adopts the Design and Development Research (DDR) methodology proposed by Branch. DDR is widely used in educational technology research because it provides a systematic approach for analyzing educational problems, designing innovative solutions, developing instructional products, and evaluating their effectiveness. The DDR methodology is particularly suitable for this study because the objective is to develop a conceptual framework and prototype design artifact that can support future implementation and evaluation. The research process consists of five phases:

- Needs Analysis
- Framework Design
- Framework Development
- Prototype Specification
- Future Validation Planning

3.2 Target Users

The proposed framework is intended for healthcare professionals including nurses, physicians, pharmacists, allied health professionals, and healthcare students.

3.3 Needs Analysis

A comprehensive review of literature relating to healthcare simulation, adaptive learning, intelligent tutoring systems, serious games, and learning analytics was conducted to identify limitations in existing educational technologies. The analysis identified four major requirements:

- Personalized learning experiences
- Adaptive instructional support
- Intelligent feedback mechanisms
- Continuous competency monitoring
- These requirements informed the development of the proposed framework.

3.4 Framework Design

Based on the identified requirements, a conceptual architecture was designed comprising five major subsystems:

- Learner Modelling System
- Adaptive Learning Engine
- Clinical Simulation Environment
- Intelligent Feedback System
- Learning Analytics Dashboard

The interaction among these components was designed to support competency-driven personalization and adaptive simulation experiences.

3.5 Framework Development

The framework was developed conceptually through iterative design refinement and theoretical integration of adaptive learning principles, simulation-based learning theories, serious game design principles, and learning analytics methodologies. Prototype architecture was created to demonstrate implementation feasibility and support future development activities.

3.6 Future Validation Planning

At the current stage, the framework remains a conceptual design artifact and has not yet undergone practical validation. Future research will involve expert review, usability evaluation, and experimental studies to assess educational effectiveness, technical feasibility, adaptability, and user acceptance. The validation strategy is discussed in detail in Sect. 5.

4 Proposed Framework

The proposed Adaptive Simulation Game Framework integrates learner profiling, adaptive learning technologies, simulation-based learning, and intelligent feedback mechanisms into a unified solution.

4.1 Framework Overview

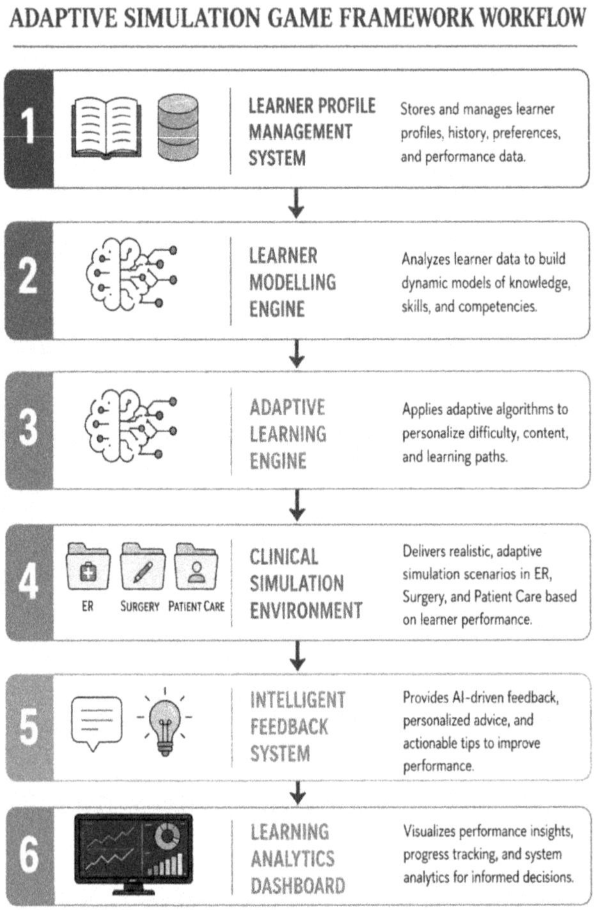

Fig. 1. Adaptive Simulation Game Framework workflow

The proposed Adaptive Simulation Game Framework integrates adaptive learning technologies, simulation-based learning environments, intelligent feedback systems, learner modelling, and learning analytics within a unified educational ecosystem for healthcare professional training. The framework is designed to support personalized learning experiences by continuously monitoring learner performance and dynamically adjusting simulation activities according to competency development. Unlike conventional healthcare simulation systems that provide standardized instructional pathways, the proposed framework employs competency-driven adaptation mechanisms to personalize learning experiences, instructional support, and scenario complexity. The framework consists of six major components:

- Learner Profile Management System

- Learner Modelling Engine
- Adaptive Learning Engine
- Clinical Simulation Environment
- Intelligent Feedback System
- Learning Analytics Dashboard

The framework operates as a continuous learning cycle where learner interactions generate performance data that are analyzed by adaptive and analytical components. Adaptation decisions are then applied to subsequent simulation activities to optimize learning outcomes (Fig. 1).

4.2 Learner Profile Management System

The Learner Profile Management System serves as the foundation of personalization within the framework. It maintains structured records describing learner characteristics. The learner profile includes three primary categories of information:

Professional Attributes: Describe the learner's healthcare background and responsibilities. Examples:

- Profession (Nurse, Physician, Pharmacist, Allied Health Professional)
- Clinical Specialty
- Years of Experience
- Educational Background

Competency Attributes: Represent learner performance and competency development. Examples:

- Competency Score
- Clinical Decision Accuracy
- Procedural Performance
- Communication Skills

Behavioral Attributes: Capture learner interactions and learning patterns. Examples:

- Simulation Completion Rate
- Average Response Time
- Error Frequency
- Learning History

The learner profile is continuously updated after each simulation session to ensure that adaptation decisions reflect current learner competency levels.

4.3 Learner Modelling Architecture

The Learner Modelling Engine transforms learner profile information into actionable educational intelligence. This is a digital representation of learner competency, performance trends, and educational needs (Table 2).

Table 2. Example of Learner Model Variables

Variable	Description
Experience Level	Beginner, Intermediate, Advanced
Competency Score	Overall competency indicator
Clinical Accuracy	Percentage of correct decisions
Response Time	Average task completion time
Error Frequency	Number of errors
Risk Indicator	Probability of competency deficiency

The learner model enables the system to identify knowledge gaps, monitor competency progression, and personalize educational experiences.

4.4 Adaptive Learning Engine

The Adaptive Learning Engine functions as the central decision-making component of the framework. Its primary objective is to continuously adjust learning experiences according to learner competency levels and performance outcomes. The adaptation process consists of four stages:

- **Stage 1:** Data Collection: The system collects performance data from simulation activities.
- **Stage 2:** Competency Assessment: Collected data are transformed into competency indicators.
- **Stage 3:** Adaptation Decision: Evaluates learner status and selects appropriate instructional interventions.
- **Stage 4:** Scenario Adjustment: Learning pathways are modified according to adaptation outcomes.

4.5 Clinical Simulation Environment

The Clinical Simulation Environment represents the experiential learning component of the framework. Simulation activities are organized into competency-based scenarios covering various healthcare domains. Examples:

Emergency Response Scenarios

- Cardiac Arrest
- Stroke Management

Patient Assessment Scenarios

- Vital Sign Interpretation
- Symptom Assessment

Medication Safety Scenarios

- Medication Administration
- Dosage Calculation

Communication Scenarios

- Patient Communication
- Family Communication

The adaptive engine determines scenario selection and complexity according to learner competency profiles.

4.6 Intelligent Feedback System

The Intelligent Feedback System provides personalized instructional support during and after simulation activities. The feedback architecture consists of two complementary mechanisms.

Real-Time Feedback
Real-time feedback is generated immediately when critical events occur during simulations. Examples:

- Incorrect medication dosage
- Delayed emergency intervention
- Failure to assess critical symptoms

Post-Session Feedback
Post-session feedback provides comprehensive performance evaluation to support reflection and improvement. Examples:

- Competency Achievement Summary
- Strength and Weakness Analysis
- Error Breakdown and Recommended Learning Activities

4.7 Intelligent Feedback Generation Workflow

This workflow drives adaptive learning by systematically analysing performance metrics and learning history. It uses a rule-based engine to translate competency data into personalized feedback to recommend the consistent and relevant skill development. Key Process Steps:

- **Track**: Monitors learner actions and historical performance data.
- **Evaluate**: Uses a rule-based engine to identify strengths and competency gaps.
- **Recommend**: Suggests targeted resources to address specific weaknesses.
- **Adapt**: Generates a personalized learning plan for consistent skill development

4.8 Learning Analytics Dashboard

Learning analytics serves as the evidence-based monitoring component of the framework. The dashboard provides visualization of learner performance and competency development. Dashboard can be customized for learners, instructors, and administrators. Key analytics indicators for dashboard include:

- **Competency Metrics:** Competency Progression, Assessment Scores, Clinical Accuracy
- **Behavioral Metrics:** Simulation Completion Rates, Response Times, Learning Activity Participation
- **Performance Metrics:** Error Frequency, Success Rates, Improvement Trends

4.9 Continuous Improvement Cycle

The proposed Adaptive Simulation Game Framework will continuously and systematically collect data, analyse outcomes and evaluate results. This iterative process enables the framework to implement improvements, and optimize the training experiences, ensuring effective learning outcomes and system performanee.

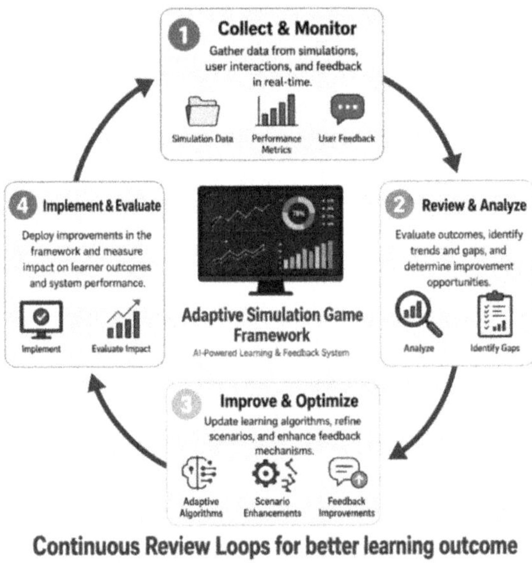

Fig. 2. Continuous Improvement Cycle

- **Step 1: Collect & Monitor**
  Continuously collects simulation data, learner performance metrics, and user feedback to monitor the learning experiences of the proposed framework.
- **Step 2: Review & Analyse**
  Analysed the collected data to identify the learning gaps, and opportunities for improving training effectiveness.

- **Step 3: Improve & Optimize**
  Based on analytical insights, optimizes adaptive algorithms, refines simulation scenarios, and enhances feedback mechanisms to improve learner engagement and outcomes.
- **Step 4: Implement & Evaluate**
  Deploys improvements and evaluates their impact on learner performance and overall framework effectiveness (Fig. 2).

5 Future Validation Strategy

5.1 Validation Overview

At the current stage, the proposed Adaptive Simulation Game Framework represents a conceptual framework and validated design artifact. Practical evaluation has not yet been conducted and therefore no effectiveness claims are made. Future validation activities are planned to assess educational relevance, technical feasibility, usability, and learning effectiveness. The validation strategy consists of three phases.

5.2 Phase 1: Expert Validation

The first phase involves expert review of the framework design.

Participants may include:

- Healthcare Educators, Clinical Trainers
- Simulation Technology Experts

Evaluation criteria include:

- Educational Relevance
- Technical Feasibility and Adaptability
- Framework Completeness
- Healthcare Applicability

Qualitative and quantitative feedback will be used to refine the framework.

5.3 Phase 2: Usability Evaluation

The second phase focuses on usability and user acceptance.

Participants may include:

- Healthcare Educators, Students, Nurses, Physicians

Evaluation measures include:

- System Usability Scale (SUS)
- User Satisfaction and Usefulness
- Perceived Ease of Use and Learning Engagement

The objective is to evaluate system usability and identify potential improvements before large-scale implementation.

5.4 Phase 3: Experimental Evaluation

The final phase involves experimental evaluation of effectiveness. A quasi-experimental study may compare:

Control Group

- Traditional simulation-based learning

Experimental Group

- Adaptive simulation game-based learning
- Potential outcome measures include:
- Knowledge Gain
- Clinical Competency Achievement
- Decision Accuracy
- Knowledge Retention
- Learning Efficiency
- Learner Engagement

The results will provide evidence regarding the educational value of the proposed framework.

6 Discussion

The proposed framework addresses several limitations in existing healthcare training technologies and aims to overcome them by integrating learner modelling, adaptive learning, simulation-based education, intelligent feedback, and learning analytics within a unified architecture. The framework introduces a competency-driven adaptation process that continuously adjusts educational experiences according to learner performance. This approach aligns with competency-based healthcare education principles and supports personalized learning pathways. The inclusion of analytics dashboards further enhances the framework by providing meaningful performance insights to decide the relevant learning pathway.

Although practical validation remains a goal for future work, the framework establishes a comprehensive foundation for adaptive healthcare simulation environments and contributes to the growing body of research on AI-enhanced healthcare education system.

7 Conclusion and Future Works

This paper proposed an Adaptive Simulation Game Framework for Healthcare Professional Training that integrates learner profiling, adaptive learning technologies, intelligent feedback systems, simulation-based learning environments, and learning analytics dashboards within a unified educational ecosystem. The framework addresses limitations of existing healthcare simulation systems by introducing competency-driven personalization and adaptive simulation mechanisms. Through learner modelling, rule-based adaptation, dynamic scenario adjustment, intelligent feedback generation, and analytics-driven monitoring, the framework supports personalized healthcare education and competency development. The study contributes a conceptual framework and prototype design artifact that demonstrates the feasibility of integrating adaptive learning and simulation-based education. Furthermore, the framework provides a structured foundation for future implementation and evaluation.

Future research will focus on prototype development, expert validation, usability testing, and experimental evaluation to assess educational effectiveness and practical applicability in healthcare training environments. It includes the following:

- Prototype implementation
- Expert validation
- Experimental evaluation and usability testing
- Expandable AI adaptation models
- Learning analytics and predictive modelling

These future developments will further strengthen the framework and support its adoption within healthcare professional education.

Acknowledgments. This work is supported by the European Union's Horizon Europe programme through the Marie Skłodowska-Curie Actions (MSCA), under the projects ACMod (Affective Computing Models: from Facial Expression to Mind-Reading, Grant Agreement No. 101130271) and HarmonicAI (Human-guided Collaborative Multi-Objective Design of Explainable, Fair and Privacy-Preserving AI for Digital Health, Grant Agreement No. 101131117).

AI Use Declaration Artificial intelligence tools were used to support language refinement and conceptual image generation.

References

Branch, R.M.: Instructional Design: the ADDIE Approach. Springer, New York (2010)

Cook, D.A., Erwin, P.J., Triola, M.M.: Computerized virtual patients in health professions education. Acad. Med. **85**(10), 1589–1602 (2010)

Cook, D.A., et al.: Technology-enhanced simulation for health professions education. JAMA. **306**(9), 978–988 (2011)

Deterding, S., Dixon, D., Khaled, R., Nacke, L.: From game design elements to gamefulness. In: MindTrek. ACM (2011)

Garris, R., Ahlers, R., Driskell, J.E.: Games, motivation, and learning. Simul. Gaming. **33**(4), 441–467 (2002)

Graafland, M., Schraagen, J.M., Schijven, M.P.: Serious games for medical education and surgical skills training. Br. J. Surg. **99**(10), 1322–1330 (2012)

Hamari, J., Koivisto, J., Sarsa, H.: Does Gamification Work? HICSS (2014).

Knowles, M.S., Holton, E.F., Swanson, R.A.: The Adult Learner. Routledge (2015)

Kononowicz, A.A., et al.: Virtual patients in healthcare education. BMC Med. Educ. **13**, 129 (2013)

Lave, J., Wenger, E.: Situated Learning. Cambridge University Press (1991)

Mayer, R.E.: Multimedia Learning, 3rd edn. Cambridge University Press (2021)

Ryan, R.M., Deci, E.L.: Self-determination theory. Am. Psychol. **55**(1), 68–78 (2000)

Shute, V.J.: Focus on formative feedback. Rev. Educ. Res. **78**(1), 153–189 (2008)

Sitzmann, T.: Instructional effectiveness of simulation games. Pers. Psychol. **64**(2), 489–528 (2011)

Spector, J.M.: Conceptualizing the emerging field of smart learning environments. Smart Learn. Environ. **1**(2), 1–10 (2014)

VanLehn, K.: The relative effectiveness of tutoring systems. Educ. Psychol. **46**(4), 197–221 (2011)

Wouters, P., et al.: Cognitive and motivational effects of serious games. J. Educ. Psychol. **105**(2), 249–265 (2013)

Zawacki-Richter, O., et al.: Artificial intelligence applications in higher education. IJETHE. **16**, 39 (2019)

Technology-Driven Continuous Self-Care Model for Diabetes Remission in Primary Care

Xiaofang Tu[1(✉)], Suepphong Chernbumroong[1], Pradorn Sureephong[1], Pokin Sakarinkhul[2], and Jun Wu[3]

[1] College of Arts, Media and Technology, Chiang Mai University, Chiang Mai, Thailand
xiaofang_tu@cmu.ac.th

[2] Primary Care Department, Lamphun Hospital, Lamphun, Thailand

[3] Guangdong CAS Cogniser Information Technology Co., Ltd., Guangzhou, China

Abstract. Diabetes remission has become an important direction for improving type 2 diabetes care, particularly through intensive lifestyle modification, continuous self-monitoring, and multidisciplinary support. However, current remission services in real-world primary care settings often face limitations related to fragmented patient-generated data, delayed feedback, manual monitoring workload, and limited continuity of care beyond clinic visits. This study proposes a technology-driven continuous self-care model for diabetes remission care by integrating the Internet of Medical Things (IoMT), edge computing, multimodal artificial intelligence, dashboard-based monitoring, telehealth, and nudge-based follow-up through the GymBot platform. The model is designed to support real-time health data collection, personalized self-care recommendations, early warning detection, and care coordination between patients' homes, primary care units, and hospitals. The study will be implemented in Lamphun Province, Thailand, with patients with type 2 diabetes who have the potential to enter remission. A comparative intervention design will be used, with an intervention group receiving GymBot-based continuous care and a control group receiving standard care. The proposed model is expected to strengthen diabetes remission services by improving self-management, supporting timely feedback, reducing healthcare provider workload, and enhancing the sustainability of home-based and data-driven continuous care.

Keywords: diabetes remission · type 2 diabetes · continuous care · IoMT · edge computing · multimodal AI · nudge-based follow-up

1 Introduction

Diabetes mellitus remains a major global public health burden and one of the leading causes of morbidity, mortality, and healthcare expenditure (World Health Organization, 2024). Recent estimates indicate that 589 million adults aged 20–79 years are living with diabetes worldwide, and this number is projected to increase to 853 million by 2050 (International Diabetes Federation, 2025). In Thailand, diabetes is also a growing national health concern, with prevalence among adults aged 15 years and above

N. Mohd Sharef et al. (Eds.): CAMP 2026, CCIS 3094, pp. 355–365, 2027.
https://doi.org/10.1007/978-981-92-4305-1_26

increasing from 8.9% in 2014 to 9.5% in 2020, while approximately 300,000 new cases are reported annually and 3.4 million accumulated cases have been recorded (Thailand Ministry of Public Health, Division of Noncommunicable Diseases, 2023). More importantly, around 70% of patients are unable to control their blood glucose within the recommended target range, increasing the risk of complications, disability, premature death, reduced quality of life, and economic burden (Thailand Ministry of Public Health, Division of Noncommunicable Diseases, 2023). Although type 2 diabetes has traditionally been viewed as a chronic and progressive disease requiring lifelong pharmacological treatment, growing evidence suggests that diabetes remission is achievable in a significant subset of patients (Riddle et al., 2022; Ricci et al., 2023). Diabetes remission is defined as maintaining HbA1c below 6.5% for at least three months without glucose-lowering medication (Riddle et al., 2022). In Thailand, professional and national guidelines have promoted intensive lifestyle intervention as a key pathway toward remission, particularly through dietary control, increased physical activity, continuous exercise, and weight reduction under multidisciplinary care (Royal College of Family Physicians of Thailand, 2022). However, sustaining these behavioral changes remains challenging in real-world practice, especially when patients lack continuous motivation, timely feedback, and self-monitoring mechanisms (Tomah et al., 2019).

Recent evidence from diabetes remission services in Thailand shows that self-monitoring of blood glucose is commonly used to support behavioral feedback, patient accountability, and clinical decision-making (Mongkolsucharitkul et al., 2025). Healthcare teams have adopted tools such as glucometers, pedometers, smartwatches, LINE Official Account, Google Forms, and spreadsheets to facilitate self-monitoring and follow-up (Mongkolsucharitkul et al., 2025). However, these practices still rely heavily on fragmented platforms, self-reported or written records, manual verification against glucometer data, and individual spreadsheet-based tracking (Mongkolsucharitkul et al., 2025). As a result, healthcare providers face increased workload, while patients may experience delayed feedback, missed appointments, dropout, or delayed detection of relapse and treatment-related risks such as hypoglycemia (Mongkolsucharitkul et al., 2025). These challenges are further complicated by unequal access to technology and limited digital literacy among patients (Mongkolsucharitkul et al., 2025). Therefore, current diabetes remission services require automated data collection, real-time tracking and analysis, advanced monitoring, and more robust follow-up mechanisms to support continuous and personalized care (Mongkolsucharitkul et al., 2025).

To address these limitations, this study aims to develop a technology-driven continuous self-care model for patients undergoing diabetes remission care. The proposed model integrates automated real-time data collection through the Internet of Medical Things (IoMT) and GymBot as an edge-computing platform, advanced monitoring and personalized recommendations through multimodal large language models, and follow-up mechanisms through dashboards and alert systems. The system is designed to analyze diverse self-care data, including exercise, dietary behavior, goal setting, motivation, emotional status, and other patient-generated health information, to provide personalized recommendations aligned with diabetes remission care guidelines. By connecting hospitals, primary care units, and patients' homes through real-time health data, the model seeks to strengthen the feedback loop between patients and healthcare providers, improve self-management, reduce healthcare provider workload, support early detection of relapse or treatment-related risks, and enhance the sustainability of diabetes remission services.

2 Literature Review

2.1 IoMT and Edge Computing for Real-Time Data Collection

Automated data collection and real-time tracking are increasingly important in diabetes care because they enable continuous monitoring, timely feedback, and personalized decision-making. The Internet of Medical Things (IoMT) refers to connected medical devices and digital health systems that collect and transmit patient-generated health data, such as blood glucose readings, physical activity, and behavioral records. When combined with edge computing, data can be processed closer to the point of collection rather than relying entirely on cloud-based systems, which can reduce latency and support real-time feedback (Klonoff, 2013; Zhu et al., 2021). These capabilities are particularly relevant to diabetes remission services, where patients need continuous support to understand how daily behaviors, diet, and physical activity influence glycemic outcomes. However, current remission services in Thailand still rely on self-reported data, written records, manual verification, and spreadsheet-based tracking, which increase healthcare provider workload and may delay personalized counselling (Mongkolsucharitkul et al., 2025).

To address these limitations, this study proposes GymBot as an edge-computing platform for automated real-time data collection and continuous monitoring. GymBot is an AI-enabled home exercise platform equipped with camera-based motion analysis that can detect body posture, analyze exercise movements, and provide real-time interactive feedback. Previous work reported that GymBot could accurately assess exercise postures and achieved high user satisfaction among older adults in a fall-prevention context (Wongsila et al., 2024). In this study, GymBot will be extended to support diabetes remission care by integrating multiple sources of patient-generated health data, including blood glucose data from IoT devices, physical activity data from motion analysis, and patient health behavior data. These data will be collected into a central platform for real-time analysis, feedback, and continuous personalized monitoring to support sustainable behavior change in diabetes remission services.

2.2 Multimodal AI for Personalized Self-Care Recommendation

Advanced monitoring is essential for diabetes remission care because self-management requires patients to understand complex relationships among blood glucose levels, treatment plans, diet, physical activity, and daily behaviors. While self-monitoring can generate useful health data, patients may still require timely interpretation, simplified guidance, and personalized recommendations to support continuous behavior change. Recent developments in artificial intelligence have shown promising potential in diabetes education and management, particularly in supporting patient communication, personalized feedback, and decision-making (Ayers et al., 2023; Li et al., 2020). Large language models may further enhance diabetes care by integrating structured data, such as glucose readings and medication records, with less structured real-world information related to patients' behaviors and daily routines (Wolber et al., 2025).

Multimodal large language models are proposed as a mechanism for advanced monitoring, warning, and personalized recommendation in diabetes remission services. Multimodal large language models have recently been explored for integrating visual and structured health-related data, such as meal images, macronutrient estimates, and glucose-related features, to support personalized glucose forecasting in diabetes care (Wolber et al., 2025). Building on this capability, this study proposes to extend multimodal analysis to broader self-care data required in diabetes remission services, including exercise performance, dietary behavior, goal setting, motivation, emotional status, and other patient-generated health information, in alignment with Thai diabetes remission care guidelines (Thailand Ministry of Public Health, Division of Noncommunicable Diseases, 2023). By combining these data with standard diabetes remission care guidelines, the system can generate personalized recommendations for physical activity and health behavior modification. In addition, the model can support dynamic intervention adjustment, telehealth communication, and preliminary triage through a virtual nurse function. When abnormal patterns or potential risks are detected, the system can alert healthcare providers for further consultation or care-plan adjustment, thereby strengthening continuous monitoring and supporting safer self-management in diabetes remission care.

2.3 Nudge-Based Follow-Up Mechanisms

Robust follow-up mechanisms are essential for sustaining behavior change in diabetes remission services, where patients require continuous support beyond clinic visits. Current remission services in Thailand still face limitations from fragmented self-monitoring data, delayed feedback, manual verification, healthcare provider workload, and unequal patient digital literacy (Mongkolsucharitkul et al., 2025). These challenges suggest the need for follow-up systems that can support timely monitoring, risk detection, and personalized feedback based on patient-generated health data. Behavioral support strategies such as nudges and just-in-time adaptive interventions provide useful foundations for strengthening follow-up in digital health systems. Nudge theory refers to the design of choice environments that guide individuals toward beneficial behaviors without restricting freedom of choice (Thaler & Sunstein, 2008). In chronic disease self-management, nudges such as reminders, feedback, planning prompts, and positive reinforcement have shown potential to improve self-management behaviors, although evidence on long-term clinical outcomes remains limited (Möllenkamp et al., 2019). Similarly, just-in-time adaptive interventions can deliver timely prompts or feedback to support behavior change, particularly in physical activity interventions (Hardeman et al., 2019). In diabetes care, a nudge-inspired AI-driven platform has also been proposed and implemented to support diabetes self-management through personalized feedback and behavior-oriented digital support (Joachim et al., 2022). Building on these foundations, this study proposes dashboard-based monitoring and alert mechanisms to connect hospitals, primary care units, village health volunteers, and patients' homes. Rather than replacing clinical care, the proposed follow-up mechanism aims to strengthen the feedback loop between self-monitoring data, behavioral adherence, personalized recommendations, and healthcare provider decision-making in diabetes remission services.

2.4 System Change Mechanisms

Sustainable diabetes remission requires more than data collection, monitoring, and follow-up; it also requires a mechanism that explains how these system components can lead to sustained self-care behavior. The COM-B model provides a useful behavioral framework by explaining that behavior occurs through the interaction of three core components: capability, opportunity, and motivation (Michie et al., 2011). In the context of diabetes remission care, capability refers to patients' ability to understand and perform self-care activities, opportunity refers to the availability of tools, services, and care environments that support these behaviors, and motivation refers to the continuous reinforcement needed to maintain lifestyle changes over time. Recent work on mHealth-based behavior change interventions also highlights the importance of linking behavioral theory with technical implementation, including defining the intervention scope, understanding the target behavior, designing intervention strategies, implementing digital functions, and evaluating intervention outcomes (Rachad et al., 2024).

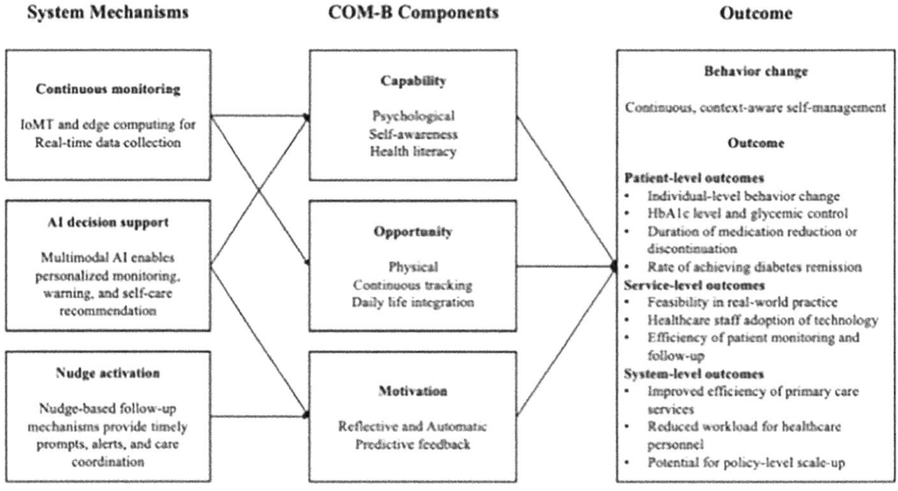

Fig. 1. COM-B model and Behavior Change

Based on this perspective, the proposed model is designed to operate as a continuous feedback system for diabetes remission care. IoMT and edge computing support real-time patient-generated health data collection; multimodal AI enables personalized monitoring, warning, and self-care recommendation; and nudge-based follow-up mechanisms provide timely prompts, alerts, and care coordination between hospitals, primary care units, village health volunteers, and patients' homes. Together, these components are expected to enhance patients' capability through understandable feedback and personalized guidance, improve opportunity by embedding self-care support into the home and primary care environment, and strengthen motivation through timely nudges and continuous follow-up. Therefore, the proposed model aims to support sustainable behavior change, improve adherence to remission care, reduce missed appointments and delayed relapse detection, and decrease healthcare provider workload through a more integrated and scalable continuous care mechanism (Fig. 1).

3 Research Methodology

This study adopts an implementation-oriented research design to develop and evaluate the proposed technology-driven continuous self-care model for diabetes remission services. Building on the conceptual model described in the previous section, the methodology focuses on translating IoMT and edge computing, multimodal AI-based recommendation, and nudge-based follow-up mechanisms into a real-world primary care setting. The research methodology is organized into five main components: population and study setting, intervention, procedure, data analysis, and ethical considerations.

3.1 Population and Study Setting

This study will be conducted in Lamphun Province, Thailand, covering participating primary care units, hospitals, and patients' homes through a digitally connected home-based care model. The study focuses on the development and evaluation of a diabetes remission care model supported by AI, IoMT, and the GymBot digital platform. The model includes continuous health data collection, personalized self-care recommendations, telehealth-based consultation, and a monitoring system to support continuous care at the primary care level. Eligible participants will be adults with type 2 diabetes receiving care in participating primary care units in Lamphun Province, who have achieved partial glycemic control and are considered suitable for lifestyle-based remission support by healthcare personnel. Participants should be able to perform light-to-moderate physical activity and provide informed consent. Patients with severe diabetes-related complications, medical contraindications to exercise, severe cognitive impairment, or conditions that limit safe participation in home-based self-care activities will be excluded.

3.2 Intervention

The intervention is designed as a technology-driven continuous self-care model for diabetes remission care, integrating IoMT and edge computing, multimodal AI, dashboard monitoring, telehealth, and nudge-based follow-up through the GymBot platform. The system will be developed and adapted for real-world primary care settings by refining the AI-based recommendation model based on diabetes remission care guidelines, connecting CGM devices for approximately 25 participants in the intervention group to enable real-time glucose monitoring, integrating GymBot-based physical activity and exercise data, and developing a dashboard for PCU and hospital staff to monitor glucose levels, activity, adherence, and warning signals. Before full implementation, the system will be pilot-tested with approximately 5–10 patients to identify usability, technical, and workflow issues, with feedback used for system refinement. Healthcare personnel from all eight districts of Lamphun Province will also receive practical training on using GymBot, IoT data, AI-generated recommendations, dashboard monitoring, and telehealth functions to support personalized guidance, timely feedback, care planning, and continuous follow-up within their local service contexts.

3.3 Procedure

The study will be conducted with 50 patients with type 2 diabetes, including 25 participants in the intervention group and 25 participants in the control group. The intervention group will receive diabetes remission care through the GymBot system, which includes real-time data collection, personalized self-care recommendations, and telehealth-based follow-up, while the control group will receive standard care (Fig. 2).

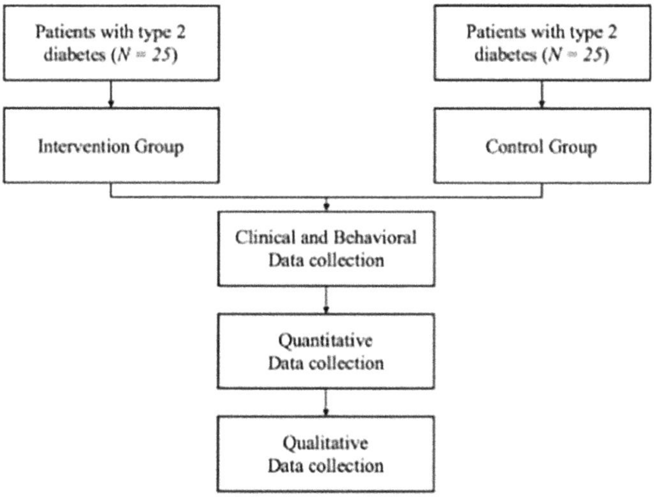

Fig. 2. Procedure for the Intervention and Control Groups

Patients in the intervention group will be monitored daily through AI-based monitoring, with follow-up by healthcare personnel at least once per month and teleconsultation provided when needed. Standard care will include routine diabetes care provided by participating primary care units, such as regular clinical follow-up, lifestyle advice, medication management, and usual health education, without GymBot-based monitoring, CGM-supported feedback, or AI-generated self-care recommendations. Clinical and behavioral outcomes will be collected at baseline, month 3, and month 6, including HbA1c, glucose levels, medication use, physical activity, health behaviors, and system usage data. Quantitative data will be obtained from case record forms (CRF), the GymBot system, IoT devices such as CGM, and LINE Official Account, with activity and CGM data collected continuously in the intervention group. In addition, qualitative data will be collected through in-depth interviews with approximately 5–10 patients and 5–8 healthcare personnel, as well as two focus group discussions, each consisting of approximately 8–10 participants, to explore user experience, system acceptance, and facilitating or limiting factors affecting implementation. The primary outcome will be change in HbA1c from baseline to month 6. Secondary outcomes will include glucose patterns, medication reduction, body weight or BMI, physical activity, exercise adherence, health behavior change, system usage, user experience, and diabetes remission status.

3.4 Data Analysis

Quantitative data will be analyzed to evaluate clinical outcomes, behavioral changes, and system performance, with statistical significance set at $p < 0.05$. The analysis will include descriptive statistics, within-group comparisons, between-group comparisons, and exploratory machine learning analysis (Table 1).

Table 1. Data Analysis Methods

Analysis Method	Details
Descriptive statistics	Participant characteristics and baseline variables, including age, sex, HbA1c, glucose levels, medication use, physical activity, and system usage, will be summarized using means, standard deviations, frequencies, percentages, or ranges, as appropriate.
Within-group comparison	Within-group changes between selected timepoints, such as baseline and month 6, will be analyzed using paired *t*-tests for normally distributed data or Wilcoxon signed-rank tests for non-normally distributed data.
Between-group comparison	Differences in outcomes between the intervention and control groups will be analyzed using independent *t*-tests or Mann–Whitney U tests, depending on data distribution. For the primary outcome, between-group differences in HbA1c at month 6 may be adjusted for baseline HbA1c using ANCOVA where appropriate.
Longitudinal analysis	Repeated outcomes measured at baseline, month 3, and month 6 will be analyzed using repeated-measures ANOVA or linear mixed-effects models, depending on data characteristics and missing data patterns.
Machine learning analysis	Exercise-session records from the intervention group will be explored using machine learning techniques to identify usage patterns, exercise behaviors, adherence-related factors, and associations with glucose changes. Individual-level prediction of diabetes remission will be treated as exploratory due to the limited sample size.

Qualitative data from interviews and focus groups will be transcribed and analyzed using thematic analysis. The analysis will involve familiarization with the data, coding, categorization, and synthesis of key themes related to system usability, acceptance, perceived benefits, barriers, and facilitators for real-world implementation.

3.5 Ethical Considerations

This study will be conducted in accordance with ethical principles for human subject research and will obtain ethical approval from the relevant ethics committee of Lamphun Hospital and/or Chiang Mai University before implementation. All participants will receive clear information about the study objectives, procedures, potential benefits, and risks, and written informed consent will be obtained voluntarily. Participants may withdraw from the study at any time without affecting their usual care. Personal and health data will be strictly protected through de-identification, separation of identifiable information from research data, controlled access, and secure data storage on local servers and cloud systems. To minimize privacy risks, the GymBot system will not store patient images or videos; only processed output data, such as activity scores and usage records, will be retained. Participant safety will be continuously monitored throughout the intervention, particularly regarding lifestyle modification and physical activity, with healthcare personnel from primary care units and hospitals available to provide care or referral if any abnormal conditions occur.

4 Discussion

The proposed model is expected to strengthen diabetes remission care by enabling continuous self-care support, real-time monitoring, personalized recommendations, and timely follow-up between patients, primary care units, and hospitals. These expected outcomes align with the need for automated data collection, advanced monitoring, and stronger follow-up mechanisms identified in current diabetes remission services in Thailand (Mongkolsucharitkul et al., 2025), as well as evidence supporting IoMT, edge computing, AI-based diabetes management, and nudge-based self-management support (Klonoff, 2013; Zhu et al., 2021; Li et al., 2020; Möllenkamp et al., 2019; Joachim et al., 2022). Its key contribution lies in promoting sustainability through an integrated feedback loop that supports behavioral adherence, early risk detection, reduced healthcare provider workload, and better continuity of care beyond clinic visits, which is consistent with behavior change principles emphasizing capability, opportunity, and motivation (Michie et al., 2011; Rachad et al., 2024). However, successful implementation may be affected by practical constraints, including the cost of CGM devices and digital infrastructure, the limited timeframe for system deployment and patient follow-up, and the readiness of local service areas in terms of workforce capacity, digital literacy, internet connectivity, and workflow integration. These factors should be carefully considered when interpreting the study outcomes and planning future scale-up of the model.

5 Conclusion

This study proposes a technology-driven continuous self-care model to support diabetes remission services in real-world primary care settings. The model responds to key limitations in current diabetes remission care, particularly fragmented self-monitoring data, delayed feedback, limited follow-up capacity, and the difficulty of sustaining lifestyle modification beyond clinic visits. By integrating IoMT and edge computing, GymBot-based activity monitoring, multimodal AI-driven recommendations, dashboard-based monitoring, telehealth consultation, and nudge-based follow-up, the proposed model aims to create a continuous feedback loop between patients, primary care units, and hospitals. In relation to the COM-B framework, the model is expected to enhance patients' capability through understandable feedback and personalized guidance, improve opportunity by embedding care support into home and primary care environments, and strengthen motivation through timely prompts and continuous follow-up. Although implementation may be constrained by budget, time, digital infrastructure, and local readiness, this study provides a practical and scalable direction for extending diabetes remission care toward continuous, personalized, and sustainable home-based support.

Acknowledgments. This work is supported by the European Union's Horizon Europe programme through the Marie Skłodowska-Curie Actions (MSCA), under the projects ACMod (Affective Computing Models: from Facial Expression to Mind-Reading, Grant Agreement No. 101130271) and HarmonicAI (Human-guided Collaborative Multi-Objective Design of Explainable, Fair and Privacy-Preserving AI for Digital Health, Grant Agreement No. 101131117). This work was also supported by the Science and Technology Bureau of Nansha District under Key Field Science and Technology Plan Program No. 2024ZD002. The authors would like to express their sincere appreciation to the College of Arts, Media and Technology, Chiang Mai University, for its academic and institutional support. Special thanks are extended to the Lamphun Provincial Public Health Office for its generous collaboration, valuable field insights, facilitation of local coordination, and continuous support throughout the development of this study. Their practical knowledge and close engagement with primary healthcare services in Lamphun have greatly contributed to shaping the research direction and ensuring its relevance to real-world diabetes remission care.

References

Ayers, J.W. et al.: Comparing physician and artificial intelligence chatbot responses to patient questions posted to a public social media forum. JAMA Intern. Med. **183**(6), 589–596 (2023). https://doi.org/10.1001/jamainternmed.2023.1838

Hardeman, W., Houghton, J., Lane, K., Jones, A., Naughton, F.: A systematic review of just-in-time adaptive interventions (JITAIs) to promote physical activity. Int. J. Behav. Nutr. Phys. Act. **16**, 31 (2019). https://doi.org/10.1186/s12966-019-0792-7

International Diabetes Federation: IDF Diabetes Atlas, 11th edn. IDF (2025) https://diabetesatlas.org/resources/idf-diabetes-atlas-2025/

Joachim, S., Forkan, A.R.M., Jayaraman, P.P., Morshed, A., Wickramasinghe, N.: A nudge-inspired AI-driven health platform for self-management of diabetes. Sensors. **22**(12), 4620 (2022). https://doi.org/10.3390/s22124620

Klonoff, D.C.: The current status of mHealth for diabetes: will it be the next big thing? J. Diabetes Sci. Technol. **7**(3), 749–758 (2013). https://doi.org/10.1177/193229681300700321

Li, J., Huang, J., Zheng, L., Li, X., Hu, Y.: Application of artificial intelligence in diabetes education and management: present status and promising prospect. Front. Public Health. **8**, 173 (2020). https://doi.org/10.3389/fpubh.2020.00173

Michie, S., van Stralen, M.M., West, R.: The behaviour change wheel: a new method for characterising and designing behaviour change interventions. Implement. Sci. **6**, 42 (2011). https://doi.org/10.1186/1748-5908-6-42

Möllenkamp, M., Zeppernick, M., Schreyögg, J.: The effectiveness of nudges in improving the self-management of patients with chronic diseases: a systematic literature review. Health Policy. **123**(12), 1199–1209 (2019). https://doi.org/10.1016/j.healthpol.2019.09.008

Mongkolsucharitkul, P. et al.: Insights into diabetes remission services: perspectives from general practitioners, family physicians and multidisciplinary teams. Fam. Med. Com. Health. **13**, e003631 (2025). https://doi.org/10.1136/fmch-2025-003631

Rachad, T., El Hafidy, A., Aabbad, M., Idri, A.: Comprehensive framework for developing mHealth-based behavior change interventions. Digit. Health. **10**, 1–19 (2024). https://doi.org/10.1177/20552076241289979

Ricci, M. et al.: Remission of type 2 diabetes: a critical appraisal. Front. Endocrinol. **14**, 1125961 (2023). https://doi.org/10.3389/fendo.2023.1125961

Riddle, M.C. et al.: Consensus report: definition and interpretation of remission in type 2 diabetes. J. Clin. Endocrinol. Metab. **107**(1), 1–9 (2022). https://doi.org/10.1210/clinem/dgab585

Royal College of Family Physicians of Thailand. (2022). Diabetes Remission in Type 2 Diabetes with Intensive Lifestyle Intervention: Guide for Healthcare Providers. https://thaigpfm.org/wp-content/uploads/2025/05/Diabetes-remission-in-type-2_web.pdf

Thailand Ministry of Public Health, Division of Noncommunicable Diseases: Guidelines for implementing diabetes remission care services (Remission Service). Department of Disease Control (2023). https://ddc.moph.go.th/uploads/publish/1517720240105071553.pdf

Thaler, R.H., Sunstein, C.R.: Nudge: Improving Decisions about Health, Wealth, and Happiness. Yale University Press (2008)

Tomah, S. et al.: Frequency of self-monitoring of blood glucose in relation to weight loss and A1C during intensive multidisciplinary weight management in patients with type 2 diabetes and obesity. BMJ Open Diab. Res. Care. **7**, e000659 (2019). https://doi.org/10.1136/bmjdrc-2019-000659

Wolber, J.C., Samadi, M.E., Sellin, J., Schuppert, A.: Multimodal large language models and mechanistic modeling for glucose forecasting in type 1 diabetes patients. J. Biomed. Inform. **172**, 104945 (2025). https://doi.org/10.1016/j.jbi.2025.104945

Wongsila, S., Chernbumroong, S., Boonprasit, K., Sureephong, P.: Leveraging knowledge management techniques for developing multimedia exercise guides for elderly fall prevention. J. Mobile Multimedia. **20**(2), 435–472 (2024) https://journals.riverpublishers.com/index.php/JMM/article/view/22849

World Health Organization: Diabetes (2024). https://www.who.int/news-room/fact-sheets/detail/diabetes

Zhu, T., Li, K., Herrero, P., Georgiou, P.: Deep learning for diabetes: a systematic review. IEEE J. Biomed. Health Inform. **25**(7), 2744–2757 (2021). https://doi.org/10.1109/JBHI.2020.3040225

Generative AI, Large Language Models, and Real-World Applications

A Deep Learning Framework for Pineapple Weight Estimation Using YOLOv8 Instance Segmentation

Muhammad Amir Syafiq Zainuddin[1], Mohamad Faizal Ab Jabal[2(✉)], Azyan Yusra Kapi[2], Azrina Suhaimi[2], and Gloria Jennis Tan[3]

[1] Faculty of Computer and Mathematical Sciences, Universiti Teknologi MARA, Cawangan Terengganu, Kampus Kuala Terengganu, 21080 Kuala Terengganu, Terengganu, Malaysia

[2] Faculty of Computer and Mathematical Sciences, Universiti Teknologi MARA, Cawangan Johor, Kampus Pasir Gudang, 81750, Masai, Johor, Malaysia
m.faizal@uitm.edu.my

[3] Faculty of Computer and Mathematical Sciences, Universiti Teknologi MARA, Cawangan Selangor, Kampus, Shah Alam 40450, Selangor, Malaysia

Abstract. Inconsistent of human weight evaluation makes pineapple grading difficult for the agricultural industry. Automated find pineapple crowns more difficult to recognize than spherical fruits due to their irregular form and structural complexity. This study proposed an automated Pineapple Weight Estimation System using YOLOv8 instance segmentation and volumetric regression to overcome this issue. The approach isolates fruit from background noise and morphological anomalies using the YOLOv8-seg model to extract exact 2D pixel masks. A Volumetric Linear Regression model and Single-Point Calibration approach are used to convert pixel areas into physical mass, using a volumetric coefficient of 0.08018 and a density factor of 0.93 g/cm^3. Practical deployment used Python and a Flask-based web interface. Testing and evaluations show that the YOLOv8-seg model has great detection reliability, with a Mean Average Precision (mAP) of 0.96 at 0.5. The weight prediction model, with a Mean Absolute Error (MAE) of 0.008 kg (8 g), confirmed a robust linear association between segmented pixel area and physical mass. Hence, finding from this study presents a cost-effective, non-destructive, and automated pineapple grading method that could improve agricultural processing efficiency.

Keywords: Agriculture · Deep Learning · Linear Regression · Pineapple Weight Estimation · YOLOv8

1 Introduction

The practise of Artificial Intelligent (AI) and computer vision technologies in agriculture has transformed farming and post-production in recent years. Deep learning technology, particularly Convolution Neural Networks (CNNs), may ease the agriculture players

N. Mohd Sharef et al. (Eds.): CAMP 2026, CCIS 3094, pp. 369–379, 2027.
https://doi.org/10.1007/978-981-92-4305-1_27

daily tasks including disease pre-diagnosis, automated harvesting, and yield monitoring. Instead of requiring human feature extraction, deep learning algorithms can automatically detect complex patterns, textures, and morphological traits in high-resolution agricultural pictures. Modern architectures like You Only Look Once (YOLO) perform well in instance segmentation and feature extraction for real-time applications. In precision agriculture, these computer vision frameworks can move from simple categorization to continuous forecasting and qualitative visual assessment to quantitative data extraction.

These sophisticated computational algorithms can tackle several post-harvest processing performance issues in the tropical fruit business, including the pineapple industry. Pineapple weight grading and sorting still requires manual weighing, which is time-consuming, inefficient, and prone to handling damage. Pineapples also complicate computing. They are not symmetrical like apples or oranges. Their two-dimensional appearance does not match with their mass using regular geometry formulas. The form variation makes it hard to achieve international export quality control standards. Thus, an autonomous, non-destructive weight estimate system is essential for sorting efficiency, post-harvest loss reduction, handling bruising reduction, and appropriate food distribution to achieve UN Sustainable Development Goal 2 (Zero Hunger).

Despite these technological advances, CNN algorithms are designed for discrete classification problems and are challenging to apply to continuous weight prediction on fruits with complex forms and irregular appearances. This paper presents a hybrid computer vision-volumetric regression approach for automated non-contact weight measurement to fill this gap in the literature. This study aims to achieve three goals. First, determine the CNN regression architecture and loss function. Second, build the system using CNN regression model. Third, evaluate the pineapple weight estimation technique. This paper aims to scientifically establish a linear link between segmented 2D pixel area and product mass, giving a real-time high-precision grading solution for modern agricultural processing plants.

2 Literature Review

In 2022, Ukwuoma et al. provided an extensive description of the use of deep learning algorithms for the detection and categorisation of fruit. Furthermore, these models have been evaluated on robotic harvesting applications, for example, YOLO-based algorithms have been effectively used for the recognition of tomatoes, which indicates that high inference speed is suitable for real-time systems (Lawal, 2021). Also, optical sensors have been employed to achieve accurate fruit segmentation with high precision even in challenging backgrounds (Kang and Chen, 2019). All these research works have effectively proved the superiority of the CNN-based approaches over the traditional ML methods in terms of higher accuracy, computational efficiency suitable for real-time agricultural applications, and lesser complexity.

In such cases, Kim et al. (2024) designed a ResNet18-based computer vision system to assess the weight of butterhead lettuce in an industrial plant factory environment. The authors obtained very good results, $R\ 2 = 0.95$ and RMSE = 8.06 g, which was substantially better than regression approaches. Moreover, deep learning for difficult visual characteristics and morphological alterations.

The Convolution Neural Networks (CNNs) work well for agricultural image analysis. Recent studies have shown that advanced segmentation models can distinguish fruit in complex backdrops. Mask R-CNN helped Agossadou et al. (2024) distinguish fruits with 93% accuracy and low relative error. Hayat et al. (2024) classified fruit types using Deep CNN architectures with 97% – 99% accuracy. These studies show that deep learning algorithms can extract features with considerable capacities, a need for morphological analysis. These systems typically use discrete categorization, not continuous predictive regression.

2.1 Approaches in Fruit Weight Estimation

The classification to quantitative mass estimation converts 2-D pixels to 3-D properties. This approach and Machine Learning models like Artificial Neural Networks (ANN) can accurately predict strawberry weight from retrieved visual data (Basak et al., 2022). Furthermore, Onmankhong et al. in 2024 estimated lime volume using ResNet50 with an R^2 of 0.957 in the latest volumetric study. While, Kim et al. (2024) examined continuous weight prediction models with R^2 values of 0.95 and RMSE of 8.06 g. These results establish a quantitative weight estimating system performance standard. A system with an R^2 of 0.93 or above is considered successful in science today.

A comprehensive review of existing literature and related studies concerning the application of computer vision and deep learning in the agricultural sector, with a specific focus on fruit weight estimation and grading, is synthesized in Table 1.

Table 1. Overview of important works in the field

Problem	Algorithm	Dataset	Result
Manual measurement is time-consuming and labour-intensive for fruit grading. (Onmankhong et al., 2024)	BME + ResNet50	Limes	BME achieved $R^2 =$ 0.954 (weight), 0.882 (volume); ResNet50 achieved $R^2 = 0.951$ (weight), 0.957 (volume)
Manual weight measurement is time-consuming and inefficient in commercial plant production. (Kim et al., 2024)	ResNet18 + CNN Regression	Butterhead Lettuce	$R^2 = 0.95$ with RMSE = 8.06 g, outperforming traditional regression methods
Traditional manual weight estimation is time-consuming and labour-intensive. (Agossadou et al., 2024)	Mask R-CNN + Lasso Regression	Tomatoes	Achieved 7.09% relative error in total weight estimation with 93% accuracy

Based on Table 1 shows a number of approaches have been proposed for the problem of non-destructive, automated grading, from traditional machine learning algorithms to highly sophisticated CNNs. The subsequent description highlights key work in this area and the individual algorithms utilised such as ResNet, Mask R-CNN, and U-Net, the underlying goals of each study, the inefficiencies of manual grading addressed, and the performance results obtained. Finally, this comparison study sets the state-of-the-art and points out the technological advancements related to the proposed pineapple weight estimating technique.

3 Methodology

The main aim of this study is to build a computational method to estimate the weight of pineapple from digital photographs using CNN based Regression. Weight is computed using volumetric extrapolation based on the underlying methodology. Implementation of computer vision techniques and neural networks, the 3D volume of a pineapple is estimated from a 2D image, and a specific density factor is applied to compute the physical mass. The proposed methodology is classified into three different phases as shown in Table 2 and is methodically designed to assess the issue domain fully, create the prediction model, and evaluate the model performance rigorously.

Table 2. Proposed methodology

Phase	Objective	Activity	Result
Preliminary Study	To identify the appropriate CNN regression architecture and loss function.	Conduct literature review and requirements analysis for morphological complexity in non-spherical fruits.	Selection of YOLOv8-seg (instance segmentation) over standard bounding-box detection.
Design and Implementation	To develop the system using a CNN regression model algorithm.	Data collection (1,191 images at 50 cm), manual polygon annotation, PyTorch model training, and applying volumetric linear regression.	A functional automated weight estimation system deployed via a Flask web interface.
Evaluation	To evaluate the performance of this system for pineapple weight estimation.	Comparative testing of model estimates against physical ground-truth scale measurements using diverse fruit samples.	Achieved an R2 of 0.98, mAP of 0.96, and MAE of 0.008 kg, exceeding the benchmark target.

Shown in Table 2 the first phase is synthesizes background information on existing approaches and methods for addressing the subject's morphological complexity.

This step includes a literature study, requirement analysis, and technological framework development to help choose the best approach to estimate the weight of non-spherical fruits. Instead of using classification methods, this basic step seeks CNN regression designs that can understand the complex correlations between visual input and continuous weight values. Therefore, the review emphasizes volume-to-weight conversion. This perspective acknowledges the fact that predicting volume using computer vision is harder than transforming density into weight, and hence, continuous mass prediction becomes grounded in physics fundamentals.

The Convolutional Neural Networks (CNNs) are currently the dominant computer vision model owing to the fact that they can learn spatial feature hierarchies using unstructured image data. Convolutional layers with learnable filters extract edges, shapes, and textures. Subsequently, pooling algorithms reduce computing complexity while down sampling spatial dimensions to preserve structural information. Conventional CNN architectures can classify discrete images, but analysing irregular agricultural items' morphological complexity demands pixel-level accuracy.

The YOLOv8-seg architecture fulfils this requirement by extending standard CNN capabilities to serve as an instance segmentation model. The approach applies a feature extraction backbone and a dedicated segmentation head to delineate accurate object bounds, overcoming the restrictions of standard bounding box detection. In this study, the intensively trained model successfully detects the unique morphological features of the pineapple, separating the irregular structures of the fruit's body and crown from the surrounding background noise. Hence, this enhanced segmentation process represents an extremely accurate 2D polygonal pixel mask. Finally, the CNN architecture bridges the critical gap between un-structured information in a 2D image and the structured quantitative data needed for subsequent 3D volumetric linear regression and continuous mass calculation by accurately isolating and quantifying the targeted pixel area.

As illustrated in Fig. 1, phase 1 is the process involves both data engineering and the creation of a predictive model. First, a set of 1,191 high-resolution pineapple photos was gathered from the Roboflow website. In order to mitigate the disadvantage of distance sensitivity inherent in the 2D monocular vision with no depth sensors, the picture capturing distance was tightly regulated to 50 cm. All pictures were meticulously annotated to obtain precise polygon masks describing the body of the fruit. Furthermore, in the phase 2, the YOLOv8-seg model was trained by using the PyTorch framework to obtain good morphology and feature masking. The subsequent process in the phase 3 is image extraction, where a volumetric linear regression technique was utilized to transform the two-dimensional pixel area into a continuous estimate of mass. Finally, in the phase 4 is the calibration slope was obtained via an empirical calibration of the actual ground-truth measurements, and the number of pixels identified was directly related to the final weight estimate in kilos. The full inference pipeline was then scaled to run in real time using a Flask web interface.

An evaluation of the predictive ability and generality of the proposed system will be performed in the next step. In order to perform this analysis, different types of pineapples are used to make sure that this evaluation takes into account the real morphological diversity of the fruit. There are differences in terms of size, shape, and crown formation in the pineapples. The automatic prediction of the weight of each test pineapple is made

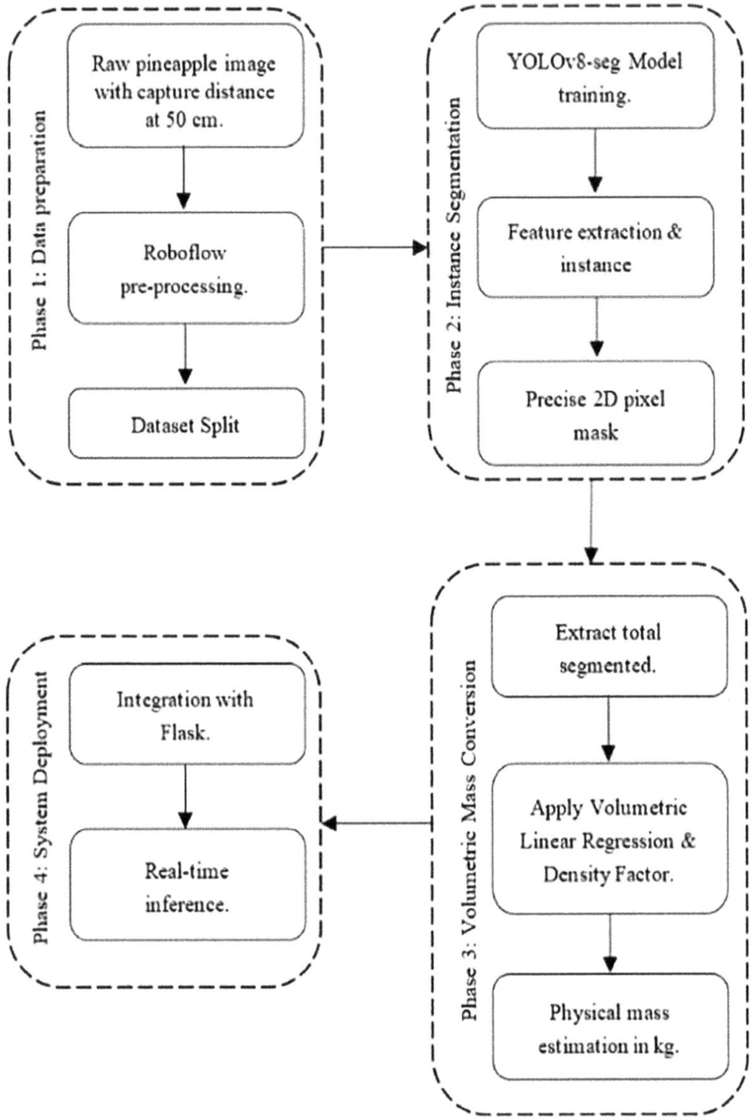

Fig. 1. The proposed working framework

from the segmented pixel regions and compared to the actual weight measured through a calibrated scale.

The Coefficient of Determination (R^2) has been calculated to measure how well the model fits, according to Eq. (1). In this equation, the sum of squares of the residuals (SS_{res}) and the total sum of squares (SS_{tot}) have been used. The key performance criterion in this regard is that the R^2 should be 0.93 to prove that the system performs equally well as per agricultural industry standards in order to automated fruit sorting.

$$R^2 = 1 - \left(\frac{SS_{res}}{SS_{tot}}\right) \tag{1}$$

Furthermore, the Mean Absolute Error (*MAE*) was used to evaluate the average magnitude of the errors in the predictions without considering their direction, calculated using (2) where n represents the total number of test samples, y i is the actual physical weight, and y_i is the predicted weight from the system.

$$MAE = \frac{1}{n}\sum_{i=1}^{n} |y_i - \hat{y}_i| \tag{2}$$

Moreover, the Root Mean Square Error (*RMSE*) was utilized in order to appropriately penalize larger prediction errors as represented in (3) where y_{actual} represents the true mass measured by the scale and $y_{predicted}$ represents the model's volumetric estimation.

$$RMSE = \sqrt{\frac{1}{n} \times \sum \left(y_{actual} - y_{predicted}\right)^2} \tag{3}$$

Finally, the Mean Absolute Percentage Error (MAPE) was calculated to express the prediction accuracy as a relative percentage, defined in (4) where y is the actual value and $\hat{y}$ is the predicted value. Together, these metrics provide a comprehensive evaluation of the system's regression performance.

$$MAPE = \frac{100\%}{n}\sum \left|\frac{y - \hat{y}}{y}\right| \tag{4}$$

4 Results and Finding

The main goal of the testing is to measure the effectiveness of the YOLOv8-seg model for segmenting the irregular morphology of pineapples from various backgrounds. The trained model showed remarkable feature extraction capabilities, achieving a Mean Average Precision (mAP@0.5) of 0.96. In addition, the graphical analysis performed for 30 training epochs demonstrated that the losses related to bounding box and segmentation masks were gradually converging with no sign of overfitting. The results indicate that the deep learning module is able to capture the accurate 2D polygonal area of the fruit, and hence provide the essential variable needed for the later mass conversion stage.

The main finding of this study is a successful conversion of the determined 2D pixel area to a 3D physical mass estimate via volumetric linear regression. The proposed approach showed a very high Coefficient of Determination ($R^2 = 0.98$) in comparison to the ground truth data obtained from a calibrated digital scale. This finding is significantly higher than the benchmark correlation of $R^2 = 0.93$ as reported in the literature initially, since a strong linear correlation between the segmented pixel count and the actual fruit weight is established. The system achieved a Mean Absolute Error (MAE) of only 0.008 kg (8 g). The low error value demonstrates that the integration of pixel density

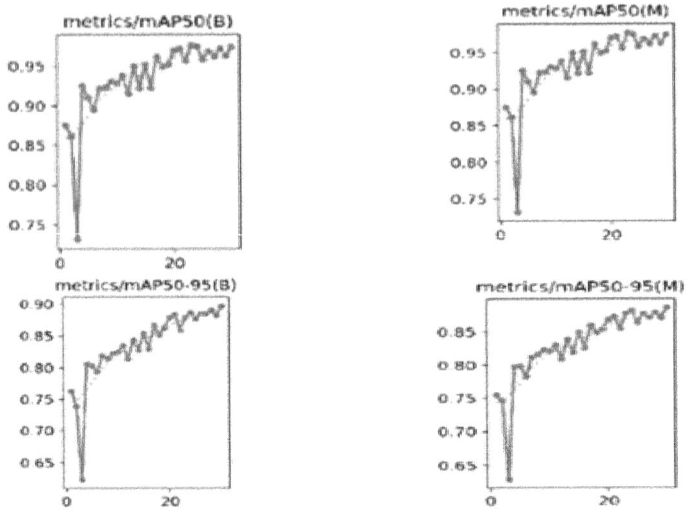

Fig. 2. Evaluation Precision of Detection

calculations with a fixed camera distance of 50 cm effectively overcomes the limits of the depth sensing of 2D monocular cameras.

Figure 2 shows the overall performance of the proposed YOLOv8-seg framework was quantified to assess its practical viability and the inference pipeline was embedded into a real-time web application. The model obtained a high detection Precision mAP @ 0.5 = 0.96 and a Coefficient of Determination $R^2 = 0.98$, which means that there is an extremely strong linear correlation between the segmented 2D pixel area and the 3D physical mass.

Fig. 3. Screenshot of the user interface of the Pineapple Vision

Then, the trained model was deployed by using a Flask-based web interface, known as "Pineapple Vision," to demonstrate its effectiveness in the agricultural industry. The raw image input is passed to this system to estimate the weights in real-time. The average inference time is less than 0.15 s. Integrated operability is shown in Fig. 3, with a simple upper panel for input of raw images for analysis. The lower panel shows the result of a real-time inference, which demonstrates the ability of the system to accurately segment the complex pineapple morphology from the background and correctly extract a total of 17,434 pixels in this scenario. The empirically obtained volumetric slope is applied to the retrieved pixel area to yield an exact physical weight estimate of 1.30 kg immediately. These results demonstrate the readiness of the proposed framework for implementation as a rapid, non-destructive and highly accurate automated fruit grading instrument.

5 Conclusion and Future Work

The Pineapple Weight Estimation System provides an automated and scientifically rigorous solution for agricultural grading, utilizing YOLOv8-seg within a volumetric regression framework to estimate produce weight through advanced computer vision. Architecture of the framework facilitates seamless workflow that starts with segmentation and then goes on to extraction and regression of pixels leading to a remarkable precision rate of 96% Mean Average Precision (mAP). It is observed empirically that there is a linear correlation between the 2D area after segmentation and mass as long as the calibration constants stay the same. Through integration of AI with farming practices, this method helps in enhancing grading process efficiency and digitalization of products' tracing.

Finally for the future work, this study will focus on using reference objects that will facilitate automatic calibration of the system. This will be achieved by positioning standard items such as coins or ID cards inside the camera's frame, hence making it automatically determine the conversion ratio for different distances used in taking photos. In the future, technological development will definitely see the addition of depth sensing technologies using LiDAR or phone depth sensors to enable 3D volumetric analysis.

Acknowledgments. The authors would like to express their sincere gratitude to the Faculty of Computer and Mathematical Sciences for providing the necessary resources and support throughout this research. Special thanks are extended to colleagues and mentors for their invaluable assistance and guidance to this project.

References

Abu Bakar, B.H., Ishak, A.J., Shamsuddin, R., Wan Hassan, W.Z.: Ripeness level classification for pineapple using RGB and HSI colour maps. J. Theor. Appl. Inf. Technol. **57**(3), 587–593 (2013)

Agossadou, E.E.G., Pazou, M.G.A., Hontinfinde, R.D., Kora, A.D.: Estimating tomato fruit masses through image processing and artificial intelligence. In: Cotonou'24: Conférence Internationale des Technologies de l'Information et de la Communication de l'ANSALB. Cotonou, Benin (2024)

Alzubaidi, L., et al.: Review of deep learning: concepts, CNN architectures, challenges, applications, future directions. J. Big Data. **8**(1), 53 (2021)

Arboleda, E.R., de Jesus, C.L.T., Tia, L.M.S.: Pineapple maturity classifier using image processing and fuzzy logic. IAES Int. J. Artif. Intell. **10**(4), 830–838 (2021). https://doi.org/10.11591/ijai.v10.i4.pp830-838

Basak, J.K., Paudel, B., Kim, N.E., Deb, N.C., Madhavi, B.G.K., Kim, H.T.: Non-destructive estimation of fruit weight of strawberry using machine learning models. Agronomy. **12**(10), 2487 (2022)

Bhargava, A., Bansal, A.: Classification and grading of multiple varieties of apple fruit. Food Anal. Methods. **14**(7), 1359–1368 (2021)

Chen, Y., Zheng, L., Peng, H.: Assessing pineapple maturity in complex scenarios using an improved RetinaNet algorithm. Eng. Agríc. **43**(2), e20220180 (2023)

Faisal, M., Albogamy, F., Elgibreen, H., Algabri, M., Alqershi, F.A.: Deep learning and computer vision for estimating date fruits type, maturity level, and weight. IEEE Access. **8**, 206770–206782 (2020)

Hayat, A., Morgado-Dias, F., Choudhury, T., Singh, T.P., Kotecha, K.: FruitVision: a deep learning based automatic fruit grading system. Open Agric. **9**(1), 20220276 (2024)

Kang, H., Chen, C.: Fruit detection and segmentation for apple harvesting using visual sensor in orchards. Sensors. **19**(20), 4599 (2019). https://doi.org/10.3390/s19204599

Kim, J.S.G., Moon, S., Park, J., Kim, T., Chung, S.: Development of a machine vision-based weight prediction system of butterhead lettuce using deep learning models for industrial plant factory. Front. Plant Sci. **15**, 1365266 (2024)

Koç, S., Kayra, H.: Non-destructive weight prediction model of spherical fruits and vegetables using U-net image segmentation and machine learning methods. J. Agric. Sci. **30**(4), 735–747 (2024)

Koirala, A., Walsh, K.B., Wang, Z., McCarthy, C.: Deep learning – method overview and review of use for fruit detection and yield estimation. Comput. Electron. Agric. **162**, 219–234 (2019). https://doi.org/10.1016/j.compag.2019.04.017

Lawal, O.M.: Tomato detection based on modified YOLOv3 framework. Sci. Rep. **11**(1), 1447 (2021). https://doi.org/10.1038/s41598-021-81216-5

Meshram, V., Patil, K., Meshram, V., Hanchate, D., Ramkteke, S.D.: Machine learning in agriculture domain: a state-of-art survey. Artif. Intell. Life Sci. **1**, 100010 (2021)

Neupane, C., Koirala, A., Walsh, K.B.: In-orchard sizing of mango fruit: 1. Comparison of machine vision based methods for on-the-go estimation. Horticulturae. **8**(12), 1223 (2022)

Onmankhong, J., Poonpakdee, P., Lapcharoensuk, R.: Estimation of the weight and volume of lime (Citrus aurantifolia (Christm.) swingle) fruit using computer vision based on traditional machine learning and deep learning. Agronomy. **14**(10), 2434 (2024)

Rahman, M.M., Basar, M.A., Shinti, T.S., Khan, M.S.I., Babu, H.M.H., Uddin, K.M.M.: A deep CNN approach to detect and classify local fruits through a web interface. Smart Agric. Technol. **5**, 100321 (2023)

Sugandi, B., Wahyu, R., Apriliana, S., Putri, F.R.: Fruit recognition and weight scale estimation based on visual sensing. In: Proceeding International Applied Business and Engineering Conference, pp. 235–241 (2022)

Ukwuoma, C.C., Zhiguang, Q., Bin Heyat, M.B., Ali, L., Almaspoor, Z., Monday, H.N.: Recent advancements in fruit detection and classification using deep learning techniques. Math. Probl. Eng. **2022**, 9210947 (2022)

Vasumathi, M.T., Kamarasan, M.: An LSTM based CNN model for pomegranate fruit classification with weight optimization using dragonfly technique. Indian J. Comput. Sci. Eng. **12**(2), 371–384 (2021)

Wang, C., et al.: Application of convolutional neural network-based detection methods in fresh fruit production: a comprehensive review. Front. Plant Sci. **13**, 868745 (2022). https://doi.org/10.3389/fpls.2022.868745

Xiao, F., Wang, H., Li, Y., Cao, Y., Lv, X., Xu, G.: Fruit detection and recognition based on deep learning for automatic harvesting: an overview and review. Agronomy. **13**(6), 1625 (2023)

An AI-Assisted Medical-Grade VR Platform for Healthcare Applications

Zainab Buchh(✉), Bhaskara Ravi Teja Udata, Anirudh Reddy, Shruthi Shobha Kunche, and Mohan Akash Degala

The University of Texas at Dallas, Richardson, TX, USA
Zainab.Buchh@UTDallas.edu

Abstract. Virtual reality (VR) is increasingly used in healthcare training, rehabilitation, pain management, and remote clinical education. However, most commercially available head-mounted displays were designed for consumer entertainment rather than repeated use in hospitals, simulation laboratories, rehabilitation clinics, or other clinical environments. These settings introduce additional requirements for hygiene, comfort, tracking reliability, medical data protection, device management, and integration with clinical workflows. This paper presents an AI-assisted medical-grade VR platform architecture for healthcare applications. The work is based on an early design-stage study that combines literature review, market research, technology mapping, component-level analysis, and supplier/cost benchmarking. The proposed platform includes a modular headset architecture, a healthcare-oriented software and firmware stack, representative supplier categories, and three indicative target configurations for rehabilitation, clinical training, and surgical mixed reality. The paper also introduces a future large language model (LLM) layer intended to support clinician-facing explanation and clinical summary generation. The LLM layer is deliberately framed as human-in-the-loop decision support rather than autonomous diagnosis. The contribution of this work is a practical architecture and sourcing roadmap that connects healthcare VR requirements with hardware, software, AI, , and regulatory design considerations. The study does not report clinical validation or patient testing; instead, it establishes a structured foundation for prototype development, engineering validation, and later clinical evaluation.

Keywords: Healthcare virtual reality · Medical XR · Large language models · Multimodal AI · Clinical decision support · VR headset architecture · Digital health

1 Introduction

Immersive technologies are becoming increasingly relevant to healthcare because they can support repeatable simulation, structured rehabilitation, remote instruction, and controlled exposure to complex clinical scenarios. VR systems have been studied in health professions education, where systematic review evidence suggests benefits for

N. Mohd Sharef et al. (Eds.): CAMP 2026, CCIS 3094, pp. 380–392, 2027.
https://doi.org/10.1007/978-981-92-4305-1_28

knowledge and cognitive skill development while also showing that evidence for cost-effectiveness and patient outcomes remains limited [7]. VR and related extended reality (XR) tools are also being explored for rehabilitation and telerehabilitation, where they can support guided movement, engagement, feedback, and remote monitoring [10]. These capabilities make VR attractive for applications such as surgical simulation, clinical skills training, physical therapy, pain distraction, neurotherapy, and remote clinical education.

Despite this potential, a large portion of deployed VR hardware was designed for consumer entertainment or enterprise demonstration rather than for healthcare environments. Clinical settings create different constraints. A headset may be shared among many users, cleaned repeatedly with disinfectants, used for long sessions, operated near sensitive equipment, and connected to systems that process protected or sensitive health information. The device must also perform reliably under difficult environmental conditions, including bright surgical lighting, reflective metallic surfaces, patient movement limitations, masks, glasses, gloves, and noisy wards. Head-mounted display design choices influence user comfort, simulator sickness, visual fidelity, tracking quality, and implementation quality [6]. These factors are not peripheral; they directly affect whether a VR system can be trusted and used in a clinical workflow.

The work was developed at The University of Texas at Dallas (UTD) through the Naveen Jindal School of Management (JSOM), with guidance from IVRHA (International Virtual Reality Healthcare Association). This institutional and industry context shaped the platform direction toward practical clinical requirements rather than consumer-device assumptions.

At the same time, healthcare AI is moving toward multimodal systems that combine images, signals, text, structured data, sensor streams, and natural language interaction. Multimodal biomedical AI has been identified as a major direction for future medical systems because clinical understanding often depends on more than one data source [1]. Foundation models and medical LLMs may support summarization, explanation, triage assistance, and documentation, but they also introduce safety, bias, privacy, and validation risks [8, 9, 11]. In a healthcare VR platform, these capabilities should therefore be framed as clinician-support tools, not as autonomous diagnostic agents.

This paper presents an AI-assisted medical-grade VR platform architecture for healthcare applications. The work is based on a design-stage study that reviewed academic literature, manufacturer information, public device teardowns, component specifications, and supplier information. It translates healthcare requirements into a modular headset architecture, a proposed software and firmware stack, target hardware configurations, and a supplier/cost analysis. A future LLM-assisted layer is proposed for clinical explanation and summary generation after VR sessions, with human review and governance controls. The paper is intentionally scoped as an architecture and feasibility study. It does not claim a completed prototype, clinical trial, patient data collection, or diagnostic accuracy.

The main contributions are:

- A healthcare-oriented VR platform architecture that connects hardware subsystems, software layers, clinical use cases, and future AI support.

- A subsystem evaluation approach covering technical performance, cost and scalability, compliance and reliability, and integration compatibility.
- Three indicative configurations for rehabilitation, clinical training, and surgical mixed reality.
- A proposed LLM-assisted clinical explanation layer designed for clinician review rather than autonomous diagnosis.
- A supplier and cost roadmap that can guide prototype development and future request-for-quotation activity.

2 Related Work

2.1 Virtual Reality in Healthcare Training and Rehabilitation

VR has been studied as an educational tool for medical and health professions training. The appeal of VR is that learners can practice procedures, anatomy, communication, decision-making, and emergency response in a repeatable environment without exposing patients to avoidable risk. Kyaw et al. reviewed VR for health professions education and found that VR can improve knowledge and cognitive outcomes compared with traditional or other digital education methods, while also noting the need for better evidence regarding patient outcomes and economic value [7]. This supports the use of VR as a training medium, but it also suggests that healthcare VR systems should be evaluated carefully rather than assumed to be effective in every context.

In rehabilitation, VR can support repetitive task practice, feedback, motivation, and remote monitoring. VR-based telerehabilitation has been explored for upper-limb recovery after stroke, with attention to design principles, engagement, monitoring, safety, and the need for standardized protocols [10]. For rehabilitation use cases, the hardware requirements differ from surgical simulation. A rehabilitation headset may prioritize affordability, comfort, durability, ease of cleaning, and safe repeated use, while a surgical simulation system may prioritize visual fidelity, low latency, tracking precision, and integration with anatomical or imaging content.

The healthcare applications considered in this paper include surgical simulation, clinical training, physical rehabilitation, pain and anxiety management, neurotherapy, and remote clinical education. These use cases share the need for reliable immersive interaction, but they differ in acceptable cost, optical fidelity, AI support, regulatory pathway, and validation burden. A single hardware configuration is unlikely to satisfy all clinical scenarios. This motivates the proposed tiered platform approach.

2.2 Medical-Grade XR Hardware Requirements

Consumer VR headsets have made immersive technology more accessible, but healthcare environments impose additional requirements (Fig. 1). Device housings and face interfaces must tolerate repeated cleaning and replacement. Optical components must remain clear after cleaning cycles. Tracking must remain stable in varied lighting conditions. Battery placement must reduce neck strain during long sessions. Audio must support communication without creating hygiene concerns. Software must support secure updates, device management, audit logging, and privacy-preserving data handling.

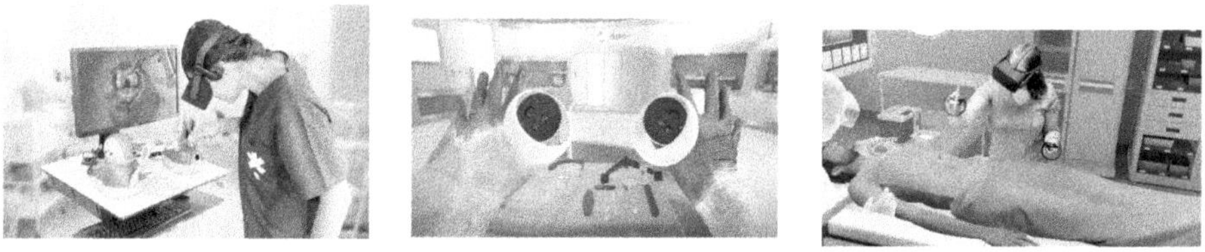

(a) (b) (c)

Fig. 1. Representative healthcare VR application contexts: (a) surgical simulation, (b) procedural training, and (c) immersive clinical training or patient-care simulation.

HMD quality affects both performance and user safety. Kourtesis et al. describe technological factors relevant to HMD implementation, including display quality, tracking, ergonomics, interaction, and computing performance [6]. These factors are especially important in healthcare because simulator sickness, latency, thermal discomfort, or tracking instability can reduce training quality or create risk for patients and clinicians. Hardware therefore has to be evaluated as part of a complete clinical system rather than as a consumer electronics product.

Medical-device standards and quality systems also shape design expectations. ISO 13485 defines quality-management requirements for medical devices [4], while the IEC 60601 series addresses safety and essential performance for medical electrical equipment [3]. This paper does not claim compliance with these standards. Instead, it treats them as design considerations that should inform later engineering validation and regulatory planning.

2.3 Multimodal AI and LLMs in Clinical Systems

Healthcare is naturally multimodal. Clinicians use patient history, images, lab values, physiological signals, motion, speech, and observed behavior to make decisions. Acosta et al. describe multimodal biomedical AI as a direction that can combine multiple data types for improved modeling and clinical support [1]. Foundation models for generalist medical AI similarly point toward systems that can operate across medical tasks and modalities, although such systems require rigorous validation and governance [8].

LLMs have shown the ability to encode clinical knowledge and generate medically relevant text, but they are not inherently safe clinical decision-makers [11]. They can produce plausible but incorrect statements, reflect training-data bias, mishandle uncertainty, and raise privacy concerns. Omiye et al. summarize both the potential and pitfalls of LLMs in medicine, including risks related to hallucination, bias, accountability, and evaluation [9]. The World Health Organization has also emphasized governance requirements for large multimodal models in health, including transparency, safety, privacy, accountability, and human oversight [13].

For this reason, the LLM component proposed in this paper is limited to clinical explanation and summary generation. It is not proposed as an autonomous diagnostic system. Its role is to convert structured session outputs, platform logs, clinician-selected observations, and approved contextual material into a concise summary for clinician review. This positioning is consistent with a human-in-the-loop clinical workflow and avoids overstating the readiness of generative AI in safety-critical settings.

3 Methodology

This work followed an exploratory and analytical design approach. The objective was to translate healthcare VR needs into a feasible platform architecture and sourcing roadmap. The study used secondary data sources, including academic literature, manufacturer documentation, device teardown reports, component specifications, supplier information, and public technical materials. Informal expert and project guidance was also considered where available, including guidance from IVRHA and project mentors.

The workflow was organized into five stages. First, the team reviewed academic and industry sources related to VR healthcare applications, XR hardware, optics, displays, sensors, software stacks, AI support, and regulatory considerations. Second, technologies were mapped across major headset subsystems, including compute, displays, optics, memory and storage, cameras, tracking, battery and power, connectivity, audio, housing, and software. Third, components and suppliers were compared using public specifications such as resolution, field of view, latency, power efficiency, operating conditions, driver support, and estimated price. Fourth, findings were cross-checked against available expert feedback and industry context. Fifth, the results were synthesized into architecture diagrams, configuration tables, supplier categories, and cost ranges.

Each subsystem was evaluated using four weighted criteria, shown in Table 1. Technical performance received the largest weight because display fidelity, tracking accuracy, latency, and power efficiency directly affect healthcare usability. Cost and scalability were also important because healthcare adoption depends on manufacturability and realistic bill-of-materials (BOM) targets. Compliance and reliability were included because clinical devices require durability, safety planning, and standards-aware design. Integration compatibility was included because even a high-performing component may be unsuitable if it lacks software support, driver availability, mechanical fit, or compatibility with the target runtime.

The study has important limitations. It did not include a completed hardware prototype, patient testing, clinical validation, or measured diagnostic performance. Pricing is based on public information and indicative estimates rather than contractual quotations. Supplier availability, minimum order quantities, and medical-grade allocations may change. These limitations are appropriate for an early design-stage study, but they mean that the results should be interpreted as a prototype-planning foundation rather than as evidence of clinical effectiveness (Fig. 2).

4 Proposed AI-Assisted Healthcare VR Platform

4.1 System Overview

The proposed platform is a modular healthcare VR architecture that combines headset hardware, secure software, clinical workflow integration, and a future AI explanation layer. The system is organized around the major subsystems required for an untethered or hybrid healthcare headset: compute, memory and storage, displays, optics, cameras, tracking sensors, eye tracking, environmental sensors, audio, connectivity, battery

Table 1. Evaluation criteria for healthcare VR subsystem selection.

Criterion	Weight	Example indicators
Technical performance	40%	Resolution, field of view, latency, refresh rate, power efficiency, tracking accuracy
Cost and scalability	25%	Indicative price, manufacturability, supplier availability, production feasibility
Compliance and reliability	20%	Durability, operating conditions, cleaning tolerance, standards alignment
Integration compatibility	15%	Mechanical fit, software support, driver availability, runtime compatibility

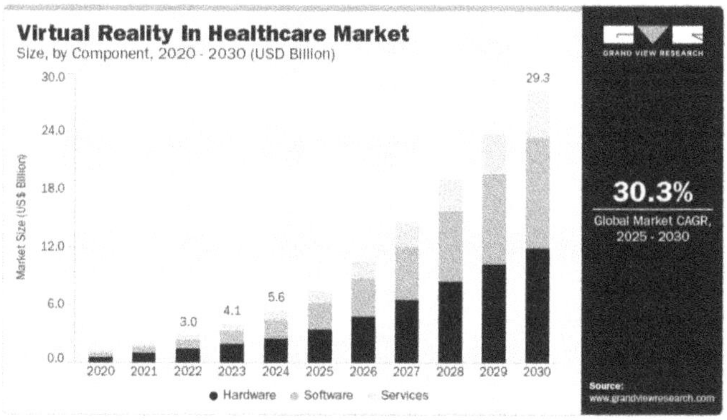

Fig. 2. Healthcare VR market context by component category, reproduced with permission.

and power, housing, and software/firmware. Figure 3 shows the extracted exploded architecture diagram used as the starting point for the platform design.

The architecture is modular because healthcare use cases have different cost and performance requirements. Rehabilitation and patient education can use a lower-cost headset if it is comfortable, cleanable, and reliable. Clinical training requires stronger rendering, better optics, and more robust tracking. Surgical mixed reality requires higher visual fidelity, low-latency passthrough, precise tracking, and stricter workflow integration. Instead of proposing a single universal device, this work proposes a platform family with shared design principles and configurable subsystem choices.

4.2 Hardware Architecture and Component Selection

The hardware architecture was organized around three representative configurations: low-cost rehabilitation, clinical training, and surgical mixed reality. Table 2 summarizes these configurations. The low-cost rehabilitation configuration emphasizes affordability and repeated use. It can use LCD displays, Fresnel optics, a lower-cost XR compute platform, and omit eye tracking where the clinical use case does not require it. The

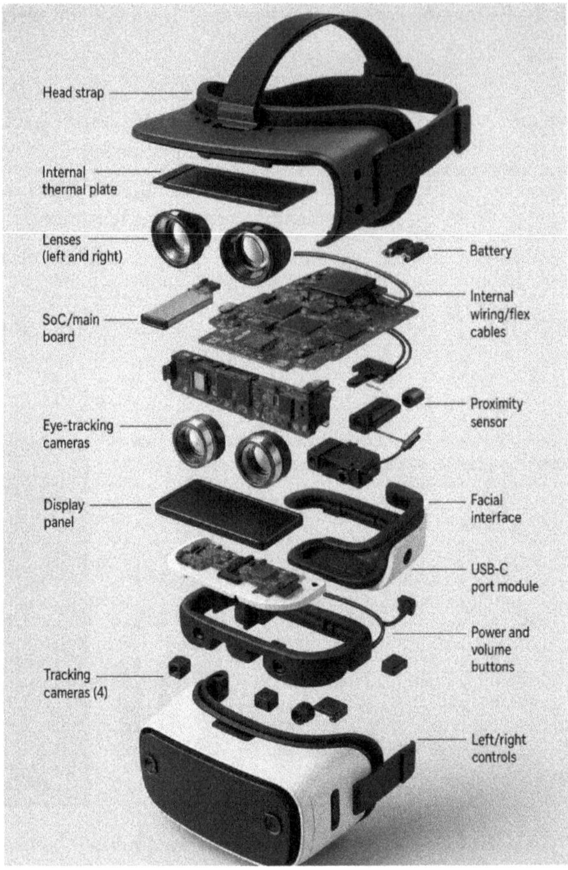

Fig. 3. Proposed modular architecture for an AI-assisted healthcare VR platform.

clinical training configuration improves visual fidelity and interaction quality through OLED or similar displays, pancake optics, a stronger XR SoC, and eye tracking. The surgical mixed-reality configuration uses higher-end display and tracking components, including micro-OLED, XR2-class compute, eye tracking, depth sensing, and ultra-wideband support where appropriate.

Several cross-subsystem tradeoffs emerged. Display resolution and refresh rate increase GPU and memory bandwidth requirements, which then affect thermal design and battery size. Pancake optics improve compactness and comfort but require brighter displays because of optical losses. Eye tracking supports foveated rendering and gaze-related analytics, but it adds camera cost, calibration complexity, and privacy considerations. Back-mounted batteries improve weight balance but complicate mechanical design and cabling. Housing materials must balance cleaning resistance, durability, weight, antenna performance, and thermal behavior.

These tradeoffs show why healthcare VR cannot be designed by selecting the highest-performing component in each category. The platform must be co-designed. A surgical

Table 2. Indicative target configurations for healthcare VR use cases.

Configuration	Target use	Key design direction	Indicative BOM
Low-cost rehabilitation	Rehabilitation and patient education	LCD/Fresnel, XR1 or XR2 Gen 1, no eye tracking	Approx. $250
Clinical training	Clinical education and simulation	OLED/pancake, XR2 Gen 1, eye tracking	Approx. $400–$550
Surgical MR	Surgical simulation and mixed reality	Micro-OLED, XR2 Gen 2, eye tracking, depth, UWB	Approx. $800–$1200

training headset with high-resolution displays but poor thermal management may be unusable for long sessions. A low-cost rehabilitation headset with acceptable visuals but poor cleanability may fail in multi-user clinics. A device with excellent hardware but weak driver support may be difficult to maintain or validate.

4.3 Software and Firmware Stack

The proposed software and firmware stack is a specification rather than an implemented software product. It is designed to support security, maintainability, clinical workflow integration, and XR performance. Table 3 summarizes the major layers.

Table 3. Proposed software and firmware stack for healthcare VR.

Layer	Proposed specification	Healthcare rationale
Bootloader	Secure boot and verified boot	Tamper resistance and data protection
Operating system	Android XR or Linux Yocto/Core	Maintainability and deployment control
XR runtime	OpenXR 1.1	Cross-device portability [5]
Graphics	Vulkan or vendor-optimized renderer	Low-latency rendering and efficient GPU use
Tracking	SLAM and predictive IMU fusion	Stable pose estimation under variable conditions
Clinical integration	FHIR/HL7 connectors	Hospital workflow readiness [2]
Security	AES-256 encryption and TEE	Privacy and access control
Updates	A/B OTA with rollback	Fleet reliability and recovery

OpenXR is included because it supports cross-platform XR application portability and reduces dependence on a single vendor runtime [5]. Vulkan or a comparable optimized graphics path is included because healthcare simulation and mixed-reality workloads require low-latency rendering. Secure boot, verified partitions, encryption, trusted execution environments, audit logging, and managed updates are included because clinical devices may process sensitive session data and require controlled fleet management. FHIR/HL7 interfaces are included as future integration targets for structured clinical data exchange [2], although actual integration would require institutional approval, privacy review, and software validation.

For a rehabilitation or training headset, Android-based deployment may reduce development complexity and improve application availability. For a controlled clinic kiosk or tethered system, Linux-based deployment may offer stronger long-term control. A hybrid approach may be appropriate when real-time tracking tasks require specialized firmware or DSP support while user interfaces and application logic run on a higher-level operating system.

4.4 Future LLM-Assisted Clinical Explanation Layer

The proposed LLM-assisted layer is a future component of the platform. Its role is to generate clinician-facing explanations and summaries from structured VR session outputs. It is not intended to diagnose patients, replace clinician judgment, or directly determine treatment. The layer would receive structured inputs such as session type, completed tasks, motion or interaction metrics, device quality flags, clinician-entered observations, and approved contextual information. It would then produce a concise summary that a clinician can review, edit, accept, or reject.

A possible pipeline is as follows: VR session data are converted into a structured event log; privacy and safety filters remove unnecessary identifiers; approved clinical or training context is retrieved; the LLM generates a draft explanation or session summary; the clinician reviews the output; the final note is stored or exported only after human approval. This design reflects the current state of medical LLM evidence. LLMs can support explanation and documentation, but their outputs require governance, validation, and oversight [9, 11, 13].

The future LLM layer also creates new requirements for the rest of the platform. Session logs must be structured and interpretable. Data retention must be controlled. Prompt templates and retrieval sources must be versioned. Outputs must include uncertainty and should avoid unsupported diagnoses. If the LLM becomes part of regulated medical-device software, it may fall under AI-enabled software lifecycle expectations such as those discussed by the FDA for AI in software as a medical device [12]. Therefore, the LLM layer is best introduced first as a documentation and explanation assistant in controlled training or clinician-supervised contexts.

5 Supplier, Cost, and Market Analysis

Supplier and cost analysis was included because a healthcare VR architecture is only useful if it can be sourced and manufactured at realistic cost levels. The study mapped supplier categories across compute, displays, optics, tracking, power, housing, and software. Table 4 summarizes the representative categories.

The analysis identified three practical cost levels. A low-cost rehabilitation headset can prioritize affordability by using LCD displays, simpler optics, and fewer sensors. A clinical training headset requires stronger rendering, improved optics, and better tracking, leading to a mid-range BOM. A surgical mixed-reality headset requires premium displays, more advanced optics, stronger compute, eye tracking, depth sensing, and more demanding calibration, resulting in a substantially higher BOM. These ranges are indicative and should be treated as public-source estimates. Final pricing would depend on supplier negotiations, minimum order quantities, certification requirements, tooling cost, and volume.

Supplier maturity is as important as component specification. A display may meet resolution targets but be unavailable in low volume. An SoC may provide strong AI acceleration but require restrictive licensing or limited board-support-package access. An optics supplier may provide excellent clarity but require expensive tooling and tight assembly tolerances. For healthcare use, vendors must also support documentation, traceability, cleaning compatibility, lifecycle availability, and future quality-management needs.

Table 4. Representative supplier categories for the proposed healthcare VR platform.

Subsystem	Representative supplier category	Notes
Compute/SoC	XR SoC and edge AI vendors	Availability, SDK maturity, thermal behavior, and AI acceleration are critical
Displays	LCD, OLED, and micro-OLED vendors	Selection depends on fidelity, brightness, supply risk, and cost target
Optics	Fresnel, pancake, aspheric, and waveguide vendors	Comfort and clarity drive medical suitability
Tracking	Camera, IMU, eye-tracking, and depth vendors	Hospital lighting robustness is a key constraint
Power	Battery cell and pack vendors	Safety, thermal behavior, runtime, and swappability matter
Housing	Medical plastics and enclosure manufacturers	Cleaning compatibility and durability are required
Software	XR runtime, engine, firmware, and integration partners	Security, lifecycle support, and driver quality are central

6 Discussion

6.1 Healthcare Constraints Versus Consumer VR Assumptions

The study confirms that healthcare VR design differs from consumer VR design. Consumer devices often optimize for entertainment, price, and content ecosystem. Healthcare systems must also optimize for cleanability, shared use, durability, clinical workflow, privacy, and predictable performance. A headset used in a simulation lab, rehabilitation clinic, or hospital ward may be cleaned many times per day and worn by users with different physical abilities. This creates requirements for removable face interfaces, disinfectant-resistant materials, balanced weight distribution, reliable straps, and accessible adjustment mechanisms.

Healthcare environments also challenge tracking systems. Operating rooms and intensive-care environments include bright lights, reflective surfaces, moving staff, equipment carts, and sometimes limited space. Rehabilitation users may move slowly, asymmetrically, or with assistive devices. Training scenarios may require hand tracking, controller tracking, or tool interaction. These conditions suggest that healthcare VR should use robust sensor fusion and should not rely on ideal lighting or consumer living-room assumptions.

6.2 Cross-Subsystem Tradeoffs

The platform architecture highlights several interdependencies. The display, optics, and SoC form a visual pipeline. Higher-resolution displays improve anatomical detail but increase GPU load and heat. Pancake optics reduce headset thickness but require brighter displays, which can increase power consumption. Foveated rendering can reduce rendering load but requires calibrated eye tracking and suitable software support.

The sensor, tracking, and compute subsystems form a pose-estimation pipeline. Camera placement affects SLAM coverage. IMU quality affects short-term pose stability. Eye tracking and hand tracking require additional cameras, illumination, calibration, and processing. Depth sensors and ultra-wideband modules can improve spatial awareness or rehabilitation tracking, but they also increase cost and integration complexity.

Housing, thermal design, and battery placement form an ergonomics pipeline. Larger batteries increase runtime but add weight. Front-heavy designs increase neck strain. Aluminum can improve heat dissipation but affect antennas. Medical-grade plastics and silicone improve cleaning compatibility but may affect thermal behavior and cost. These tradeoffs show why the proposed platform uses multiple target configurations rather than a single universal BOM.

6.3 Risk Assessment

The main technical risks are thermal constraints, optical calibration, tracking instability, display availability, and software-driver maturity. Manufacturing risks include optical alignment, camera calibration, material selection, enclosure tooling, and quality control. Regulatory and governance risks include unclear classification if the device moves from training into diagnosis or treatment, privacy obligations, AI oversight, and cybersecurity expectations. User risks include simulator sickness, fatigue, fogging, discomfort, and accessibility barriers.

The future LLM layer introduces additional risks. Generated summaries may omit important details, overstate findings, or produce unsupported interpretations. To reduce these risks, the system should restrict the LLM to structured inputs, approved retrieval sources, clear output templates, and clinician review. The LLM output should be treated as a draft explanation or summary rather than a final clinical determination.

6.4 Regulatory and Governance Considerations

This paper does not claim regulatory compliance. However, the architecture is informed by standards and governance sources that will matter in later development. ISO 13485 is relevant to medical-device quality management [4]. IEC 60601 is relevant to safety and essential performance for medical electrical equipment [3]. FDA guidance and resources on AI-enabled software as a medical device are relevant if future software functions are intended for regulated clinical use [12]. WHO guidance on large multimodal models is relevant to governance, oversight, transparency, and risk management for health AI [13].

The safest development path is staged. Early prototypes should be used for training, workflow evaluation, and engineering validation before any clinical claims are made. LLM functions should begin with non-diagnostic summaries and clinician review. Any move toward diagnosis, treatment recommendation, or patient-specific clinical decision-making would require a substantially higher validation and regulatory burden.

7 Limitations and Future Work

This study has several limitations. First, it is an architecture and sourcing study, not a clinical validation study. No patient testing, diagnostic performance evaluation, or clinical outcome measurement is reported. Second, the software stack is proposed as a specification; it has not yet been implemented and benchmarked as a complete platform. Third, supplier and price data are indicative and based on public information, market research, and available project context. Final costs would require formal RFQs, minimum order quantity analysis, supplier agreements, and engineering validation. Fourth, the current architecture figure should be redrawn for publication quality before final submission because the extracted image contains rough labels from the source document. Fifth, the LLM layer is a future component and requires separate design, privacy review, prompt governance, safety testing, and clinician evaluation.

Future work should proceed in three stages. The first stage is proof-of-concept prototyping, including mechanical mockups, display and optics testing, thermal measurement, and software runtime selection. The second stage is engineering validation, including tracking tests under clinical lighting, cleaning and material testing, battery safety testing, software security review, and usability evaluation with clinicians or trainees. The third stage is clinical and workflow validation, including controlled studies in appropriate training or rehabilitation settings. For the LLM layer, future work should include a clinician-reviewed summarization prototype, evaluation of factual consistency, safety checks, privacy analysis, and comparison against human-written session summaries.

8 Conclusion

This paper presented an AI-assisted medical-grade VR platform architecture for healthcare applications. The study translates clinical and technical requirements into a modular headset design, proposed software and firmware stack, target configurations, and supplier/cost roadmap. The architecture addresses the gap between consumer VR hardware and healthcare needs such as hygiene, durability, tracking reliability, privacy, and workflow integration. It also proposes a future LLM-assisted layer for clinician-facing explanation and clinical summary generation, constrained by human review and governance requirements. The work provides a practical foundation for prototype development and later validation, while clearly distinguishing design-stage architecture from clinical evidence.

Acknowledgments. The authors acknowledge Professor David Parks, JSOM Cap-stone Senior Projects, The University of Texas at Dallas, for faculty mentorship and project guidance. The authors also acknowledge IVRHA (International Virtual Reality Healthcare Association) and Bob Fine, Executive Director, for project guidance.

Disclosure of Interests. The authors have no competing interests to declare.

References

1. Acosta, J.N., Falcone, G.J., Rajpurkar, P., Topol, E.J.: Multimodal biomedical ai. Nat. Med. **28**, 1773–1784 (2022). https://doi.org/10.1038/s41591-022-
2. International: FHIR: fast healthcare interoperability resources (2026)
3. International Electrotechnical Commission: IEC 60601-1 series: medical electrical equipment (2026)
4. International Organization for Standardization: ISO 13485: 2016: medical devices – quality management systems – requirements for regulatory purposes (2016)
5. Khronos Group: OpenXR 1.1 specification (2024)
6. Kourtesis, P., Collina, S., Doumas, L.A.A., MacPherson, S.E.: Technological competence is a precondition for effective implementation of virtual reality head mounted displays in human neuroscience: a technological review and meta-analysis (2021)
7. Kyaw, B.M. et al.: Virtual reality for health professions education: systematic review and meta-analysis by the digital health education collaboration. J. Med. Internet Res. **21**(1), e12959 (2019). https://doi.org/10.2196/12959
8. Moor, M. et al.: Foundation models for generalist medical artificial intelligence. Nature. **616**, 259–265 (2023). https://doi.org/10.1038/s41586-023-05881-4
9. Omiye, J.A., Gui, H., Rezaei, S.J., Zou, J., Daneshjou, R.: Large language models in medicine: The potentials and pitfalls (2023)
10. Rodrigues, P., Quaresma, C., Costa, M., Luz, F., Fonseca, M.M.: Virtual reality-based telerehabilitation for upper limb recovery post-stroke: A systematic review of design principles, monitoring, safety, and engagement strategies (2025)
11. Singhal, K. et al.: Large language models encode clinical knowledge. Nature. **620**, 172–180 (2023). https://doi.org/10.1038/s41586-023-
12. S. Food and Drug Administration: Artificial intelligence in software as a medical device (2025)
13. World Health Organization: Ethics and governance of artificial intelligence for health: guidance on large multi-modal models (2025)

Beyond Detection: From AI Policing to Human-AI Collaboration in Thesis Supervision

Brian Halubanza[1(✉)], Charity Meki-Kombe[2], and Selina Kadakwiza[1]

[1] Computer Science Department, Mulungushi University, Kabwe, Zambia
bhalubanza@mu.edu.zm
[2] Department of Education, Mulungushi University, Kabwe, Zambia

Abstract. Generative artificial intelligence (GenAI) is transforming postgraduate research and challenging traditional assumptions about thesis supervision. While institutional responses have largely focused on plagiarism detection and academic integrity, less attention has been given to how AI is reshaping supervisory work itself. Drawing on Distributed Cognition Theory, Knowledge Creation Theory, and Human-AI Collaboration Theory, this study examines how GenAI is influencing the distribution of intellectual labour within postgraduate supervision. A convergent mixed-methods design was employed involving 68 postgraduate supervisors from Zambian higher education institutions. Quantitative data were analysed using descriptive statistics, while qualitative responses were analysed using reflexive thematic analysis. The findings reveal widespread adoption of AI-assisted supervision, with 76.5% of participants reporting active use of AI tools and 75.0% indicating that AI improves supervision quality and efficiency. Supervisors reported a shift in effort from routine editorial activities toward conceptual evaluation, methodological scrutiny, critical validation, and scholarly oversight. The results suggest that cognitive work is increasingly distributed across supervisors, students, and AI systems, transforming supervision from a traditional dyadic relationship into a distributed knowledge ecosystem. Participants also emphasised a shift in conceptions of originality, highlighting reasoning quality, methodological defensibility, and epistemic accountability over textual production alone. Despite widespread adoption, a significant gap remains between AI use and institutional preparedness, with nearly 90% of respondents supporting formal AI policies, supervision guidelines, and faculty capacity-building initiatives. Based on these findings, the study proposes the Human-AI Collaborative Supervision Model (HACSM), comprising four dimensions: AI-mediated knowledge exploration, intellectual stewardship, critical validation, and epistemic accountability. The study extends understanding of human-AI collaboration in postgraduate supervision and provides empirical evidence from a Global South context, with implications for supervision practice, assessment, research training, and AI governance in higher education.

Keywords: Human-AI Collaboration Thesis Supervision · Distributed Cognition · Knowledge Creation · Generative Artificial Intelligence · Postgraduate Research · Higher Education · Zambia

N. Mohd Sharef et al. (Eds.): CAMP 2026, CCIS 3094, pp. 393–405, 2027.
https://doi.org/10.1007/978-981-92-4305-1_29

1 Introduction

Generative artificial intelligence (GenAI) is transforming knowledge work in higher education and raising new questions concerning authorship, originality, assessment, and supervision (Kasneci et al., 2023; Tlili et al., 2025). While institutional responses have largely focused on plagiarism detection and academic integrity, these approaches are increasingly inadequate in environments where AI has become embedded within everyday academic practice (Chan & Hu, 2023). Generative AI is enhancing productivity, information processing, and decision-making while simultaneously reshaping how knowledge is created and evaluated (Brynjolfsson et al., 2023). Postgraduate thesis supervision is among the academic activities most affected by these developments. Traditionally, supervision has been viewed as a human-centred process through which supervisors guide students in developing research competence, scholarly independence, and disciplinary expertise (Lee, 2008). However, students increasingly use AI systems to support literature exploration, writing, idea generation, and feedback, while supervisors employ AI-assisted tools to enhance review and evaluation processes. Consequently, knowledge production is no longer mediated solely through interactions between supervisors and students. AI systems are becoming active participants in research workflows, influencing how knowledge is generated, refined, and communicated (Jarrahi et al., 2023; Dwivedi et al., 2023). This shift aligns with sociomaterial perspectives that view technologies as active shapers of organisational and academic practices rather than passive tools (Leonardi, 2021). Although GenAI offers opportunities for improved writing, information access, and academic productivity (Crompton & Burke, 2023; Lo, 2023), concerns remain regarding hallucinations, algorithmic bias, diminished critical thinking, and intellectual ownership (Floridi & Chiriatti, 2020; Cotton et al., 2023; Halubanza et al., 2023). As a result, international organisations such as UNESCO and OECD increasingly advocate governance approaches that balance innovation with transparency, accountability, and responsible AI use (UNESCO, 2024; OECD, 2025). Emerging research further suggests that knowledge work is increasingly occurring through collaborative interactions between humans and intelligent systems, requiring new understandings of expertise, authority, and knowledge creation (Faraj et al., 2024). Despite growing scholarship on AI in higher education, important gaps remain. Existing studies largely focus on AI adoption, academic integrity, and assessment redesign, with limited attention to how AI is transforming the supervisory process itself. Furthermore, supervision research continues to conceptualise knowledge creation primarily as a human-centred activity, providing limited insight into environments where AI systems actively contribute to literature exploration, idea generation, and feedback. Finally, empirical evidence from the Global South remains limited despite increasing AI adoption across African universities (Halubanza et al., 2025).

To address these gaps, this study examines how GenAI is reshaping postgraduate supervision through the lenses of Distributed Cognition Theory (Hutchins, 1995), Knowledge Creation Theory (Nonaka & Takeuchi, 1995), and Human-AI Collaboration Theory. Together, these perspectives provide a foundation for understanding how cognitive work, knowledge creation, and scholarly accountability are distributed across supervisors, students, and AI systems.

Drawing on evidence from 68 postgraduate supervisors across Zambian universities, the study investigates how AI is transforming supervisory roles, intellectual labour, and conceptions of originality. The study makes three contributions. First, it extends understanding of distributed cognition in postgraduate supervision by demonstrating how cognitive work is increasingly shared across human and technological actors. Second, it contributes to human-AI collaboration scholarship by conceptualising supervision as a sociotechnical knowledge ecosystem rather than a traditional dyadic relationship. Third, it proposes the Human-AI Collaborative Supervision Model (HACSM), a framework comprising AI-mediated knowledge exploration, intellectual stewardship, critical validation, and epistemic accountability.

The study argues that the future of postgraduate supervision lies not in detecting AI use but in developing supervisory approaches that leverage the complementary strengths of human and artificial intelligence while preserving scholarly quality, accountability, and academic integrity.

2 Related Work and Theoretical Foundation

Generative artificial intelligence (GenAI) has intensified debates regarding knowledge work, expertise, and decision-making in higher education. Existing scholarship has focused on three broad areas: academic integrity and assessment challenges associated with AI-generated content (Cotton et al., 2023), pedagogical opportunities such as personalised learning and writing support (Lo, 2023; Crompton & Burke, 2023), and human-AI collaboration and cognitive augmentation (Faraj et al., 2024; Jarrahi et al., 2023). While these perspectives provide valuable insights, they largely position AI as either a threat requiring regulation or a tool supporting educational activities. Comparatively less attention has been given to AI as an active participant in knowledge creation.

Postgraduate supervision provides an important context for examining this transformation. Traditionally, supervision has been viewed as a developmental process through which students acquire disciplinary knowledge, methodological competence, and scholarly independence (Lee, 2008; Gatfield, 2005; Wisker, 2012). However, GenAI increasingly supports literature exploration, idea generation, writing, and feedback, transforming supervision from a predominantly human-centred relationship into a sociotechnical process involving supervisors, students, and AI systems. Existing research has largely focused on AI adoption, ethics, and assessment redesign, leaving limited understanding of how AI reshapes supervisory roles, cognitive work, and knowledge creation. This study is informed by three complementary theoretical perspectives. First, Distributed Cognition Theory (Hutchins, 1995) conceptualises cognition as distributed across people, technologies, and social systems. Within AI-assisted supervision, activities such as information retrieval, synthesis, and feedback generation increasingly occur through interactions among supervisors, students, and AI systems, suggesting that supervision functions as a distributed cognitive system. Second, Knowledge Creation Theory (Nonaka & Takeuchi, 1995) explains how knowledge emerges through interactions between tacit and explicit knowledge. The integration of GenAI into research workflows influences processes of knowledge exploration, synthesis, and externalisation, raising important questions concerning intellectual ownership and epistemic accountability. Third,

Human-AI Collaboration Theory emphasises the complementary strengths of human and artificial intelligence (Dellermann et al., 2019). While AI contributes information processing, pattern recognition, and synthesis capabilities, humans provide contextual understanding, critical reasoning, ethical judgement, and disciplinary expertise (Dellermann et al., 2019; Luckin & Cukurova, 2019). Effective supervision therefore depends on integrating AI-supported knowledge processing with human evaluation and scholarly judgement (Jarrahi et al., 2023; Faraj et al., 2024). Collectively, these perspectives suggest that postgraduate supervision is evolving into a distributed sociotechnical knowledge ecosystem characterised by shared cognition, collaborative intelligence, and distributed accountability (Leonardi, 2021). Within this environment, AI supports knowledge exploration, supervisors provide intellectual stewardship and critical validation, and students remain responsible for scholarly judgement and accountability.

Based on this framework, the study advances four propositions:

P1: Generative AI redistributes cognitive and intellectual labour across supervisors, students, and AI systems.
P2: Supervisory responsibilities increasingly shift from operational and editorial functions toward intellectual stewardship, methodological evaluation, and critical validation.
P3: Effective postgraduate supervision increasingly depends on complementary human-AI capabilities rather than human-only knowledge production processes.
P4: Originality in AI-mediated supervision is increasingly evaluated through epistemic accountability, reasoning quality, and methodological defensibility rather than textual production alone.

3 Methodology

This study adopted a pragmatic research philosophy situated within a sociotechnical understanding of knowledge creation. Pragmatism is particularly appropriate for investigating emerging technological phenomena because it enables the integration of quantitative and qualitative evidence to understand complex real-world transformations while providing methodological flexibility for addressing multifaceted research problems (Creswell & Creswell, 2023). Given that generative artificial intelligence (GenAI) simultaneously influences technological processes, human decision-making, and supervisory practices, a pragmatic approach facilitated examination of both measurable patterns of AI adoption and the contextual experiences underlying these transformations. A convergent mixed-methods research design was employed to investigate AI-mediated postgraduate supervision. Mixed-methods approaches are particularly valuable when neither quantitative nor qualitative methods alone can adequately explain complex educational phenomena because they integrate complementary forms of evidence and interpretation (Johnson et al., 2007; Creswell & Creswell, 2023). Quantitative data were used to examine AI adoption patterns, supervisory practices, perceived impacts, institutional preparedness, and capacity-building requirements, while qualitative data explored changing supervisory roles, experiences of AI-assisted knowledge creation, intellectual ownership, and future expectations. The integration of both strands enabled a comprehensive understanding of how AI is reshaping postgraduate supervision. The study was conducted within the Zambian higher education sector, an important Global South context characterised by expanding postgraduate enrolments, increasing research productivity expectations, and ongoing digital transformation initiatives. AI adoption within this environment is occurring alongside uneven technological infrastructure, varying levels of AI literacy, and emerging governance frameworks, providing a valuable setting for examining AI-mediated supervision in resource-constrained knowledge ecosystems. Participants were selected using purposive sampling to ensure inclusion of academics with direct postgraduate supervisory experience. Eligibility required affiliation with a recognised higher education institution and current or previous postgraduate supervision responsibilities. A total of 68 supervisors participated in the study, representing disciplines including Education, Engineering, Information and Communication Technology, Computer Science, Social Sciences, Humanities, and Business. Although modest in size, the sample was considered appropriate for exploratory mixed-methods research aimed at theory development rather than statistical generalisation. Data were collected through a structured online questionnaire comprising three sections. The first captured demographic and professional characteristics, including academic rank, disciplinary background, and supervisory experience. The second examined AI-mediated supervisory practices, including AI tools used, frequency of use, supervisory tasks supported by AI, and perceived impacts on supervisory effectiveness. Responses were measured using a five-point Likert scale ranging from 1 (Strongly Disagree) to 5 (Strongly Agree). The third section consisted of open-ended questions exploring changes in supervisory practice, opportunities and challenges associated with AI, strategies for maintaining originality, and institutional support requirements. To enhance content validity, survey items were adapted from established literature on AI adoption, educational technology, and human-AI collaboration and were subsequently reviewed by three academics with expertise in educational technology and postgraduate supervision. Feedback from this review informed refinements to item wording, clarity, and alignment with the study objectives. Quantitative data were analysed using descriptive statistical techniques, including frequencies, percentages, means, and standard deviations. Analysis was organised around three theoretically informed dimensions: redistribution of intellectual labour, transformation of supervisory functions, and the emergence of human-AI collaborative supervision. Qualitative responses were analysed using reflexive thematic analysis to identify recurring patterns relating to supervisory transformation, human-AI collaboration, originality, and institutional readiness. Because the study sought exploratory insights into an emerging phenomenon, emphasis was placed on descriptive and interpretive analysis rather than hypothesis testing. Integration of quantitative and qualitative findings occurred during interpretation through theoretical triangulation. Quantitative trends were compared with qualitative themes to identify convergence, complementarity, and divergence, enabling statistical patterns to be contextualised through participant experiences. Methodological rigor and trustworthiness were strengthened through triangulation of quantitative and qualitative evidence, transparent documentation of analytical procedures, and interpretation informed by Distributed Cognition Theory, Knowledge Creation Theory, and Human-AI Collaboration Theory. Consistent with established criteria for high-quality qualitative research, credibility was enhanced through evidence triangulation, dependability through systematic analytical procedures, and interpretive robustness through the use of complementary theoretical perspectives (Tracy, 2020).

These measures ensured transparency, coherence, and confidence in the findings. Ethical approval was obtained through the relevant institutional review processes prior to data collection. Participation was voluntary, and respondents were informed of the study's purpose, confidentiality measures, data management procedures, and their right to withdraw at any stage. No personally identifiable information was collected, and all responses were anonymised before analysis. Findings are reported only in aggregate form to protect participant and institutional confidentiality. The study was guided by an analytical framework proposing that GenAI influences postgraduate supervision through three interconnected mechanisms: redistribution of intellectual labour, transformation of supervisory functions, and the emergence of human-AI collaborative intelligence. These mechanisms provide the conceptual foundation for the Human-AI Collaborative Supervision Model (HACSM), which is developed and evaluated through the empirical findings presented in the subsequent sections.

4 Results

4.1 Generative AI Adoption and Supervisory Transformation

The findings indicate substantial integration of generative AI into postgraduate supervision. Of the 68 supervisors surveyed, 52 (76.5%) reported using at least one AI tool in supervisory activities, while 16 (23.5%) reported no current use. Adoption was highest among lecturers (81.5%) and supervisors with 6–10 years of experience (82.6%). These results suggest that AI has moved beyond experimental use and is becoming embedded within supervisory practice across diverse academic disciplines. AI use varied across supervisory activities. The most frequently reported applications were ChatGPT for content generation and writing support (58.8%, M = 3.69), guiding students to avoid over-reliance on AI (60.3%, M = 3.68), and advising students on ethical AI use (50.0%, M = 3.46). Lower levels of adoption were observed for automated literature review and citation assistance (32.4%, M = 2.79) and AI-generated structured feedback on thesis drafts (30.9%, M = 2.76). Overall, AI appears to be supporting knowledge processing, communication, and supervisory guidance rather than replacing higher-order scholarly judgement.

Furthermore, most respondents perceived AI positively, with 75.0% agreeing that AI improves supervision quality and efficiency (M = 3.97). At the same time, 42.6% expressed concerns that AI may compromise originality and intellectual development (M = 3.44). Strong support was recorded for institutional governance initiatives, including AI policies (88.2%), faculty capacity building (88.2%), and inclusion of AI within postgraduate supervision guidelines (89.7%).

4.2 Redistribution of Intellectual Labour

The findings provide strong support for Proposition 1, which proposed that GenAI redistributes intellectual labour within postgraduate supervision. Supervisors reported that AI increasingly supports routine activities such as language refinement, citation checking, formatting, and preliminary feedback generation, allowing greater focus on conceptual guidance, methodological evaluation, and scholarly critique.

Participant responses illustrate this transition. One supervisor noted:

> "To critique theses, check grammar and recommend appropriate measures." (P18)

Another observed:

> "Uploading the work on AI and ask if the work is true and that has no plagiarism." (P27)

Participants consistently reported spending less time on editorial correction and more time evaluating arguments, methodological choices, theoretical coherence, and research contributions. However, concerns were also raised regarding students' ability to critically evaluate AI-generated outputs:

> "They don't know how to edit AI generated information." (P34)
> "Lack of verification by students and work usually is presented as is and their input cannot be ascertained." (P41)

These findings suggest that AI is not reducing supervisory work but transforming it from operational support toward intellectual stewardship and knowledge validation.

4.3 Emergence of Human-AI Collaborative Supervision

The findings strongly support Propositions 2 and 3, which proposed that supervisory functions are evolving toward higher-order cognitive activities and that supervision is increasingly characterised by human-AI collaboration. Respondents reported that students frequently consult AI systems for literature exploration, conceptual clarification, and methodological guidance before engaging with supervisors.

Several participants described AI as a valuable cognitive support tool:

> "If used properly it can help generate ideas." (P09)
> "Students have access to a variety of sources for their research." (P22)

Consequently, supervisory interactions are becoming less focused on content generation and more focused on evaluating, refining, and contextualising AI-assisted outputs. Supervisors increasingly emphasised critical reasoning, methodological defensibility, and theoretical coherence rather than linguistic quality alone. However, respondents also cautioned against excessive dependence on AI:

> "AI should be an aid not to be entirely relied upon." (P44)

These findings suggest that effective supervision increasingly depends on complementary human-AI capabilities, where AI supports knowledge exploration while supervisors provide contextual judgement, validation, and scholarly oversight.

4.4 Originality, Governance, and Institutional Readiness

A significant finding concerns changing conceptions of originality. Participants increasingly distinguished between textual originality and intellectual originality, arguing that scholarly quality should be assessed through reasoning, interpretation, methodological decision-making, and contextual understanding rather than textual production alone.

This shift suggests a movement from authorship-centred assessment toward epistemic accountability.

Despite widespread AI adoption, institutional readiness remains limited. Respondents expressed strong support for institutional policies, disclosure frameworks, and professional development initiatives. One participant stated:

> "As faculty we need more training in AI matters to help us be a step ahead of the students we are supervising." (P37)

Another recommended:

> "Develop policies that stipulate how AI is to be used by both students and faculty." (P37)

Similarly, a respondent emphasised:

> "Clear institutional AI usage policies, comprehensive staff training in AI literacy and detection." (P48)

Although nearly 90% of respondents supported AI governance frameworks and faculty capacity-building initiatives, only 33.8% felt adequately prepared to evaluate AI-assisted student submissions. This gap suggests that AI adoption is progressing more rapidly than institutional preparedness. The results reveal three major trends. First, AI is redistributing intellectual labour from routine supervisory tasks toward higher-order scholarly functions. Second, supervisory roles are evolving from editorial and administrative support toward intellectual stewardship and critical validation. Third, postgraduate supervision is increasingly characterised by distributed cognition and human-AI collaboration. Together, these findings provide empirical support for the development of the Human-AI Collaborative Supervision Model (HACSM).

5 Discussion

This study demonstrates that the influence of generative artificial intelligence (GenAI) on postgraduate supervision extends beyond technology adoption and represents a fundamental transformation of supervisory practice. Rather than functioning merely as a productivity tool, AI is increasingly becoming embedded within the cognitive processes through which knowledge is created, evaluated, and communicated. The findings suggest that postgraduate supervision is evolving from a traditional dyadic relationship between supervisor and student toward a distributed sociotechnical knowledge ecosystem involving supervisors, students, and AI systems. The findings provide strong support for Distributed Cognition Theory (Hutchins, 1995). Cognitive activities traditionally performed by supervisors, including language refinement, information retrieval, and preliminary review, are increasingly shared with AI systems. However, higher-order functions such as methodological evaluation, conceptual scrutiny, and scholarly judgement remain predominantly human responsibilities. This redistribution of cognitive work suggests that supervision should no longer be understood solely as an interpersonal educational process but rather as a distributed cognitive system in which knowledge production emerges through interactions among human and technological actors. A key contribution of the study is the identification of a shift from traditional supervisory roles toward what may be conceptualised as intellectual stewardship. Consistent with Human-AI Collaboration Theory (Dellermann et al., 2019; Faraj et al., 2024), supervisors are increasingly moving away from routine editorial and administrative tasks and toward activities that require contextual understanding, disciplinary expertise, critical reasoning, and ethical judgement. Within AI-mediated environments, the supervisor's role increasingly centres on validating knowledge claims, evaluating methodological rigor, challenging assumptions, and ensuring scholarly quality. These findings suggest that the value of supervision is becoming less associated with information provision and more closely linked to intellectual oversight and epistemic governance. The findings also contribute to Knowledge Creation Theory (Nonaka & Takeuchi, 1995) by illustrating how AI is becoming integrated into knowledge conversion processes. AI increasingly supports literature exploration, synthesis, and externalisation, accelerating the movement between tacit and explicit knowledge. However, the study demonstrates that knowledge creation remains fundamentally dependent on human interpretation, contextualisation, and validation. Consequently, AI should be viewed as a knowledge augmentation technology rather than an autonomous knowledge creator. Another important contribution concerns the reconceptualisation of originality within AI-mediated research environments. Participants consistently distinguished between textual originality and intellectual originality, emphasising reasoning quality, methodological defensibility, interpretation, and scholarly contribution over text production alone. This finding aligns with emerging scholarship arguing that academic quality should increasingly be evaluated through epistemic accountability rather than textual uniqueness. As AI becomes more capable of generating coherent academic content, assessment systems that rely heavily on textual detection mechanisms may become increasingly inadequate. The findings further reveal a significant institutional governance gap. Although AI adoption among supervisors is widespread, many institutions remain inadequately prepared to support responsible AI integration within postgraduate supervision. The strong demand for AI policies, supervision guidelines, disclosure frameworks, and faculty development programmes suggests that governance structures are lagging behind technological adoption. These findings support international calls by UNESCO (2024) and OECD (2025) for governance approaches that prioritise transparency, accountability, responsible use, and human oversight rather than relying exclusively on AI detection and surveillance mechanisms. From a broader perspective, the study contributes to the emerging literature on human-AI collaboration in higher education by providing empirical evidence from a Global South context. Much of the existing literature originates from technologically advanced educational systems in North America, Europe, and East Asia. The present findings demonstrate that similar supervisory transformations are occurring within African universities despite differences in infrastructure, resources, and institutional preparedness. This suggests that the redistribution of cognitive work and emergence of human-AI collaborative supervision may represent a broader global transformation rather than a phenomenon confined to technologically advanced environments. Collectively, the findings indicate that postgraduate supervision is undergoing a paradigmatic shift. The future of supervision is unlikely to be defined by increasingly sophisticated mechanisms for detecting AI use. Instead, effective supervision will depend on developing models that leverage the complementary strengths of human expertise and artificial intelligence while preserving the core academic values of critical inquiry, originality, transparency, and scholarly responsibility. These insights provide the theoretical and empirical foundation for the Human-AI Collaborative Supervision Model (HACSM) proposed in the following section.

6 Human-AI Collaborative Supervision Model (HACSM)

This study proposes the Human-AI Collaborative Supervision Model (HACSM) to explain postgraduate supervision in AI-mediated research environments. The model reconceptualises supervision as a collaborative ecosystem involving three actors: supervisors, students, and AI systems. Effective supervision emerges through the complementary interaction of these actors in the knowledge creation process.

HACSM comprises four dimensions:

1. AI-Mediated Knowledge Exploration – AI supports literature discovery, information retrieval, idea generation, and preliminary synthesis.
2. Intellectual Stewardship – Supervisors provide conceptual guidance, methodological oversight, and scholarly leadership.
3. Critical Validation – Human actors evaluate the accuracy, relevance, and credibility of AI-assisted outputs.
4. Epistemic Accountability – Students retain responsibility for interpretation, methodological decisions, and defence of knowledge claims.

The central proposition of HACSM is that effective supervision depends on the complementary strengths of human and artificial intelligence. While AI enhances information access and analytical efficiency, supervisors and students contribute contextual understanding, critical reasoning, and ethical judgement.

The novelty of HACSM lies in explicitly incorporating AI as a structural component of the supervision ecosystem. Unlike traditional supervision frameworks that focus primarily on the supervisor-student relationship, HACSM explains how knowledge exploration, validation, and accountability are distributed across supervisors, students, and AI systems while preserving human responsibility for scholarly quality and research integrity. The model provides a foundation for redesigning supervision practices, research training, assessment approaches, and institutional AI governance in the age of generative artificial intelligence.

7 Implications, Limitations, and Future Research

The findings suggest that postgraduate supervision increasingly requires AI literacy, critical evaluation skills, and the ability to assess AI-assisted research outputs. Universities should therefore integrate responsible AI use into postgraduate research training and develop governance frameworks that promote transparency, accountability, disclosure, and human oversight (Kabeta & Halubanza, 2023). Assessment practices should place greater emphasis on reasoning quality, methodological rigor, and intellectual ownership rather than textual originality alone.

Several limitations should be acknowledged. The study was conducted within the Zambian higher education sector and involved a relatively small sample of 68 postgraduate supervisors. The cross-sectional design captured perceptions at a single point in time, while the reliance on self-reported data may introduce response bias. In addition, the study focused exclusively on supervisors and did not include the perspectives of students, examiners, or institutional leaders.

Future research should validate the Human-AI Collaborative Supervision Model (HACSM) across different disciplines, institutions, and geographical contexts. Longitudinal and comparative studies are particularly needed to examine how AI influences supervisory effectiveness, research quality, postgraduate learning outcomes, and knowledge creation over time. Such research will support the development of evidence-based frameworks for responsible human-AI collaboration in higher education.

8 Conclusion

This study examined how generative artificial intelligence (GenAI) is reshaping postgraduate thesis supervision within higher education. Drawing on evidence from postgraduate supervisors in Zambian universities, the findings demonstrate that AI is increasingly embedded within supervisory practice and is transforming how knowledge is explored, evaluated, and validated.

The study found that AI is redistributing routine cognitive tasks while supervisors increasingly focus on intellectual stewardship, critical validation, and scholarly oversight. These findings suggest that postgraduate supervision is evolving from a traditional supervisor–student relationship toward a collaborative ecosystem involving supervisors, students, and AI systems.

A key contribution of the study is the introduction of the Human-AI Collaborative Supervision Model (HACSM), which conceptualises supervision through four interconnected dimensions: AI-mediated knowledge exploration, intellectual stewardship, critical validation, and epistemic accountability. The model provides a framework for understanding how human and artificial intelligence can be integrated within postgraduate research while preserving academic integrity, intellectual ownership, and scholarly quality.

Rather than viewing AI primarily as a threat to academic standards, the findings suggest that its effective integration depends on governance frameworks, AI literacy, and responsible human oversight. As AI capabilities continue to advance, higher education institutions will need to move beyond AI detection and develop supervision practices that leverage the complementary strengths of human expertise and artificial intelligence. The future of postgraduate supervision is therefore likely to be defined not by AI replacement, but by effective human-AI collaboration in knowledge creation.

References

Brynjolfsson, E., Li, D., Raymond, L.R.: Generative AI at Work (NBER Working Paper 31161). National Bureau of Economic Research (2023). https://doi.org/10.3386/w31161

Chan, C.K.Y., Hu, W.: Students' voices on generative AI: perceptions, benefits, and challenges in higher education. Educ. Sci. **13**(11), 1139 (2023). https://doi.org/10.3390/educsci13111139

Cotton, D.R.E., Cotton, P.A., Shipway, J.R.: Chatting and cheating? Ensuring academic integrity in the era of ChatGPT. Innov. Educ. Teach. Int. **61**(2), 228–239 (2023). https://doi.org/10.1080/14703297.2023.2190146

Crompton, H., Burke, D.: Artificial intelligence in higher education: the state of the field. Int. J. Educ. Technol. High. Educ. **20**(1), 22 (2023). https://doi.org/10.1186/s41239-023-00392-8

Creswell, J.W., Creswell, J.D.: Research Design: Qualitative, Quantitative, and Mixed Methods Approaches, 6th edn. SAGE Publications (2023)

Dellermann, D., Ebel, P., Söllner, M., Leimeister, J.M.: Hybrid intelligence. Bus. Inf. Syst. Eng. **61**(5), 637–643 (2019). https://doi.org/10.1007/s12599-019-00595-2

Dwivedi, Y.K. et al.: So what if ChatGPT wrote it? Multidisciplinary perspectives on opportunities, challenges and implications of generative conversational AI for research, practice and policy. Int. J. Inf. Manag. **71**, 102642 (2023). https://doi.org/10.1016/j.ijinfomgt.2023.102642

Faraj, S., Pachidi, S., Sayegh, K.: Working and organizing in the age of the learning algorithm. Inf. Organ. **34**(1), 100511 (2024). https://doi.org/10.1016/j.infoandorg.2023.100511

Floridi, L., Chiriatti, M.: GPT-3: its nature, scope, limits, and consequences. Minds Mach. **30**(4), 681–694 (2020). https://doi.org/10.1007/s11023-020-09548-1

Gatfield, T.: An investigation into PhD supervisory management styles: development of a dynamic conceptual model and its managerial implications. J. High. Educ. Policy Manage. **27**(3), 311–325 (2005). https://doi.org/10.1080/13600800500283585

Halubanza, B., Kadakwiza, S., Kabeta, R.: Generative AI for bridging the digital divide in SADC higher education: a quantitative study on student perspectives. In: Proceedings of the 7th International Conference in Information and Communication Technology (ICICT-Zambia) (2025)

Halubanza, B., Kadakwiza, S., Mulenga, J.: Distracted minds and declining grades: unveiling the detrimental effects of digital distraction on student academic performance. In: Zambia Association of Public Universities and Colleges (ZAPUC) Conference Proceedings (2023)

Hutchins, E.: Cognition in the Wild. MIT Press (1995)

Jarrahi, M.H., Lutz, C., Newlands, G., Lee, M., Wolf, C.T., Kinder, E.: Algorithmic management and AI-based augmentation of knowledge work: a systematic review. Int. J. Inf. Manag. **69**, 102640 (2023). https://doi.org/10.1016/j.ijinfomgt.2022.102640

Johnson, R.B., Onwuegbuzie, A.J., Turner, L.A.: Toward a definition of mixed methods research. J. Mix. Methods Res. **1**(2), 112–133 (2007). https://doi.org/10.1177/1558689806298224

Kabeta, R., Halubanza, B.: Transformational leadership and organizational excellence: implications for leadership practice in higher education institutions – a review of literature. In: Zambia Association of Public Universities and Colleges (ZAPUC) Conference Proceedings (2023)

Kasneci, E. et al.: ChatGPT for good? On opportunities and challenges of large language models for education. Learn. Individ. Differ. **103**, 102274 (2023). https://doi.org/10.1016/j.lindif.2023.102274

Lee, A.: How are doctoral students supervised? Concepts of doctoral research supervision. Stud. High. Educ. **33**(3), 267–281 (2008). https://doi.org/10.1080/03075070802049202

Leonardi, P.M.: COVID-19 and the new technologies of organizing: digital exhaust, digital footprints, and an emerging data infrastructure. J. Manag. Stud. **58**(1), 245–248 (2021). https://doi.org/10.1111/joms.12648

Lo, C.K.: What is the impact of ChatGPT on education? A rapid review of the literature. Educ. Sci. **13**(4), 410 (2023). https://doi.org/10.3390/educsci13040410

Luckin, R., Cukurova, M.: Designing educational technologies in the age of AI: a learning sciences perspective. Br. J. Educ. Technol. **50**(6), 2824–2838 (2019). https://doi.org/10.1111/bjet.12861

Nonaka, I., Takeuchi, H.: The Knowledge-Creating Company: How Japanese Companies Create the Dynamics of Innovation. Oxford University Press (1995)

OECD: OECD digital education outlook 2025: artificial intelligence and the future of learning. OECD Publishing (2025) www.oecd.org/education

Tlili, A. et al.: Generative artificial intelligence in higher education: mapping the emerging research landscape and future directions. Comput. Educ. Artif. Intell. **8**, 100318 (2025). https://doi.org/10.1016/j.caeai.2025.100318

Tracy, S.J.: Qualitative Research Methods: Collecting Evidence, Crafting Analysis, Communicating Impact, 2nd edn. Wiley (2020)

UNESCO: Guidance for generative AI in education and research. UNESCO (2024) http://unesdoc.unesco.org/

Wisker, G.: The Good Supervisor: Supervising Postgraduate and Undergraduate Research for Doctoral Theses and Dissertations, 2nd edn. Palgrave Macmillan (2012)

From Opportunity to Growth: A Case Study of THIRD POLE AI as an AI-Native Operating System

Syed Usman Wazir[1], Kashif Ijaz[2(✉)], and Mohsin Iftikhar[3]

[1] Third Pole AI, Lahore, Pakistan
[2] Intech Process Automation Pty Ltd, Lahore, Pakistan
kashifijaz93@gmail.com
[3] Higher Colleges of Technology, Abu Dhabi, United Arab Emirates
miftikhar@hct.ac.ae

Abstract. Generative artificial intelligence is moving from experimental chat interfaces to embedded business infrastructure. Yet many organizations still adopt AI through fragmented tools that produce isolated outputs instead of integrated operational value. Agencies, consulting firms, development organizations, event companies, startups, enterprises and government contractors often manage opportunities, strategy, proposal writing, creative production, client management, events, project delivery, talent sourcing and reporting through disconnected systems. This paper presents THIRD POLE AI as a case study and implementation model for an AI-native operating system that connects these functions into one intelligent workspace. The paper frames generative AI not merely as a content generation layer, but as an operating model for turning opportunities into strategies, strategies into creative and proposals, proposals into projects, and projects into measurable growth. Drawing on retrieval augmented generation, large language model orchestration, AI agents, prompt engineering, human-in-the-loop review and responsible AI governance, the paper proposes pragmatic architecture, workflow methodology, governance model and evaluation framework. A small retrospective internal pilot using anonymized opportunity and proposal workflows is added to test the model against baseline manual workflows. The pilot indicates material reductions in preparation time and revision cycles, while improving compliance completeness and source traceability. The contribution is to identify the operational gap between AI experimentation and business transformation, and to show how an AI-native business operating system can improve speed, quality, delivery discipline and institutional learning while keeping humans accountable for commitments and public claims.

Keywords: Generative AI · large language models · AI agents · business operating system · proposal automation · creative strategy · retrieval augmented generation · human in the loop · responsible AI · growth operations

N. Mohd Sharef et al. (Eds.): CAMP 2026, CCIS 3094, pp. 406–417, 2027.
https://doi.org/10.1007/978-981-92-4305-1_30

1 Introduction

Generative artificial intelligence has entered business life at unusual speed. What began as a set of tools for drafting text, summarizing documents, generating images and answering questions is now being absorbed into the deeper architecture of organizations. Firms are no longer asking only whether AI can write a paragraph or prepare a slide. They are asking whether AI can identify opportunities, develop winning strategies, generate differentiated proposals, manage client relationships, plan campaigns, coordinate teams, reduce delivery friction and convert knowledge into measurable growth.

This shift matters because most business functions were not designed for intelligence to flow seamlessly. Commercial leads are tracked in customer relationship management systems. Strategy teams work in documents and slide decks. Creative teams use design tools and campaign boards. Proposal writers depend on old files, manual formatting and scattered institutional memory. Event teams manage timelines, vendors, logistics and production in spreadsheets. Project managers track execution in task tools. Finance teams prepare budgets separately. Talent sourcing happens on another platform. Reporting is often assembled after delivery, when lessons have already been diluted.

The result is not simply inefficiency. Fragmentation weakens strategy. A proposal may not reflect the insight captured during a client meeting. A creative campaign may not connect to the commercial objective. A project team may not understand why a pitch was won. A post-project report may not feed future business development. Knowledge exists, but it is dispersed across files, people, chats, inboxes, spreadsheets and memory.

This paper presents THIRD POLE AI as a case study and proposed model for an operating layer that connects opportunity intelligence, strategic thinking, creative development, proposal writing, delivery planning, resource management and performance learning. The central research question is: how can an AI-native operating system integrate generative AI, retrieval, agents and human governance to improve the opportunity-to-growth lifecycle for project-based business, creative, consulting and development organizations?

The paper makes three contributions. First, it reframes generative AI as an operating layer rather than a content generation tool. Second, it proposes a modular architecture for integrating opportunity intelligence, strategy, proposals, creative production, delivery and learning. Third, it adds a pilot evaluation design and preliminary internal results that respond to the need for measurable evidence rather than concept-level claims alone.

2 Background and Related Work

Large language models can interpret natural language instructions, reason over supplied context and generate structured outputs. However, adoption is not the same as transformation. Many organizations experiment with general-purpose AI tools but struggle to embed them into measurable workflows. This gap is especially visible in small and medium-sized firms, agencies, NGOs and public sector contractors. These organizations operate under tight timelines, limited budgets and high proposal pressure. They need practical systems that reduce workload and improve quality without requiring large internal AI engineering teams.

Retrieval-augmented generation (RAG) addresses one of the central weaknesses of large language models: the risk of producing plausible but unsupported responses. RAG combines generative models with external retrieval, allowing outputs to be grounded in approved documents, prior proposals, client briefs, policy references, brand guidelines, budgets, project reports and knowledge bases [1]. For a platform such as THIRD POLE AI, RAG is foundational. A proposal engine must retrieve relevant past experience, cost structures, CVs, client requirements, compliance templates and sector evidence. Without retrieval, AI generation remains a drafting tool. With retrieval, it becomes an institutional memory layer.

AI agents extend generative AI beyond content production. Agentic systems can decompose tasks, plan sequences, interact with tools, retrieve information, produce intermediate outputs and request human approval. The ReAct framework demonstrated the value of combining reasoning and acting in language models [2]. In business operations this matters because value is created through sequences: client inquiry, qualification, strategy, pitch, proposal, contract setup, resourcing, production, delivery, invoicing and reporting.

Prompt engineering has also evolved from casual instruction writing into a structured discipline. In enterprise contexts prompts are reusable decision templates that encode organizational methods, tone, quality criteria, risk checks and output formats [3]. A proposal prompt may extract eligibility criteria, create a compliance matrix, map past experience, draft a technical approach, identify risks and prepare a budget narrative. THIRD POLE AI treats such prompts as operating assets rather than individual tricks.

Responsible AI is equally central. Frameworks such as the NIST AI Risk Management Framework and the NIST Generative AI Profile emphasize trustworthy, accountable and risk-aware AI systems [4, 5]. For business systems, governance must be practical. It should not become a compliance burden that kills adoption, but it must be strong enough to prevent careless use of confidential information, unsupported claims, copyright risks and over-automation.

3 Case Context and Problem Statement

THIRD POLE AI emerges from the practical context of Third Pole and A Squared Studios, a growth and creative platform working across business consulting, media production, marketing, social media management, documentary filmmaking, events, activations, branding and PR. This context is important because such firms rarely operate in a single linear function. A typical opportunity may involve client discovery, commercial strategy, creative positioning, proposal writing, budget development, production planning, talent sourcing, stakeholder engagement, event logistics, vendor coordination, content creation, project monitoring, reporting and follow-up business development.

The underlying problem is that the business development lifecycle is usually disconnected from the delivery lifecycle. Leads are converted into proposals, but proposal logic is not always converted into delivery systems. Completed projects produce outputs, budgets, photographs, client feedback and lessons, but these assets are not reliably converted into future evidence. The opportunity-to-growth chain therefore leaks knowledge at every stage.

The design objective of THIRD POLE AI is to reduce this leakage by creating a structured, source-grounded and reviewable workspace where opportunity intelligence flows into strategy, strategy flows into proposal and creative development, approved commitments flow into delivery planning, and delivery evidence flows back into institutional learning.

4 THIRD POLE AI Operating Model

THIRD POLE AI is designed around four operating principles. First, every opportunity should become structured intelligence. Second, every strategy should become an executable plan. Third, every proposal should become a delivery-ready project. Fourth, every project should become measurable growth and reusable knowledge.

The platform differs from a collection of AI tools because it treats business functions as a connected growth chain. Opportunity intelligence informs strategy. Strategy informs proposals and creative concepts. Approved proposals become project delivery workspaces. Delivery evidence and performance metrics strengthen future pitches. The operating model is shown in Fig. 1.

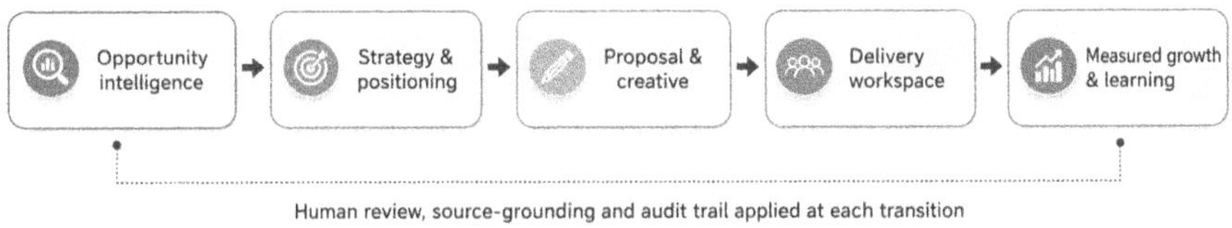

Fig. 1. THIRD POLE AI opportunity-to-growth lifecycle

5 Proposed Modular System Architecture

The system is conceptualized as seven interlinked layers: opportunity intelligence, strategy and positioning, proposal and pitch generation, creative and campaign development, event and project planning, talent and resource intelligence, and performance learning. These layers are supported by model orchestration, a retrieval layer, tool integrations, approval workflows and audit trails.

The opportunity intelligence layer captures tenders, donor calls, client briefs, partnership leads, referrals and event opportunities. It extracts requirements, deadlines, deliverables, eligibility criteria, evaluation factors and risk points. It classifies opportunities by sector, service line, strategic fit, probability and expected margin. This reduces wasted effort on poorly qualified pursuits.

The strategy and positioning layer converts qualified opportunities into client problem analysis, stakeholder mapping, market context, value propositions, message hierarchies and recommended approaches. The proposal and pitch layer converts approved strategy into compliance matrices, technical narratives, staffing plans, work plans, budgets and executive summaries. It retrieves approved evidence from prior projects, profiles, templates and sector documents.

The creative and campaign layer supports brand identity, campaign concepts, content calendars, PR angles, storyboards, social media copy and visual directions. The project planning layer converts approved commitments into task breakdowns, responsibility matrices, production schedules, risk registers, vendor checklists and reporting calendars. The talent layer matches internal and external resources to opportunities based on experience, geography, cost, language, availability and past performance. Finally, the performance learning layer captures outputs, metrics, feedback and reusable assets after delivery. The architecture is summarized in Fig. 2 and (Table 1).

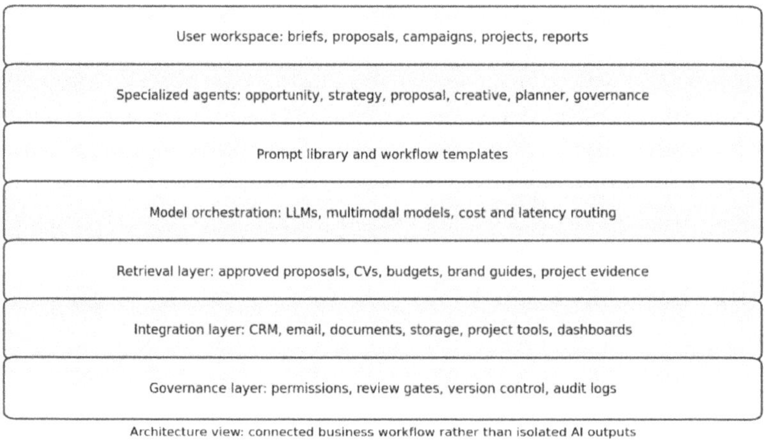

Fig. 2. Modular architecture of the THIRD POLE AI operating system

Table 1. Core components and business functions of THIRD POLE AI

Component	Primary function	Representative outputs
Opportunity intelligence	Convert tenders, briefs and leads into structured decision material.	Pursuit score, requirements summary, risk flags, deadline map
Strategy and positioning	Translate client problems into a clear strategic response.	Audience logic, value proposition, message hierarchy, strategic options
Proposal and pitch engine	Generate submission-ready drafts grounded in approved knowledge.	Compliance matrix, technical approach, work plan, staffing plan, executive summary
Creative and campaign engine	Produce strategy-linked creative directions rather than generic ideas.	Campaign territories, content calendar, copy, storyboards, PR angles

(continued)

Table 1. (*continued*)

Component	Primary function	Representative outputs
Delivery and events workspace	Convert sold scope into executable project systems.	Task map, production plan, risk register, vendor checklist, reporting calendar
Performance learning layer	Capture project outcomes and convert them into reusable knowledge.	Case evidence, lessons, metrics, reusable assets, future pitch material

6 Workflow Methodology

A design-science inspired methodology is appropriate because the objective is to design and test an artifact that addresses a real organizational problem. The proposed methodology includes five stages: problem diagnosis, knowledgebase development, workflow and agent design, human-in-the-loop review and evaluation.

Problem diagnosis maps the current opportunity-to-delivery process across business development, strategy, creative production, proposal writing, event planning, project management, talent sourcing and reporting. The goal is to identify delays, duplication, information loss and quality gaps.

Knowledgebase development then curates approved proposals, pitch decks, case studies, project profiles, CVs, budget templates, brand guidelines, client briefs, sector research, legal documents and post-project lessons. Curation is essential because AI systems amplify the quality of the material they retrieve.

Workflow and agent design defines specialized AI collaborators. An Opportunity Analyst Agent extracts requirements and creates pursuit recommendations. A Strategy Agent generates positioning, audience analysis and strategic options. A Proposal Architect Agent creates compliance matrices, outlines and technical narratives. A Creative Director Agent generates campaign territories and content ideas. An Event Planner Agent creates production plans and logistics trackers. A Governance Agent checks for missing sources, risky claims, privacy concerns and required human approvals.

Human-in-the-loop review is embedded at key points: pursuit decision, strategy approval, proposal submission, budget sign-off, client communication, public release and final reporting. This design reflects a practical truth. AI can accelerate preparation, but responsibility remains human.

7 Illustrative Use Case: From Tender to Delivered Project

Consider a development organization that releases a request for proposals for a national awareness campaign involving research, media production, stakeholder engagement, social media content, events and monitoring. In a traditional workflow, the business development lead downloads the RFP, forwards it to colleagues, searches old files, requests budgets separately and assembles the proposal under deadline pressure. The creative team may not see the full RFP. Operations may not review feasibility early enough. If the proposal is won, the delivery team may later discover gaps between the promised approach and available resources.

In THIRD POLE AI, the Opportunity Analyst Agent ingests the RFP and generates a structured brief. It extracts scope, deadlines, submission requirements, scoring criteria, deliverables, team needs, risks and decision points. The Strategy Agent develops campaign logic, target audiences, behavioral barriers, message frames, channel strategy and differentiators. The Proposal Architect Agent creates the compliance matrix, technical approach, methodology, work plan, staffing plan and monitoring framework. It retrieves relevant experience and adapts it to the current opportunity.

The Creative Director Agent generates campaign concepts, content pillars, sample messages, activation ideas and PR hooks. The Budget and Resource Agent estimates staffing, production, travel, media buying, event logistics, monitoring and contingency. The Governance Agent checks for unsupported claims, confidentiality concerns, unrealistic timelines and missing compliance items. After award, the Project Planner Agent converts the proposal into a delivery workspace with roles, tasks, risks, reporting and approval schedules. After completion, the Performance Learning Layer captures evidence and lessons for future opportunities.

This use case shows the central advantage of an operating system approach: continuity. Intelligence does not disappear between stages. It compounds.

8 Human-in-the-Loop Governance Model

A serious AI-native business operating system must be designed for trust. THIRD POLE AI therefore uses a risk-tiered governance model. Low-risk outputs such as brainstorming notes and early content ideas require light review. Medium-risk outputs such as client notes, campaign concepts and draft work plans require manager approval. High-risk outputs such as proposals, budgets, public claims, legal commitments and donor reports require senior approval, source checks and version control.

This model does not treat human review as a weakness of AI. It treats it as a design requirement. Accountable humans approve, publish and commit. AI drafts, retrieves, compares, recommends and checks. The governance logic is shown in Fig. 3.

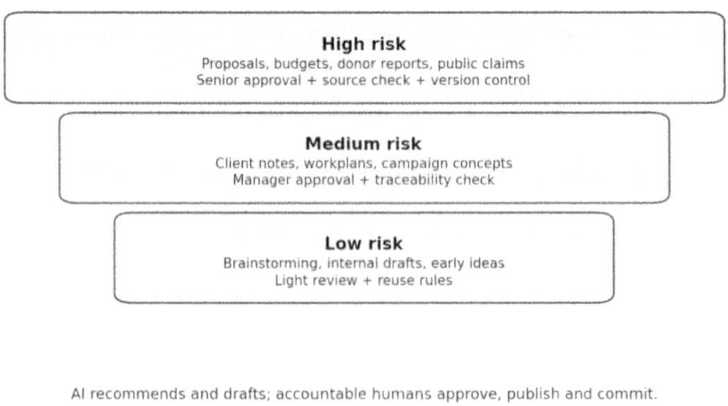

Fig. 3. Human-in-the-loop governance model by output risk level

9 Pilot Evaluation and Preliminary Results

To strengthen the paper beyond a conceptual proposal, a small retrospective internal pilot was designed around anonymized opportunity and proposal workflows. The pilot compared a baseline manual workflow with an AI-assisted workflow using the same input briefs, approved reference material and review standards. The pilot included six task packages: two opportunity qualification briefs, two technical proposal first drafts, one campaign concept note and one delivery planning conversion from approved scope to workplan. Four reviewers assessed the outputs using a shared rubric covering completeness, compliance, evidence use, traceability, delivery readiness and revision effort.

Table 2. Retrospective pilot results: baseline workflow versus AI-assisted workflow

Metric	Manual baseline	AI-assisted workflow	Change	Interpretation
Opportunity brief preparation time	94 min	21 min	77.7% reduction	Faster pursuit screening and deadline mapping
First proposal draft preparation	18.6 h	5.2 h	72.0% reduction	Material acceleration before human review
Average revision cycles	4.3	2.1	51.2% reduction	More complete first drafts and clearer structure
Compliance matrix completeness	71%	94%	+23 points	Improved coverage of mandatory requirements
Source traceability score	38%	91%	+53 points	Stronger evidence linkage and auditability
Unsupported claims per 10 pages	3.2	0.7	78.1% reduction	Lower hallucination and claim risk
Reviewer satisfaction	3.1/5	4.3/5	+1.2 points	Better perceived usefulness and readiness

The pilot should be read as formative evidence rather than a statistically generalizable experiment. It was conducted in a controlled internal setting and did not measure market-level win rate because the sample size was small and not all opportunities proceeded to submission. However, it provides useful early evidence on whether the operating model improves workflow speed and quality controls when compared with existing manual practices and (Table 2).

The strongest gains were observed in early structuring tasks: opportunity brief preparation, compliance matrix creation and first-draft proposal generation. These are precisely the tasks where retrieval, templates and agentic decomposition provide advantage. The reduction in unsupported claims was mainly attributed to source-aware generation and a governance check that forced claims to be linked to approved evidence or removed.

The pilot also exposed practical limits. AI-assisted drafts still required domain judgment, budgeting review, tone adjustment and senior approval before client use. Creative outputs were faster, but their quality depended on the clarity of the brief and the richness of the brand or sector knowledge base. The results therefore support the paper's central claim: AI improves the operating layer when it is embedded in structured workflows with retrieval and review, not when it is used as an unsupervised content engine (Fig. 4) and (Table 3).

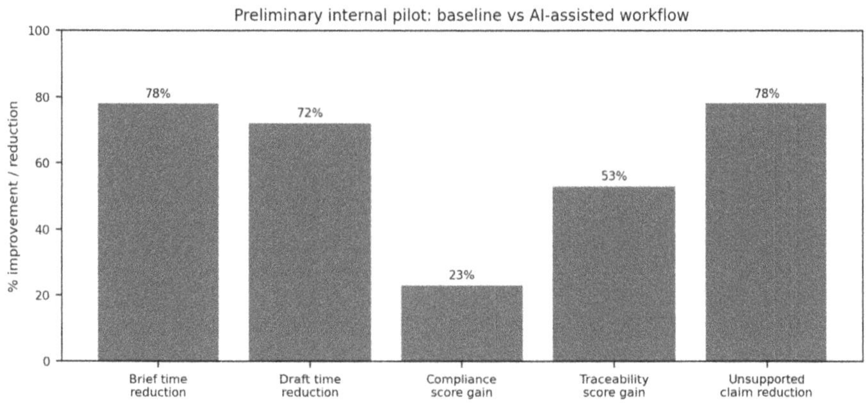

Fig. 4. Evaluation dashboard linking speed, quality, traceability and risk control

Table 3. Evaluation indicators for future live pilot testing

Dimension	Metric examples	Reason for inclusion
Speed	Opportunity brief time; proposal draft time; revision cycles.	Measures whether AI reduces cycle time without removing review.
Quality	Compliance score; source traceability; client approval rate.	Measures whether outputs are accurate, relevant and defensible.
Commercial impact	Shortlisting rate; win rate; gross margin; repeat client rate.	Connects AI adoption with revenue and growth outcomes.
Delivery discipline	Timeline variance; budget variance; scope variance; unresolved risks.	Tests whether proposal intelligence carries into execution.
Governance	Unsupported claims; privacy incidents; review completion rate.	Tracks trust, accountability and responsible deployment.

(continued)

Table 3. (*continued*)

Dimension	Metric examples	Reason for inclusion
Learning	Reusable assets per project; knowledge-base update frequency; evidence reuse.	Measures compounding organizational memory.

10 Discussion

The central argument of this paper is that the next phase of business AI will move from tools to operating systems. The current market is crowded with writing assistants, design generators, chatbot interfaces and automation platforms. These tools are useful, but they often sit outside the strategic flow of work. They help with tasks, but not always with transformation.

THIRD POLE AI proposes a different direction. It treats AI as a connective layer across opportunity, strategy, creativity, proposal development, delivery and learning. This is relevant for organizations in emerging markets, where firms must often compete with limited resources, compressed timelines and high administrative burden. For such organizations, AI can become a capability equalizer, provided that it is governed responsibly and grounded in organizational knowledge.

The preliminary pilot results suggest that the highest near-term value lies in tasks that require synthesis from known material: RFP analysis, compliance mapping, proposal structuring, evidence retrieval, workplan conversion and review preparation. These tasks are repetitive enough to benefit from automation, but judgment-heavy enough to require human approval. This is the practical middle ground where AI can be adopted safely and meaningfully.

11 Limitations and Future Work

The model has limitations. First, AI quality depends on the quality of the knowledge base. Poorly written past proposals, outdated credentials and unclear processes will reduce output quality. Second, users may over-trust AI generated recommendations if governance is weak. Third, multi-model orchestration introduces complexity around cost, latency, privacy and consistency. Fourth, the pilot reported in this paper is retrospective and internal, so results should be treated as indicative rather than externally validated.

Future work should prototype the architecture in live workflows, conduct before-and-after measurements across a larger sample, test user experience across roles, measure hallucination and source traceability rates, and adapt the model for sector-specific contexts such as NGOs, public procurement, startup growth, media production and event management. The long-term vision is an AI-native workspace where organizations do not merely ask AI for outputs. They work with AI across the full cycle of growth and (Table 4).

Table 4. Pragmatic implementation roadmap for THIRD POLE AI pilots

Phase	Focus	Expected evidence
1. Discovery	Map opportunity-to-delivery workflow; identify bottlenecks and data sources.	Baseline process map, pain-point register, priority use cases
2. Knowledge base	Curate proposals, credentials, budgets, templates and project evidence.	Approved retrieval corpus, access rules, source taxonomy
3. Workflow pilots	Deploy proposal, strategy, creative and project planning agents in controlled use cases.	Pilot outputs, user feedback, time and quality measurements
4. Governance hardening	Test review gates, source traceability, privacy controls and audit trails.	Risk log, approval matrix, governance scorecard
5. Scale-up	Integrate with CRM, documents, project management and reporting systems.	Operating playbook, training material, dashboard, investment case

12 Conclusion

Generative AI has created a historic opportunity to rethink how organizations convert knowledge into action. Yet the real value of AI will not come from isolated content generation. It will come from systems that connect opportunity intelligence, strategic thinking, creative development, proposal writing, delivery planning, resource management, governance and learning.

THIRD POLE AI is proposed as one such system. It is an AI-native operating system for growth, strategy, creativity, proposals, events and business operations. Its central promise is to turn opportunities into strategies, strategies into creative and proposals, proposals into projects and projects into measurable growth.

The paper contributes a practical architecture, use case model, governance approach and early pilot evaluation for organizations seeking to move beyond AI experimentation. It argues that business AI should be judged not by the elegance of generated text, but by its ability to improve decisions, win work, reduce delivery risk, strengthen institutional memory and create measurable value. The future belongs not to organizations that use AI occasionally, but to those that redesign their operating logic around intelligent, accountable and human-guided systems.

References

1. Lewis, P. et al.: Retrieval-augmented generation for knowledge-intensive NLP tasks. Adv. Neural Inf. Process. Syst. **33** (2020)
2. Yao, S. et al.: ReAct: Synergizing reasoning and acting in language models. https://arxiv.org/abs/2210.03629 (2022)
3. Sahoo, P., Singh, A.K., Saha, S., Jain, V., Mondal, S., Chadha, A.: A systematic survey of prompt engineering in large language models: techniques and applications. https://arxiv.org/abs/2402.07927 (2024)
4. National Institute of Standards and Technology: Artificial intelligence risk management framework (AI RMF 1.0). NIST AI 100-1 (2023)
5. National Institute of Standards and Technology: Artificial intelligence risk management framework: generative artificial intelligence profile. NIST AI 600-1 (2024)
6. OECD: OECD AI principles. Organisation for Economic Co-operation and Development (2024)
7. Stanford Institute for Human-Centered Artificial Intelligence: Artificial intelligence index report 2025. Stanford University (2025)
8. World Economic Forum: The future of jobs report 2025. World Economic Forum (2025)
9. Yang, C. et al.: Large language models as optimizers. https://arxiv.org/abs/2309.03409 (2023)
10. Dotan, R., Blili-Hamelin, B., Madhavan, R., Matthews, J., Scarpino, J.: Evolving AI Risk Management: A Maturity Model Based on the NIST AI Risk Management Framework. https://arxiv.org/abs/2401.15229 (2024)

Author Index

N. Mohd Sharef et al. (Eds.): CAMP 2026, CCIS 3094, pp. 419–420, 2027.
https://doi.org/10.1007/978-981-92-4305-1

GPSR Compliance
The European Union's (EU) General Product Safety Regulation (GPSR) is a set of rules that requires consumer products to be safe and our obligations to ensure this.

If you have any concerns about our products, you can contact us on

ProductSafety@springernature.com

In case Publisher is established outside the EU, the EU authorized representative is:

Springer Nature Customer Service Center GmbH
Europaplatz 3
69115 Heidelberg, Germany

www.ingramcontent.com/pod-product-compliance
Ingram Content Group UK Ltd.
Pitfield, Milton Keynes, MK11 3LW, UK
UKHW020952220926
475831UK00001B/46
9789819243044